Energy-Efficient Fault-Tolerant Systems

Jimson Mathew • Rishad A. Shafik
Dhiraj K. Pradhan
Editors

Energy-Efficient Fault-Tolerant Systems

Springer

Editors
Jimson Mathew
Department of Computer Science
University of Bristol
Bristol, UK

Rishad A. Shafik
Department of Computer Science
University of Bristol
Bristol, UK

Dhiraj K. Pradhan
Department of Computer Science
University of Bristol
Bristol, UK

ISBN 978-1-4614-4192-2 ISBN 978-1-4614-4193-9 (eBook)
DOI 10.1007/978-1-4614-4193-9
Springer New York Heidelberg Dordrecht London

Library of Congress Control Number: 2013944573

Printed on acid-free paper

Springer is part of Springer Science+Business Media (www.springer.com)

Dedicated to our inspirations

Raisa

Alan

Linda

Babita

Sonia

Foreword

Energy efficiency is a prime design objective for embedded systems. Over the years, various techniques have been proposed highlighting energy-efficient systems design. These techniques have shown effective ways to reduce energy, while satisfying the system performance needs. However, due to continued device miniaturization and technology scaling, reliability in the presence of ever-increasing number of faults is emerging as an additional design challenge. Incorporating fault tolerance and energy efficiency is not trivial since they are two conflicting design objectives. Currently no book exists that gives a complete account of the state-of-the-art and emerging techniques for design, analysis, and testing of energy-efficient, fault-tolerant electronic systems. The book at hand is expected to appropriately fill this gap and will be highly useful for students, researchers, and practitioners working in relevant areas.

Chalmers University of Technology
Gothenburg, Sweden

Georgi N. Gaydadjiev

Preface

Energy efficiency is a widely acknowledged design objective in electronic systems research and development. With continued technology scaling reliability is an emerging concern in these systems due to increased device-level vulnerability in the presence of electromagnetic inductions and process variability. Achieving high reliability and energy efficiency objectives jointly can, however, be challenging as energy minimization techniques exacerbate the reliability further. Hence, energy-efficient fault-tolerant (i.e. reliable) design is currently of high interest to both industry and academia to address these challenges effectively. The aim of this book is to introduce state-of-the-art and emerging issues in energy-efficient fault-tolerant design techniques as applied or prototyped for current or future generations of system-on-chip (SoC). Although necessary brief background has been provided in each chapter, fundamental theories have been omitted or avoided. The reader is expected to have preliminary background of electronic circuits and systems design and synthesis.

The book consists of nine chapters; five chapters (Chaps. 1–5) are dedicated towards systems modeling and design techniques, while four other chapters (Chaps. 6–9) delve into emerging systems architectures and design challenges. The individual chapter outlines are given below.

Chapter 1 (Introduction to Energy-Efficient Fault-Tolerant Systems) provides the necessary background relevant to this book. Low-power design aspects are briefly revisited, and concepts related to faults and reliability are reviewed. Underpinning these basics, challenges of energy-efficient fault-tolerant systems design are highlighted showing various techniques.

Chapter 2 (Reliability Evaluation Techniques) gives a comprehensive review of the reliability models. In particular, methods for evaluating system reliability with stochastic modeling and other combinatorial methods are discussed in details.

Chapter 3 (Energy-Efficient Design Techniques) presents an overview of power minimization and power management techniques. The impact of energy-efficient design techniques on the system reliability is also investigated. Extensive case studies are presented addressing energy-efficient fault-tolerant design techniques.

Chapter 4 (Error Correction Coding for Electronic Circuits) discusses various existing and emerging error detection and correction (EDAC) codes mainly used in sequential circuits. Implementation and suitability of various EDAC codes are detailed with examples, followed by their analyses in terms of fault detection and tolerance capabilities, overheads, and circuit complexities.

Chapter 5 (System-Level Design Methodology) introduces system-level design methodology, particularly suitable for multiprocessor system-on-chip (MPSoC). Various MPSoC design challenges are discussed, highlighting ways to effectively address them using system-level design methodology. Later, an extensive system-level design optimization case study is presented investigating into the impact of application task mapping on MPSoC applications.

Chapter 6 (Fault-Tolerant Reconfigurable On-Chip Network) presents fault-tolerant reconfigurable on-chip networks, which are infrastructure communication architectures for future computing platforms. The impact of workload variations on the fault-tolerance of reconfigurable on-chip network is also studied in details through a number of case studies.

Chapter 7 (Bio-Inspired Online Fault Detection in NoC Interconnect) deals with bio-inspired online and low-cost fault detection capabilities for emerging scalable network-on-chip-based multiprocessor systems. In particular, a novel real-time strategy for detecting faults in NoC interconnect is demonstrated that uses biological synapses and neurons to detect temporal and spatial faults effectively. Analysis of different fault scenarios and results from real-time experiments on an example FPGA implementation are also provided.

Chapters 8 (Power-Efficient Fault-Tolerant Finite Field Multiplier) and 9 (Low-Cost C-Testable Finite Field Multiplier Architectures) investigate into emerging low-complexity techniques for secure and fault-tolerant hardware design. To solve the problem of an attacker injecting faults into the hardware and then analyzing the incorrect outputs, different detection and correction mechanism are studied in detail. The aim is to ensure that the effect of injected errors is not visible at the outputs for the attacker to be able to intercept and compromise the security and fault-tolerance of the system. Moreover, new methods for secure error detection and correction using various type of codes are investigated for scan-based attacks during testing.

We would like to acknowledge the outstanding contributions from our chapter contributors. Without their valued efforts, this book would not have been possible. We would also like to gratefully appreciate the efforts of our expert panel of reviewers, who helped with their useful feedback to improve the overall quality and coherence of the book.

Bristol, UK

Jimson Mathew
Rishad A. Shafik
Dhiraj K. Pradhan

Contents

1 Introduction to Energy-Efficient Fault-Tolerant Systems 1
Rishad A. Shafik, Jimson Mathew, and Dhiraj K. Pradhan

2 Reliability Evaluation Techniques 11
Rong-Tsorng Wang

3 Energy-Efficient Design Techniques 99
Rong Ye and Qiang Xu

4 Error Correction Coding for Electronic Circuits 137
Juan A. Maestro, Pedro Reviriego, and Mark F. Flanagan

5 System-Level Design Methodology 169
Rishad A. Shafik, Bashir M. Al-Hashimi, and Krishnendu Chakrabarty

6 Fault-Tolerant Reconfigurable On-Chip-Network 211
Mohammad Hosseinabady and Jose L. Nunez-Yanez

7 Bio-Inspired Online Fault Detection in NoC Interconnect 241
Malachy McElholm, Jim Harkin, Liam McDaid, and Snaider Carrillo

8 Power-Efficient Fault-Tolerant Finite Field Multiplier 269
Jimson Mathew, A.M. Jabir, R.A. Shafik, and D.K. Pradhan

9 Low Cost C-Testable Finite Field Multiplier Architectures 307
Jimson Mathew, H. Rahaman, and D.K. Pradhan

Index 333

Acronyms

ABB	Adaptive base bias
AHB	Advanced high-performance bus
AMBA	Advanced microprocessor bus architecture
APB	Advanced peripheral bus
ASA	Adaptive simulated annealing
ASB	Advanced system bus
ASIC	Application-specific integrated circuit
AXI	Advanced extensible interface
BILP	Binary integer linear programming
DCT	Discrete cosine transformation
DRAM	Dynamic random access memory
DSP	Digital signal processor
DTU	Data transaction unit
DUT	Device under test
DVS	Dynamic voltage scaling
EDA	Electronic design automation
ESL	Electronic system level
FER	Frame error ratio/rate
FI	Fault injection
FIFO	First-in first-out
FIR	Fault injection rate
FIT	Failures in time
FLIT	Flow control unit
FPGA	Field programmable gate array
HDL	Hardware description language
HW	Hardware
HW/SW	Hardware/software
IDCT	Inverse discrete cosine transformation
ILP	Integer linear programming
ISQ	Inverse scan and quantisation
MC	Motion compensation

MCU	Microcontroller unit
MPARM	Multiprocessor ARM
MPEG	Moving picture experts group
MPSoC	Multiprocessor system-on-chip
MTBF	Mean time between failure
MTTF	Mean time to failure
MTTR	Mean time to repair
NI	Network interface
NoC	Network-on-chip
OSCI	Open SystemC initiative
P2P	Point-to-Point
PE	Processing element
PSNR	Peak signal-to-noise ratio
RAM	Random access memory
RHEL	Red Hat Enterprise Linux
RTEMS	Real-time executive for embedded systems
RTOS	Real-time operating system
RTL	Register transfer level
SA	Simulated annealing
SER	Soft error rate
SEU	Single-event upset
SoC	System-on-chip
SRAM	Static random access memory
SW	Software
VLD	Variable length decoder
VLSI	Very large-scale integration

Chapter 1
Introduction to Energy-Efficient Fault-Tolerant Systems

Rishad A. Shafik, Jimson Mathew, and Dhiraj K. Pradhan

Embedded systems are making their way into more and more devices, from hand-held gadgets to household appliances, and from mobile devices to cars. The current trend is that this growth will continue and the market is expected to experience a three-fold rise in the demand from 2013 to 2018 [20]. This growth has been possible due to continued technological advancements in terms of device miniaturization, feature richness, design cost control and performance improvement, originally described by the Moore's Law [32].

Since a vast majority of today's embedded systems are battery powered, a prime design objective is to minimize power consumption. Minimized power consumption can extend the battery operating lifetime of a system with a given energy budget. Although technology scaling has enabled the fabrication of faster and more power-efficient devices than their predecessors due to smaller geometry and device capacitance [3], increased computational demand and packing density has caused a diminishing effect on the overall power consumption at system- and application-level. In fact, the overall power consumption of a system-on-chip (SoC) is now increasing beyond available maximum power density budget at chip-level [12]. This has necessitated efficient and low power design techniques that have been studied extensively by researchers in the indsutry and the academia [4, 15, 36].

A major challenge for modern SoC design is the increasing number of hardware faults, such as those caused by imperfect lithographic pattering during manufacturing and induced electromagnetic induction during operational lifetime [5, 12]. These faults manifest themselves as logic upsets at circuit-level and can affect the signal transfers and stored values leading to incorrect execution in embedded systems. According to ITRS, 1 out of every 100 chips will experience a fault per day during operational lifetime, while manufacturing defect rate will reach the level of approximately 1,000 defects/m^2 in the next few years [20]. Indeed, under the

R.A. Shafik (✉) • J. Mathew • D.K. Pradhan
University of Bristol, Bristol, BS8 1UB, UK
e-mail: rishad.shafik@bristol.ac.uk; jimson@cs.bris.ac.uk; pradhan@cs.bris.ac.uk

J. Mathew et al. (eds.), *Energy-Efficient Fault-Tolerant Systems*,
DOI 10.1007/978-1-4614-4193-9_1, © Springer Science+Business Media New York 2014

circumstances, reliable design and testing of current and future generations of SoCs is of critical importance, in particular for high availability, safety-critical systems etc. [24]. However, designing energy-efficient reliable systems is highly challenging due to conflicting design trade-offs between power consumption and reliability objectives [36]. This is because, reliable design and testing techniques generally introduce redundant hardware or software resources that can increase the overall power consumption of the system [15].

The rest of this chapter is outlined as follows. Section 1.1 gives brief introduction to energy-efficient design. Section 1.2 outlines necessary background on faults and reliability, highlighting the challenges of energy-efficient reliable design.

1.1 Energy-Efficient Design

The total power dissipated in a CMOS circuit is formed of two major components: dynamic (P_{dyn}) and static power (P_{stat}) dissipation, i.e.

$$P_{total} = P_{dyn} + P_{stat}. \qquad (1.1)$$

Dynamic power is mainly caused by circuit activity. The main contribution of dynamic power dissipation in (1.1) is incurred by capacitive switching current that charges and discharges the circuit capacitive loads during various logic changes, given by

$$P_{dyn} = \alpha \, C_L \, V_{dd}^2 \, f, \qquad (1.2)$$

where C_L is the average load capacitance per cycle (generally constant for a given circuit), V_{dd} is the supply voltage, f is the circuit operating frequency and α is the switching activity factor (i.e. the ratio of switching activity over a given time). Another contributor of dynamic power is short-circuit current that flows in the circuit due to direct current (DC) path between the supply and ground when the output transition is taking place. Compared to switching capacitive switching power, short-circuit power is small and often it is ignored in total dynamic power dissipation. Static power, on the other hand, is incurred even without any circuit activity. The main contribution of static power is from sub-threshold gate leakage currents. Sub-threshold current arises from the inversion charges that exist at the gate voltages below the normal circuit threshold voltage. A simpler static power model of leakage power (P_{leak}) can be given by

$$P_{leak} = V_{dd} N k I_{leak}, \qquad (1.3)$$

where N is the number of transistors and and I_{leak} is the leakage current. The leakage current (I_{leak}) depends on technology parameters like threshold voltage, while k is circuit constant that depends on the number of transistors operating at

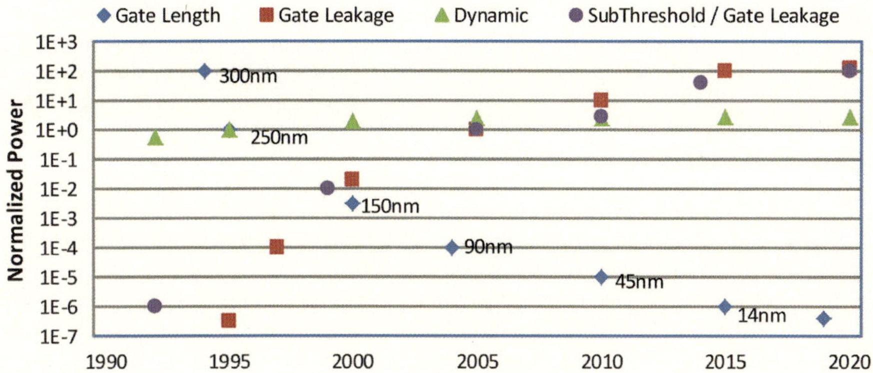

Fig. 1.1 Trends of power dissipations with technology scaling

a given time. Similar to dynamic power, as can be seen from (1.3) leakage power depends on the supply voltage (V_{dd}) and the number of transistors (N), which is increasing with technology scaling.

Figure 1.1 shows the normalized power dissipation trends with continued technology scaling over a span of 30 years (from 1990 to 2020) [34]. As can be seen, with previous technology nodes dynamic power dissipation dominated the total power consumption in CMOS circuits. From (1.2) it is evident that the most effective means of lowering total (dynamic and also static) power dissipation (P_{dyn}) for these technology nodes is to reduce the operating voltage (V_{dd}). However, lowering V_{dd} increases circuit propagation delay. This delay eventually restricts the circuit operating frequency, which requires the operating clock frequency to be lowereed to tolerate the propagation delay [39]. Dynamic voltage and frequency scaling (DVFS) is an effective power minimization technique that reduces dynamic power through lowering V_{dd} and f during runtime [11, 13]. The main working principle of DVFS is to lower V_{dd} and f during slack times (i.e. time between early completion of the previous computational task or late starting of the next computational task) [42]. Over the last decade, power minimization using DVFS-enabled SoCs has been extensively investigated considering its effects on system performance [4, 15].

However, with continued technology scaling, static power dissipation is emerging as a major concern for systems designers (as shown in Fig. 1.1). To reduce static power dissipation dynamic power management (DPM) is an effective technique. The main strategy employed in DPM is to control the operational times of supply voltages in system components. For example, power supply can be shut down for components within an MPSoC when they are idle and can be switched on when they are operational (otherwise known as power gating) or supply clock can be gated off when certain components in a circuit are not active. However, shutting down supply power or clock for these components can result in delay in retaining fully operational mode of the circuit. Hence, DPM technique needs to carefully take into consideration this delay effect to achieve power minimization without compromising the system performance [7]. Often, today's MPSoCs include both DVFS and DPM techniques to minimize dynamic and leakage power consumptions.

1.2 Faults and Reliable Design

An emerging challenge in today's electronic system design is reliability of the system when it is subjected to different errors or faults [23, 30]. In fact due to technolgy scaling and aggressive voltage scaling, the number of these faults occuring in a circuit is increasing exponentially. These faults manifest themselves as temporary logic upsets, such as single-event upsets (SEUs), and can affect the signal transfers and stored values leading to incorrect or undesired outcomes in circuit and systems. Several academic and industrial studies, such as [18, 31, 35, 43, 48], have investigated the presence and increase of these faults highlighting the impact of operating environments.

Faults in electronic systems can be classified in two types: permanent and transient. Permanent faults are related to irreversible physical defects in the circuit. Major causes of permanent faults include improper lithographic process or systematic aberrations during manufacturing or post-manufacturing burn-in, or even burn-in caused by electro-migration and negative bias temperature instability (NBTI) during operational lifetime etc. Since permanent faults can cause persistent failures in the circuit, faulty chips are discarded or repaired after post-manufacturing tests. Transient faults, also known as soft errors, can appear during the operation of a circuit. Unlike permanent faults, transient faults do not represent a physical defect in the circuit. Major causes of transient faults include cross-talk, power supply noise and neutron or alpha radiations during operational lifetime. Radiation induced faults are generally considered major source of transient faults as they take place during operational lifetime when a single ionizing radiation event produces a burst of hole-electron pairs in a transistor that is large enough to cause the circuit to change state.

To evaluate the rate and effect of fault occurrence, different parameters have been reported to date. Major parameters are briefly defined below:

Fault Density is the measure of number of faults found in a device per unit of data. For memories, this is generally expressed as the number of faults per megabyte or gigabyte data. This parameter is used for permanent faults or defects only [9].

Fault injection time (FIT) is the rate at which the faults take place per unit of time in an electronic component. It is generally denoted as λ and expressed in unit of number of fault per million of operating hours (fault/10^9 operating hours).

Mean time-to-failure (MTTF) is an estimate of the mean time expected until the first fault occurs in a component of an electronic system. MTTF is a statistical value and is meant to be the mean over a long period of time and large number of units. It is usually expressed in unit of millions of hours. For constant failure rate systems, MTTF is the inverse of FIT (i.e. $MTTF = \frac{1}{\lambda}$) [44].

Mean time-to-repair (MTTR) is described as the time elapsed between a fault occurrence and its repair is completed, i.e. the system return back to its operational mode.

Mean time-between-failures (MTBF) is described as the time elapsed before a component in an electronic system experiences another fault. Unlike MTTF, the time elapsed in MTBF includes the time required to recover from a fault, i.e. $MTBF = MTTF + MTTR$.

Soft error rate (SER) is the rate at which the soft errors take place per unit time and per unit data. It is generally used to describe the severity of an operating environment and is expressed as number of soft errors per bit per cycle or number of soft errors per cycle per chip.

Reliability is the ability of a system to perform a required function under given conditions for a given time interval. Hence, with a given FIT (λ) and time interval, reliability can be expressed as $R = e^{-\lambda t}$.

Availability is the ability of a system to be in a state to perform a required function at a given instant of time or at any instant of time within a given time interval, assuming that the external resources, if required, are provided. In other words, it can also be expressed as the ratio of up time of a system to the total operating time (up and down time). Since faults can cause failures and down time of a system, availability can get affected.

Appropriate fault modeling is crucial for systems designers as it describes how a physical fault in the underlying device affects the circuit-level behavior. Depending on the nature and impact of occurrence of faults, fault models can be classified in the following major ones:

Stuck-at Fault Model The most common fault model is the single stuck-at model. In this model the defects behave like the given circuit line is permanently connected to ground (stuck-at-0) or to power supply (stuck-at-1); and only a single fault is present in a circuit at anytime. The stuck-at model has several advantages that include simplicity, logical behaviour, tractability and measurability [2]. The single stuck-at fault model remains the most commonly used, even though multiple technologies and numerous process shrinks.

Bridging fault An extension of the stuck-at model is the bridging fault model and this models the short between two lines. This model is appealing because shorts are generally considered the most common fault in CMOS circuits. Inductive analysis has shown that the most commonly occurring type of fault resulting from fabrication defects, modeled as dust particles of various sizes on photo-masks, is the bridging fault [17]. Much of the earlier work in bridging faults claimed that either wired-and or wired-or resulted when two nodes were bridged [1,29]. Randomly placed bridging faults are complicated and can not be modeled by a single fault model. As technology advance to smaller geometries, more metal layers, reduced design margins and higher frequencies, the effect of these defects will grow in complexity, increasing the variety of bridging behaviour we need to detect.

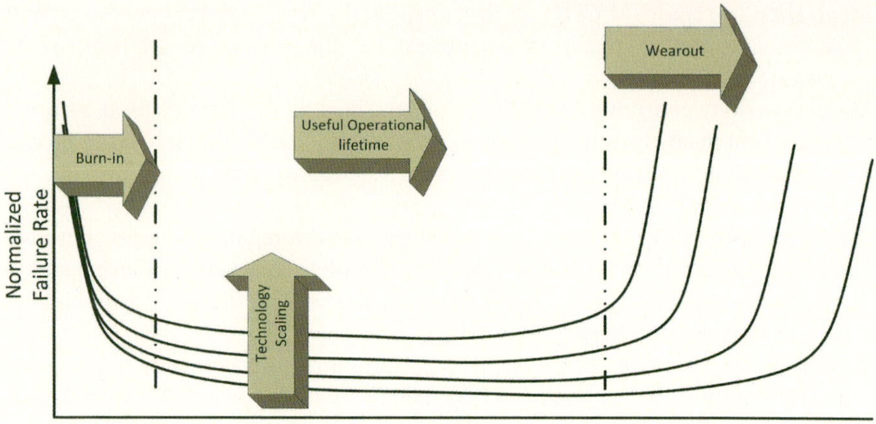

Fig. 1.2 Trends of power dissipations with technology scaling

Stuck Open Fault While shorts remain the most common type of defects in most CMOS processes, open fault also cause concern. As the number of wiring levels in circuits increases, the number of vias proliferates. We do not well understood the effects of missing vias, partial vias and resistive vias on circuit operation [2]. In some cases, the circuit still functions correctly but at a lower speed. The best known open model is the stuck-open fault model [2]. In this model, the gate to a given transistor is fixed at the open or 'off' value. As a result transistor cannot pull the cell output to its voltage. The length of time that the output remains at the previous value depends on the length of time required to discharge the load capacitance. A stuck-open fault requires a two-pattern test [22]. The first pattern sets up the fault pulling the circuit output to a known state and the second pattern activates the fault.

Delay Fault Failures causing logic circuits to malfunction at desired clock rates, or not meeting timing specifications are modeled by what are called delay faults. A change in the timing behaviour of the circuit, causes incorrect operation only at certain frequencies. The two broad classes of delay faults are gate delay fault and path delay faults. Gate delay fault models defects local to individual circuit elements; whereas path delay fault models distributed failures caused by statistical process variations and imprecise modeling [37, 41]. Because delays refers to differences in behaviour over time, delay fault models focus on transitions in logic values rather than logic values. A test for a delay fault consists of an initialization pattern, which sets the stage for a transition and a propagation pattern, which creates a desired transition and propagates it to observable points.

Figure 1.2 shows the trends of device failure rates over their lifetimes, highlighting the impact of technology scaling [34]. As can be seen, due to manufacturing silicon or metal defects many devices experience early burn-in and get discarded from the shipment line. The good circuits that are delivered to customers also experience failures due to mainly transient faults. Note that failure rates during useful operating lifetime are more or less constant as failures during this period is dominated by the environmentally induced faults. However, as the devices experience wearouts due to electromigration effects etc., the number of failures start to increase. As can be seen, with technology scaling the normalized device failure rates are also increasing. Moreover, useful operating lifetime is also shortening as technology scaling increases the rate or device wearouts substantially. These issues pose serious challenges to systems designers to ensure reliability in the presence of different faults.

To mitigate the effect of faults, different techniques have been employed over the years [33, 42]. These techniques are briefly outlined in the following.

Hardware Redundancy such as [28], is an effective technique, which employs extra hardware and incorporates voting from multiple outputs to a single output to mitigate the effect of transient or permanent faults. Due to usage of extra hardware resources this technique incurs area and power overheads, while achieving desired reliability. The trade-off between reliability and system overheads in hardware redundancy techniques has been extensively studied [28]. An example implementation is Maxwell's SCS750 supercomputer used in space applications, which employs triple modular redundancy (TMR) technique for its IBM's PowerPC processors [27]. Due to such redundancy, SCS750 achieved fault-tolerance with more than 180% overall hardware overheads.

Time Redundancy techniques require minimal hardware resources to detect transient faults. Upon detection of faults, task replication or re-execution is carried out during idle/slack times. Due to software-dependent nature of time redundancy techniques, the hardware overhead is generally much lower than hardware redundancy techniques. But this greatly depends on the availability of idle or slack times during computation. For example, in [6] it is demonstrated that the fault-tolerance can be improved without any impact on the execution time by utilizing idle processors for duplicating some of the computations of the active processors. However, several studies reported that these techniques cause overheads in terms of computation and communication performance [25]. Application check-pointing is another effective time redundancy technique. The fault-tolerance using such technique is achieved by selectively repeating (or rolling back) an execution interval of an application to realise fault-tolerant design. However, fault-tolerance using this technique is achieved at the cost of high complexity of application design. Examples of effective application check-pointing techniques highlighting such increased cost are, adaptive and non-uniform online application check-pointing proposed in [47], offline application check-pointing shown in [19].

Information Redundancy is another effective technique for fault detection and correction, particularly for memory devices. Using this technique fault detection and correction (EDAC) codes are integrated with the original logic circuit data [45]. These extra information codes are generated from the original logic data to effectively identify the presence of one or more transient or permanent faults and possibly correct them. Recently, Intel introduced dual-core Xeon-7100 system with several EDAC features to incorporate fault detection and correction [14].

Combination of time and information redundancy techniques, such as [16], can also be highly effective in achieving desired fault-tolerance at low cost.

System-level Techniques are highly effective as they can design hardware and software with combinations of various redundancy techniques to achieve effective fault detection and tolerance [8]. A number of system-level techniques have been proposed in past few years showing different fault-tolerance techniques to extract maximum benefit in terms of fault-tolerance and low power consumption. For example, pre-emptive online scheduling [46] of failed tasks has been demonstrated as an effective system-level fault-tolerance technique. However, the effectiveness of such technique depends upon predictability of slack times. Fault-tolerance-based optimization of cost-constrained distributed real-time systems has been proposed [21]. Fault-tolerance in [21] is achieved through system-level mapping and assignment of fault-tolerance policies to processing cores. Another approach to fault-tolerant design using process re-execution and scheduling in available hardware computing resources in an multiprocessor SoC (MPSoC) application has been proposed in [38]. Highlighting the impact of faults at application-level, various other researchers [10, 26] have shown low-cost fault-tolerance techniques. The basic principle of these works is that the faults manifested at architectural-level do not always lead to faults at application level. Exploiting the relaxed requirements of application-level correctness, low cost and energy-efficient fault-tolerance techniques have been proposed.

Indeed, energy-efficient fault-tolerant systems design is highly challenging [40]. The following chapters present state-of-the-art and also emerging issues in energy-efficient fault-tolerant systems design. Various case studies have also been illustrated, where necessary, to demonstrate effective means of addressing the design challenges.

References

1. M. Abramovici, P. Menon, A practical approach to fault simulation and test generation for bridging faults. IEEE Trans. Comptut. **C-34**, 658–663 (1985)
2. R.C. Aitken, Nanometer technology effects on fault models for ic testing. Computer **32**, 46–51 (1999)
3. B.M. Al-Hashimi (ed.), *System-on-Chip: Next Generation Electronics* (IEE, London, 2006). Ch. 17
4. F. Angiolini, P. Meloni, S.M. Carta, L. Raffo, L. Benini, A layout-aware analysis of networks-on-chip and traditional interconnects for MPSoCs. IEEE Trans. Comput. Aided Des. Integr. Circuits Syst. **26**(3), 421–434 (2007)
5. R.C. Baumann, Soft errors in advanced semiconductor devices-part i: three radiation sources. IEEE Trans. Device Mater. Reliab. **1**, 17–22 (2001)
6. H. Beitollahi, S.G. Miremadi, G. Deconinck, Fault-tolerant earliest-deadline-first scheduling algorithm, in *Proceedings of the IEEE International Parallel and Distributed Processing Symposium IPDPS*, Long Beach, 26–30 Mar 2007, pp. 1–6
7. L. Benini, G.D. Micheli, *Dynamic Power Management: Design Techniques and CAD Tools* (Springer, USA, 1997)

8. L. Benini, G. Micheli, System-level power optimization: techniques and tools. ACM Trans. Des. Autom. Electron. Syst. **5**, 115–192 (2000)
9. M.A. Breuer, Multimedia applications and imprecise computation, in *Proceedings of 8th Euromicro Conference on Digital System Design*, 2005, Porto, pp. 2–7
10. M.A. Breuer, H.H. Zhu, Error tolerance and multimedia, in *Proceedings of International Conference on Intelligent Information Hiding and Multimedia Signal Processing*, Pasadena, Dec 2006, pp. 521–524
11. T. Burd, T. Pering, A. Stratakos, R. Broderson, A dynamic voltage scaled microprocessor system. IEEE J. Solid-State Circuits **35**(11), 1571–1580 (2000)
12. S. Campagna, M. Violante, An hybrid architecture to detect transient faults in microprocessors: an experimental validation, in *Proceedings of the Design, Automation & Test in Europe Conference & Exhibition (DATE)*, Dresden, 2012, pp. 1433–1438
13. A. Chandrakasan, S. Sheng, R. Brodersen, Low power CMOS digital design. IEEE J. Solid-State Circuits **27**(4), 473–484 (1992)
14. J. Chang, M. Huang, J. Shoemaker, J. Benoit, S.-L. Chen, W. Chen, S. Chiu, R. Ganesan, G. Leong, V. Lukka, S. Rusu, D. Srivastava, The 65-nm 16-MB shared on-die l3 cache for the dual-core Intel Xeon processor 7100 series. IEEE J. Solid-State Circuits **42**(4), 846–852 (2007)
15. F. Dabiri, A. Nahapetian, M. Potkonjak, M. Sarrafzadeh, *Lecture Notes in Computer Science: Power and Timing Modeling, Optimization and Simulation*, vol. 4644, Ch. Soft Error-Aware Power Optimization Using Gate Sizing (Springer, Berlin/New York, 2007), pp. 255–267
16. A. Ejlali, B.M. Al-Hashimi, M.T. Schmitz, P. Rosinger, S.G. Miremadi, Combined time and information redundancy for SEU-tolerance in energy-efficient real-time systems. IEEE Trans. Very Larg. Scale Integr. (VLSI) Syst. **14**(4), 323–335 (2006)
17. F.J. Ferguson, J.P. Shen, Extraction and simulation of realistic cmos faults using inductive fault analysis, in *Proceedings of International Test Conference*, Washington, DC, 1988, pp. 475–484
18. C.J. Gelderloos, R.J. Peterson, M.E. Nelson, J.F. Ziegler, Pion induced soft upsets in 16 mbit dram chips. IEEE Trans. Nucl. Sci. **44**, 2237–2242 (1997)
19. J. Han, Q. Li, Dynamic power-aware scheduling algorithms for real-time task sets with fault tolerance in parallel and distributed computing environment, in *International Parallel and Distributed Processing Symposium*, Denver, 2005, pp. 6–16
20. ITRS, International technology roadmap for semiconductors, http://www.itrs.net/
21. V. Izosimov, P. Pop, P. Eles, Z. Peng, Design optimization of time-and cost-constrained fault-tolerant distributed embedded systems, in *DATE '05: Proceedings of the conference on Design, Automation and Test in Europe*, Munich (IEEE, Washington, DC, 2005) pp. 864–869
22. B.W. Johnson, *Design and Analysis of Fault-Tolerant Digital Systems* (Addison-Wesley Publishing, Reading, 1989)
23. I. Koren, C.M. Krishna, *Fault-Tolerant Systems* (Morgan-Kaufman, San Francisco, 2007)
24. P. Kudva, J. Rivers, Balancing new reliability challenges and system performance at the architecture level. IEEE Des. Test Comput. **66**, 1 (2012 in press)
25. S.-C. Lai, S.-L. Lu, K. Lai, J.-K. Peir, DITTO processor, in *Proceedings of the International Conference on Dependable Systems and Networks, DSN*, Washington, DC, 2002, pp. 525–534
26. X. Li, D. Yeung, Exploiting application-level correctness for low-cost fault tolerance. J. Instr.-Lev. Parallel. **10**, 1–28 (2008)
27. L. Longden, R. Thibodeau, C. Hillman, P. Layton, M. Dowd, Designing a single board computers for space using the most advanced processor and mitigation technologies. Eur. Space Agency Publ. ESA-SP **507**, 313–316 (2002)
28. A. Maheshwari, W. Burleson, R. Tessier, Trading off transient fault tolerance and power consumption in deep submicron (DSM) VLSI circuits. IEEE Trans. Very Larg. Scale Integr. (VLSI) Syst. **12**(3), 299–311 (2004)
29. K.C.Y. Mei, Bridging and stuck-at faults. IEEE Trans. Comptut. **C-35**, 720–727 (1974)
30. S. Mitra, N. Seifert, M. Zhang, Q. Shi, K. Kim, Robust system design with built-in soft error resilience. IEEE Comput. **38**, 43–52 (2005)

31. S. Mitra, P. Sanda, N. Seifert, Soft errors: technology trends, system effects, and protection techniques, in *Proceedings of 13th IEEE International On-Line Testing Symposium (IOLTS)*, Heraklion, 2007, p. 4
32. G. Moore, Cramming more components onto integrated circuits. Electronics **38**(8), (1965). Ex-Director, Fairchild Semiconductors
33. S. Mukherjee, *Architecture Design for Soft Errors* (Morgan Kaufmann, San Diego, 2008)
34. NASA, National aeronautics and space administration. *Scaled CMOS Technology Reliability Users Guide*. Technical report, Mar 2008
35. M. Olmos, Radiation Results of the SER Test of Actel, Xilinx and Altera FPGA instances. Technical Report, iRoC, Oct 2004
36. C. Piguet, C. Schuster, J.-L. Nagel, *Static and Dynamic Power Reduction by Architecture Selection*, vol. 4148 (Springer, Berlin/Heidelberg, 2006)
37. I. Pomeranz, S.M. Reddy, Design-for-testability for path delay fautls in large combinational circuits using test points. IEEE Trans. Comput.-Aided Des. Integr. Circuits Syst. **17**, 333–343 (1998)
38. P. Pop, K. Poulsen, V. Izosimov, P. Eles, Scheduling and voltage scaling for energy/reliability trade-offs in fault-tolerant time-triggered embedded systems, in *Proceedings of International Conference on Hardware/Software Codesign and System Synthesis (CODES+ISSS)*, Salzburg, 2007, pp. 233–238
39. J. Pouwelse, K. Langendoen, H. Sips, Dynamic voltage scaling on a low-power microprocessor, in *Proceedings of the 7th Annual International Conference on Mobile Computing and Networking*, Rome, July 2001, pp. 251–259
40. D.K. Pradhan, *Fault-Tolerant Computer System Design* (Prentice-Hall, Upper Saddle River, 1996)
41. A.K. Pramanik, S.M. Reddy, On unified delay fault testing, in *Proceedings of the International Conference on VLSI Design*, Bombay, 1993, pp. 265–268
42. M.T. Schmitz, B.M. Al-Hashimi, P. Eles, *System-Level Design Techniques for Energy-Efficient Embedded Systems* (Kluwer, Dordrecht/Boston, 2004)
43. R.D. Schrimpf, D.M. Fleetwood, *Radiation Effects and Soft Errors in Integrated Circuits and Electronic Device* (World Scientific, Singapore, 2004)
44. Soft Error in Electronic Memory – A White Paper. Tezzaron Semiconductor, Jan 2004, http://www.tezzaron.com/about/papers/
45. J. Sosnowski, Transient fault tolerance in digital systems. IEEE Micro **14**(1), 24–35 (1994)
46. Y. Zhang, K. Chakrabarty, Energy-aware adaptive checkpointing in embedded real-time systems, in *DATE '03: Proceedings of the conference on Design, Automation and Test in Europe* (IEEE, Washington, DC, 2003), pp. 10918
47. Y. Zhang, K. Chakrabarty, A unified approach for fault tolerance and dynamic power management in fixed-priority real-time embedded systems. IEEE Trans. Comput.-Aided Des. Integr. Circuits Syst. **25**(1), 111–125 (2006)
48. J.F. Ziegler, Terrestrial cosmic rays. IBM J. Res. Dev. **40**(1), 19–39 (1996). IBM Corp., USA

Chapter 2
Reliability Evaluation Techniques

Rong-Tsorng Wang

2.1 Introduction

From commercial to life-critical applications, the proliferation of computing systems in everyday life has substantially increased our dependence on them. Failures in air traffic control systems, nuclear reactors, or hospital patient monitoring systems can bring catastrophic consequences. In order to enhance the dependability of computing systems, an effective evaluation of their reliability is desired. This chapter presents methods for evaluating system reliability, and indicates that stochastic modeling has provided an effective and unified framework for analyzing various aspects of reliability. The content of this chapter is devoted to combinatorial methods, Markov models, and software models, and equally important is techniques based on the theory of stochastic processes.

Section 2.1.1 reviews some basic notation and terminology of reliability. In Sect. 2.2.1, we discuss combinatorial methods used in reliability analysis, including basic system structures, reliability block diagram, and fault tree analysis. When applying redundancy to improve system reliability, it is commonly assumed that components work independently and their intrinsic failure time distributions have no interdependence. In practical situations, however, the reliability of the system could be significantly affected by components (or subsystems) with dependency, which often occurs in three ways among others: components subjected to the same stresses, components sharing the same load, or components failing with common causes [11]. In each case, the random variables of interest tend to act similarly – degrade the performance and promote the system failure. For example, parallel systems such as the pair of engines of an aeroplane have components whose lifetimes are dependent. Thus, it is more realistic to assume certain forms of positive dependence among components.

R.-T. Wang (✉)
Department of Statistics, Tunghai University, Taichung, Taiwan
e-mail: rtwang@thu.edu.tw

J. Mathew et al. (eds.), *Energy-Efficient Fault-Tolerant Systems*,
DOI 10.1007/978-1-4614-4193-9_2,

In system reliability analysis, dependence concepts of multivariate models for nonrepairable system have been studied by many authors, e.g., Barlow and Proschan [11], Joe [43], and Balakrishnan and Lai [10]. For bivariate cases, for example, the emphasis is either on the relationship between the joint reliability function $P(T_1 > t_1, T_2 > t_2)$ and the product of marginal reliability functions $P(T_1 > t_1)P(T_2 > t_2)$, or on the conditional reliability functions $P(T_2 > t_2 | T_1 > t_1)$. Balakrishnan and Lai [10] investigate various notions of positive dependence between two random variables. In particular, they study dependence concepts that are stronger (or weaker) than *positive quadrant dependence* (PQD), which appears as a straightforward concept of positive dependence and is given by $P(T_1 \geq t_1, T_2 \geq t_2) \geq P(T_1 \geq t_1) \cdot P(T_2 \geq t_2)$.

One technique for characterizing a system having dependent component failure rates is to apply the multi-version of exponential distributions. The most cited bivariate exponential distribution is the one proposed by Marshall and Olkin [67], known as the BVE distribution. The BVE distribution enable us to model common-cause failures, and it is obtained by considering a model in which two components fail separately or simultaneously upon receiving a shock that is governed by a homogeneous Poisson process. Freund [29] proposed a bivariate extension of the exponential distribution by allowing the failure rate of the surviving component to be affected after the failure of another component. Freund's model is one of the first to study bivariate distributions from reliability considerations, and it can be used to model load-sharing systems. In Sect. 2.2.2, we extend the BVE and Freund's model to n-components redundant systems for applying redundant system with common-cause and load-sharing failures.

Fault tree analysis has been used in analyzing safety systems in nuclear power plants, aerospace, and defense. However, dependencies of various types that do occur in practice are not easily captured by fault tree models. In contrast, Markov model is powerful in that it can solve system with dynamic and dependent behaviors, although Markov model has the significant disadvantage that its size grows exponentially as the size of the system increases. From the counting process point of view, the birth process which counts the number of failed components provides a simple representation for the failure process of a system. This approach, however, does not help us to analyze the condition when component lifetimes are dependent. Thus a more general model is required.

In Sect. 2.3, we discuss Markov chains, Poisson processes and nonhomogeneous Poisson processes, and birth-death processes. Continuous-time Markov chains can be used to model both nonrepairable and repairable system. By definition a repairable system is a system which, after failing to perform one or more of its functions satisfactorily, can be restored to fully satisfactory performance by a method other than replacement of the entire system [8]. Commonly used models for a repairable system are renewal processes (perfect repair) and nonhomogeneous Poisson processes (minimal repair) [54].

In Sect. 2.4, we discuss the software reliability models. We aim to extend common assumptions used in statistical software reliability, classify existing software reliability models, and propose mathematical models for describing software reliability. We begin with the assumptions deduced from Jelinski and Moranda model [42]. Then, we show relations between the transition failure rates of the failure process and the conditional failure rates of times between failures, and specify several common types of software reliability models. A self-exciting and mixture model is then proposed as a generalization of perfect debugging models.

Section 2.5 discusses three examples. In Sect. 2.5.1, we consider a load-sharing system with common-cause failures. Specifically, we formulate reliability properties by introducing a continuous-time Markov chain to characterize systems in which component lifetimes are dependent in two ways: common-cause failures and increased failure rates for surviving components. In Sect. 2.5.2 we discuss a model for a shared-load k-out-of-n:G repairable system studied by Shao and Lamberson [97]. Specifically, a continuous-time Markov chain is used to create a set of differential equations. Section 2.5.3 reviews a bivariate birth-death model for a N-version programming fault-tolerant system proposed by Wang [115].

Effective and optimal techniques may be addressed according to system structures (e.g., hardware or software or integrated systems, binary-state or multi-state systems, parallel or k-out-of-n systems, individual or network systems), failure mechanisms (e.g., independent or common-cause failures), and repair patterns (e.g., repairable or non-repairable, perfect or minimal or general repair) [11, 22, 48, 64, 69, 71, 76, 78, 84, 85, 92, 100, 111]. Dependence concepts of Markov dependence, stationarity, and Martingales have been applied for system failure characterizations. Recent researches indicate that the modern theory of stochastic processes allows for developing a general reliability model that incorporates time dynamics and different information levels. More advanced issues can be found in for examples, Singpurwalla [99], Peña [82], and Aalen et al. [1].

2.1.1 *Definitions of Reliability*

The lifetime of a component is usually unknown and is characterized by a non-negative random variable (r.v.), say T, representing the time to failure of a system since its inception. Usually T is measured from time $t = 0$. By definition, *reliability* is the probability of a device performing its purpose adequately for the period of time intended under the operating conditions encountered [11]. (Failures can be defined in terms of degradation reaching a certain level, see, e.g., Meeker and Escobar [69].) Let $F(t) = P(T \leq t)$ be the cumulative distribution function (c.d.f.) of T such that $F(0-) = 0$ and $F(+\infty) = 1$. The *reliability function* (or survival function) of T is defined by

$$R(t) := \bar{F}(t) = P(T > t), \quad \forall t \geq 0. \tag{2.1}$$

For example, the reliability required for aircraft control systems has been specified as $R(t) > 1 - 10^{-9}$ for $t = 1$ h. The corresponding conditional reliability function (or conditional survival function) of T of age x is given by

$$R(t|x) := \frac{\bar{F}(x+t)}{\bar{F}(x)}, \quad \forall t, x \geq 0,$$

provided that $\bar{F}(x) > 0$. The expected time of T, also known as the *mean time to failure* (MTTF), is given by

$$E[T] = \int_0^\infty \bar{F}(t)dt = \int_0^\infty R(t)dt. \tag{2.2}$$

In addition, the second moment of T is given by

$$E[T^2] = 2\int_0^\infty t \cdot R(t)dt.$$

The *mean residual life* of T of age x is given by

$$E[T - x|T > x] = \frac{\int_x^\infty \bar{F}(t)dt}{\bar{F}(x)} = \int_0^\infty \frac{\bar{F}(t+x)dt}{\bar{F}(x)} = \int_0^\infty R(t|x)dt.$$

Equation (2.1) may or may not continuous in t. Suppose that Eq. (2.1) is everywhere differentiable, then the probability density function (p.d.f.) $f(t)$ of T exists for all $t \geq 0$, and

$$f(t) \cdot \Delta t \approx \Pr(t \leq T < t + \Delta t),$$

where Δt is a small increment of t. The *failure rate* (or conditional failure rate) of the r.v. T at time t is defined as

$$h(t) := \lim_{\Delta t \to 0+} \frac{1}{\Delta t} \Pr(t < T \leq t + \Delta t | T > t), \qquad t \geq 0. \tag{2.3}$$

$h(t)$ gives the probability that a component of age t will almost immediately fail. Since, for those t where $h(t)$ is continuous,

$$h(t) = \frac{f(t)}{\bar{F}(t)}, \quad \forall t \geq 0. \tag{2.4}$$

$h(t)$ is also known as the hazard rate, the force of mortality, and the intensity rate. The functions $f(t)$, $F(t)$, $\bar{F}(t)$, and $h(t)$ are mathematically equivalent, since

$$\int_0^t h(x)dx = -\ln \bar{F}(t),$$

$$\bar{F}(t) = \exp\left\{-\int_0^t h(x)dx\right\}.$$

The function $\int_0^t h(x)dx$ is called the *cumulative hazard rate* function. Many non-normal univariate distributions have been used in reliability, such as exponential, Gamma, Weibull and so forth. In particular, these distributions are applied to three phases of the so-called bathtub curve: the infant mortality phase, the normal operating phase, and the wear-out phase. The most commonly used distribution is the exponential distribution.

Example 2.1. If the failure rate is given by $h(t) = \beta^2 t/(1 + \beta t)$, where $\beta > 0$, $t > 0$. Then

$$\bar{F}(t) = \exp\left\{-\int_0^t h(x)dx\right\} = \exp\left(-\int_0^t \left[\beta - \frac{\beta}{1+\beta x}\right] dx\right) = (1 + \beta t)e^{-\beta t}.$$

Observed that T has a gamma distribution. The MTTF is

$$E[T] = \int_0^\infty \bar{F}(t)dt = \int_0^\infty (1 + \beta t)e^{-\beta t} = \frac{2}{\beta}.$$

Example 2.2. If T has a Weibull distribution with *shape* parameter $\beta > 0$ and *scale* parameter $\eta > 0$, whose c.d.f. is given by

$$F(t) = 1 - \exp\left[-\left(\frac{t}{\eta}\right)^\beta\right], \quad t > 0.$$

The failure rate is

$$h(t) = \frac{d}{dt}\left[-\ln \bar{F}(t)\right] = \frac{\beta}{\eta}\left(\frac{t}{\eta}\right)^{\beta-1}.$$

The hazard rate is increasing for $\beta > 1$, decreasing for $0 < \beta < 1$, and constant for $\beta = 1$.

2.2 Combinatorial Methods

System failure is modeled in terms of the failures of the components of the system. Both the system and its components are often allowed to take only two possible states: a working state and a failed state. The linking of component failures to system failures can be understood in several ways. Among these are the *reliability block diagram* (success-oriented) and the *fault tree analysis* (failure-oriented). In some cases, it is possible to convert the fault tree to a reliability block diagram, and vice versa. In Sect. 2.2.1, we review basic system structures, reliability block diagram, and fault tree analysis.

Parallel redundancy is a common method to increase system reliability and MTTF. Usually components in a reliability structure are dependent as they may share the same load or may be failed with common-cause failures. Bivariate and multivariate lifetime distributions play important roles in modeling these dependencies. Many bivariate and multivariate exponential distributions have been proposed [10], however, only a certain class of multivariate distributions is applicable for reliability systems. The BVE distribution of Marshall and Olkin [67] is suited for modeling common-cause failures. Freund's model [29] can be applied to the situation that the failed component increases the stress on the surviving component and consequently increases the other component's tendency of failure. In Sect. 2.2.2, we extend the BVE and Freund's distributions to n-components redundant systems, and discuss the reliability functions of series, parallel, and k-out-of-n:F structures.

2.2.1 *Independent Components*

2.2.1.1 Three Basic Structures

Consider a system with n independent components, and the system reliability $R(t)$ can be determined by components' reliability $R_i(t)$. We write $R(t)$ a function of $R_1(t), \ldots, R_n(t)$ as

$$R(t) = \psi(R_1(t), \ldots, R_n(t)).$$

The function ψ is decided by the structure of the system. We often use *reliability block diagram* or *network diagram* to describe the system reliability structure. Let $T_1, \ldots, T_n$ be the lifetimes of components. Assume that $T_1, \ldots, T_n$ are independent, there are three common structures:

(1) series: $$R(t) = \prod_{i=1}^{n} R_i(t), \tag{2.5}$$

(2) parallel: $$R(t) = 1 - \prod_{i=1}^{n}(1 - R_i(t)), \tag{2.6}$$

(3A) k-out-of-n:F: $$R(t) = \sum_{i=0}^{k-1} \binom{n}{i} [R_1(t)]^{n-i}[1 - R_1(t)]^i, \quad \text{if } R_i(t) = R_1(t), \tag{2.7}$$

(3B) k-out-of-n:G: $$R(t) = \sum_{i=k}^{n} \binom{n}{i} [R_1(t)]^{i}[1 - R_1(t)]^{n-i}, \quad \text{if } R_i(t) = R_1(t). \tag{2.8}$$

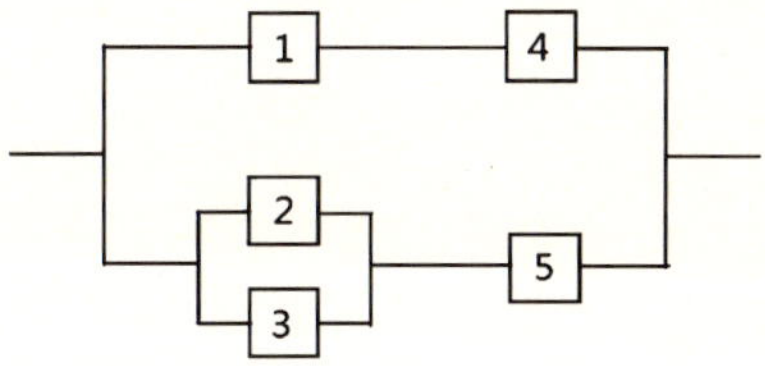

Fig. 2.1 A reliability block diagram of a complex system

A series system functions if and only if each component functions, and a parallel system functions if and only if at least one component functions. A system is said to be a k-out-of-n:F system if it fails if and only if at least k of the n components fail. A dual concept called k-out-of-n:G system is defined as that it is good if and only if at least k of n components are good. A system is a k-out-of-n:F system if and only if it is a $(n-k+1)$-out-of-n:G system. Likewise, a system is a k-out-of-n:G system if and only if it is a $(n-k+1)$-out-of-n:F system. Apparently,

$$\text{a system is 'series'} \Longleftrightarrow \text{it is a 1-out-of-}n\text{:F system (or } n\text{-out-of-}n\text{:G)},$$

$$\text{a system is 'parallel'} \Longleftrightarrow \text{it is an } n\text{-out-of-}n\text{:F system (or 1-out-of-}n\text{:G)}.$$

The reliability of a k-out-of-n:F system in (2.7) with independently and identically distributed (i.i.d.) components is equal to the probability that the number of failing components is less than or equal to $k-1$. As a k-out-of-n:F system is equivalent to a $(n-k+1)$-out-of-n:G system, (2.7) is equivalent to

$$\sum_{j=n-k+1}^{n} \binom{n}{j} [R_1(t)]^j [1 - R_1(t)]^{n-j}.$$

If we denote $R_G(n,k)$ the reliability of a k-out-of-n:G system and $R_F(n,k)$ the reliability of a k-out-of-n:F system, we have

$$R_G(n,k) = R_F(n, n-k+1).$$

2.2.1.2 Reliability Block Diagram

A reliability block diagram is a success-oriented diagram describing a specified function of a system. Each rectangular block in the diagram represents a function of a component. If the function is available, we have connection through the block, and if the function is failed, there is no connection through the block.

Consider the reliability block diagram in Fig. 2.1. The system is a parallel structure of two independent modules: the structure comprising components 1 and 4,

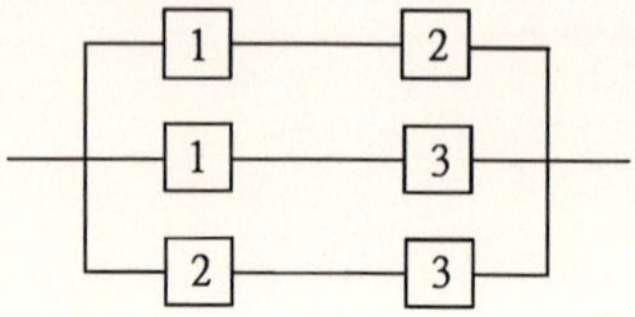

Fig. 2.2 A reliability block diagram for a 2-out-of-3:G system

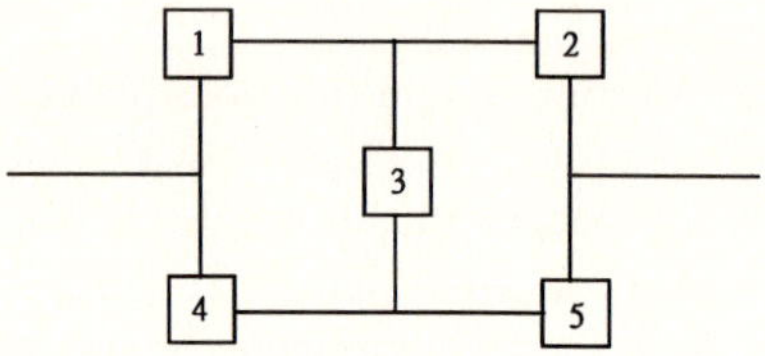

Fig. 2.3 A bridge structure

and the structure comprising components 2, 3, and 5. From (2.5) and (2.6), the reliability of the former structure equals $R_1 R_4$, whereas the reliability of the latter equals $[1 - (1 - R_2)(1 - R_3)]R_5$. Thus the system reliability is given by

$$R_{system} = 1 - (1 - R_1 R_4)\{1 - [1 - (1 - R_2)(1 - R_3)]R_5\}.$$

A reliability block diagram of 2-out-of-3:G system is illustrated in Fig. 2.2.

2.2.1.3 A Bridge Structure

The reliability block diagram of a bridge configuration is given in Fig. 2.3. Each block in the diagram denotes a component. For independent components, the system reliability of the bridge structure can be obtained by using *minimal path* or *minimal cut sets* [11] and is expressed by

$$\begin{aligned} R_{system} &= R_1 R_2 + R_4 R_5 + R_1 R_3 R_5 + R_2 R_3 R_4 \\ &\quad - R_1 R_2 R_3 R_4 - R_1 R_2 R_3 R_5 - R_1 R_2 R_4 R_5 - R_1 R_3 R_4 R_5 \\ &\quad - R_2 R_3 R_4 R_5 + 2 R_1 R_2 R_3 R_4 R_5. \end{aligned}$$

The minimal path sets of the bridge structure are $\{1,2\}$, $\{4,5\}$, $\{1,3,5\}$, and $\{2,3,4\}$. The minimal path sets may represent the bridge as a parallel-series structure. The minimal cut sets are $\{1,4\}$, $\{2,5\}$, $\{1,3,5\}$, and $\{2,3,4\}$. The minimal cut sets may represent the bridge as a series-parallel structure. Alternative method to obtain the system reliability of a bridge structure is the *pivot-decomposition method* (or factoring method) [123]. The minimal path sets and minimal cut sets for the bridge structure are given in Fig. 2.4.

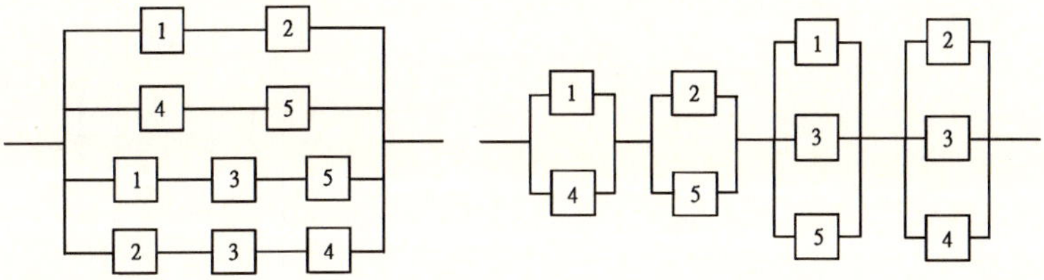

Fig. 2.4 Minimal path and minimal cut representations for bridge

2.2.1.4 A Non-i.i.d. *k*-out-of-*n*:F System

The reliability of a non-i.i.d. k-out-of-n:F system is given by

$$R(t) = \sum_{i=0}^{k-1} \sum_{j=1}^{\binom{n}{i}} \prod_{m=1}^{n} R_m^{\delta_{m,j}} (1 - R_m)^{\bar{\delta}_{m,j}}, \quad k = 1, \ldots, n,$$

where $\delta_{m,j}$ and $\bar{\delta}_{m,j}$ are complementary indicator functions for which

$$\sum_{m=1}^{n} \delta_{m,j} = n - k, \qquad \sum_{m=1}^{n} \bar{\delta}_{m,j} = k,$$

and

$$\delta_{m,j} = \begin{cases} 1 & \text{if the } m\text{th component functions} \\ 0 & \text{if the } m\text{th component fails.} \end{cases}$$

Myers [77] has studied extensively about k-out-of-n:G systems of this type.

Example 2.3. The reliability of a non-i.i.d. 3-out-of-4:F system is given by

$$\begin{aligned} R &= \sum_{i=0}^{2} \sum_{j=1}^{\binom{4}{i}} \prod_{m=1}^{4} R_m^{\delta_{m,j}} (1 - R_m)^{\bar{\delta}_{m,j}} \\ &= R_1R_2R_3R_4 + R_1R_2R_3(1-R_4) + R_1R_2R_4(1-R_3) + R_1R_3R_4(1-R_2) \\ &\quad + R_2R_3R_4(1-R_1) + R_1R_2(1-R_3)(1-R_4) + R_1R_3(1-R_2)(1-R_4) \\ &\quad + R_1R_4(1-R_2)(1-R_3) + R_2R_3(1-R_1)(1-R_4) \\ &\quad + R_2R_4(1-R_1)(1-R_3) + R_3R_4(1-R_1)(1-R_2). \end{aligned}$$

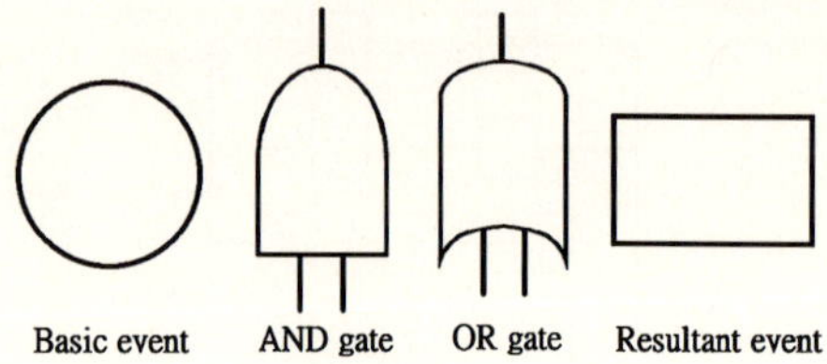

Fig. 2.5 Common fault tree symbols

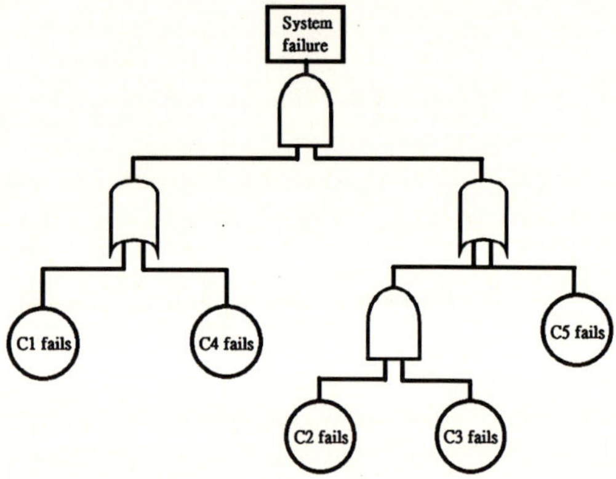

Fig. 2.6 A fault tree diagram

2.2.1.5 Fault Tree Analysis

Fault tree analysis is a widely used method in the industrial sector to perform reliability analysis of systems during their design and development, particularly in system safety analysis. A *fault tree diagram* is the underlying top-down graphical model in fault tree analysis. The starting point of a fault tree is a system failure, known as a *top event*. Fault events that can cause the occurrence of the top event are generated and connected by logic operators such as ‘AND’ and ‘OR’. The status of output/top event can be derived by the status of input events and the connections of the logical gates. The logical meaning of AND(OR) gate is the same as that of the intersection(union) of events. That is, the AND gate provides a True output (fault) if all its inputs are True (fault) and the OR gate provides a True output (fault) if one or more of its inputs are True (fault). In other words, an AND gate in a fault tree is logically equivalent to a parallel reliability block diagram, and an OR gate corresponds to a series reliability block diagram. The fault tree is developed by repeatedly asking which combinations of the component failures will result in a system failure. Fault events are generated successively until the fault events need not be developed any further. These lowest level causes are called *basic events*.

Four commonly used symbols in the construction of fault trees are shown in Fig. 2.5. A fault tree diagram corresponding to Fig. 2.1 is shown in Fig. 2.6.

2.2.1.6 Order Statistics

A r.v. T with d.f. F is said to be exponential distributed with parameter λ, denoted by $T \sim \text{Exp}(\lambda)$ or $F \sim \text{Exp}(\lambda)$, if the reliability function is given by $R(t) = e^{-\lambda t}$ for $t > 0, \lambda > 0$.

Suppose that $T_1, \ldots, T_n$ are i.i.d. r.v.'s with d.f. F. We index the failure times chronologically by $0 \leq T_{1:n} \leq \cdots \leq T_{n:n}$, where $T_{k:n}$ denotes the occurrence time of the kth failure, for $k = 1, \ldots, n$. That is, $T_{1:n}, \ldots, T_{n:n}$ are the corresponding *order statistics* of $T_1, \ldots, T_n$. We also denote the *spacings* by, for $k = 1, \ldots, n$, $D_k = T_{k:n} - T_{k-1:n}$, where $T_{0:n} = 0$. That is, D_k denotes the time between the $(k-1)$th and the kth failure. The reliability of a system can then be studied through the order statistics and spacings as an *general order statistics* model [90]. In particular, if $F \sim \text{Exp}(\lambda)$ the model is called an i.i.d. *exponential order statistics* [70]. Moreover the kth order statistic $T_{k:n}$ can be considered as the time of failure of a k-out-of-n:F system [11]. In the case that $T_1, \ldots, T_n$ are i.i.d. r.v.'s with d.f. F, the reliability function of a k-out-of-n:F system at time t is given by

$$P(T_{k:n} > t) = \sum_{i=0}^{k-1} \binom{n}{i} [F(t)]^i [1 - F(t)]^{n-i}. \tag{2.9}$$

Example 2.4. Let $T_1, \ldots, T_n$ be i.i.d. r.v.'s with d.f. F. If $F \sim Exp(\lambda)$, then the spacings D_k are independent exponential distributions with parameters $(n-k+1)\lambda$, for $k = 1, \ldots, n$ [11]. Since $T_{k:n} = \sum_{i=1}^{k} D_i$, the p.d.f. of $T_{k:n}$ is given by

$$f_{T_{k:n}}(t) = \beta_1 \ldots \beta_k \sum_{i=1}^{k} \left[\Psi_{i,k} e^{-\beta_i t}\right], \tag{2.10}$$

where

$$\begin{aligned}
\beta_i &= (n - i + 1)\lambda, \\
\Psi_{i,k}^{-1} &= (\beta_1 - \beta_i) \ldots (\beta_{i-1} - \beta_i)(\beta_{i+1} - \beta_i) \ldots (\beta_k - \beta_i).
\end{aligned}$$

The distributions of $T_{k:n}$ are Erlang distributions. The reliability functions of $T_{k:n}$ are more conveniently obtained from Eq. (2.9) than from the p.d.f. (2.10). Namely, for $k = 1, \ldots, n$,

$$P(T_{k:n} > t) = \sum_{i=0}^{k-1} \binom{n}{i} (1 - e^{-\lambda t})^i (e^{-\lambda t})^{n-i}.$$

In particular, the reliability functions of series and parallel systems are, respectively,

$$P(T_{1:n} > t) = e^{-n\lambda t} \quad \text{and} \quad P(T_{n:n} > t) = 1 - (1 - e^{-\lambda t})^n.$$

2.2.1.7 Linear Combinations of Reliability of Series Systems

Consider a parallel system of n components. Let T and T_i be the lifetimes of the parallel system and the component i, respectively, and consider the events $A = \{T > t\}$ and $A_i = \{T_i > t\}$. Then $A = \bigcup_{i=1}^n A_i$. Apparently, the sum of the component reliabilities gives an upper bound of the parallel system reliability

$$P\left(\bigcup_{i=1}^n A_i\right) \le \sum_{i=1}^n P(A_i).$$

To evaluate $P(A)$, apart from (2.6), we consider the inclusion-exclusion identity:

$$P(A) = \sum_{i=1}^n P(A_i) - \sum_{i<j} P(A_i \cap A_j) + \sum_{i<j<k} P(A_i \cap A_j \cap A_k) - \cdots + (-1)^{n+1} P\left(\bigcap_{i=1}^n A_i\right). \quad (2.11)$$

Here

$$P(A_{i_1} \cap A_{i_2} \cap \cdots \cap A_{i_k}) = P(T_{i_1} > t, T_{i_2} > t, \ldots, T_{i_k} > t) = P(U > t),$$

where U is the lifetime of the series system of components $i_1, i_2, \ldots, i_k$. Note that the series system reliability has a lower bound

$$P\left(\bigcap_{i=1}^n A_i\right) \ge \sum_{i=1}^n P(A_i) - (n-1),$$

known as *Bonferroni's inequality*.

Equation (2.11) gives that the reliability of a parallel system can be expressed as a linear combination of the reliability of all possible series systems. Taking integration in (2.11) and applying (2.2), we have that the MTTF of parallel systems can be found as a linear combination of the MTTF of all possible series systems. Rade [89] gives the following theorem.

Theorem 2.1. *Let μ be the MTTF of a parallel system of n components and let $\mu_{i_1 i_2 \ldots i_k}$ be the MTTF of the series system of components $i_1, i_2, \ldots, i_k$, assuming the joint distribution of the components' lifetime is the same as in the parallel system. Then*

$$\mu = \sum_{i=1}^n \mu_i - \sum_{i<j} \mu_{ij} + \sum_{i<j<k} \mu_{ijk} - \cdots + (-1)^{n+1} \mu_{12\ldots n}. \quad (2.12)$$

If all series systems of k components have the same MTTF $\mu_{(k)}$, then

$$\mu = \binom{n}{1}\mu_{(1)} - \binom{n}{2}\mu_{(2)} + \binom{n}{3}\mu_{(3)} - \cdots + (-1)^{n+1}\binom{n}{n}\mu_{(n)}. \tag{2.13}$$

Rade [89] also generalizes Theorem 2.1 to a series-parallel system and a k-out-of-n:G system.

Example 2.5. Let the lifetimes $T_1, T_2, \ldots, T_n$ of the components be independent and exponentially distributed with parameters $\lambda_1, \lambda_2, \ldots, \lambda_n$. Then (2.12) gives

$$\mu = \sum_{i=1}^{n} \frac{1}{\lambda_i} - \sum_{i<j} \frac{1}{\lambda_i + \lambda_j} + \cdots + (-1)^{n+1} \frac{1}{\lambda_1 + \cdots + \lambda_n}.$$

In this case, the lifetime of the series system of components with intensities $\lambda_1, \lambda_2, \ldots, \lambda_k$ has an exponential distribution with parameter $\lambda_1 + \cdots + \lambda_k$ and mean $1/(\sum_{j=1}^{k} \lambda_j)$. If, furthermore, all λ_i are equal to λ, the formula (2.13) gives

$$\mu = \frac{1}{\lambda} \sum_{j=1}^{n} \frac{(-1)^{j+1}\binom{n}{j}}{j} = \frac{1}{\lambda} \sum_{j=1}^{n} \frac{1}{j}.$$

A direct calculation from (2.6) yields

$$\mu = \int_0^\infty \left[1 - (1 - e^{-\lambda t})^n\right] dt = \frac{1}{\lambda} \sum_{j=1}^{n} \frac{1}{j}.$$

For large n, μ is approximately $\frac{1}{\lambda}(\ln n + \gamma)$, where $\gamma = 0.577215664\ldots$ is Euler's constant. If the system is a k-out-of-n:G system, the mean time to failure of the system is

$$\int_0^\infty \left[\sum_{j=k}^{n} \binom{n}{j} e^{-j\lambda t}(1 - e^{-\lambda t})^{n-j} \right] dt = \frac{1}{\lambda} \sum_{j=k}^{n} \frac{1}{j}.$$

2.2.2 *Dependent Components*

2.2.2.1 The Model of Marshall and Olkin

Marshall and Olkin [67] introduced a bivariate exponential distribution (BVE) by considering a reliability model in which two components fail separately or simultaneously upon receiving a shock that is governed by a homogeneous Poisson process. They derived the BVE in several ways: the bivariate lack of memory property, shock models, a random sum model, and a minima model. The BVE distribution is the most important bivariate exponential distribution. It contains many properties which inspire intensive investigations of the bivariate exponential distributions.

We say that (T_1, T_2) has the BVE distribution if the joint reliability function is given by, for $t_1 \geq 0$ and $t_2 \geq 0$,

$$\bar{F}(t_1, t_2) = e^{-\lambda_1 t_1 - \lambda_2 t_2 - \lambda_{12} \max(t_1, t_2)},$$

where $\lambda_1, \lambda_2, \lambda_{12}$ are non-negative parameters such that $\lambda_1 + \lambda_{12} > 0$, $\lambda_2 + \lambda_{12} > 0$. T_1 and T_2 are independent when $\lambda_{12} = 0$. For $0 < t_1 < t_2$ and $0 < t_2 < t_1$ the density can be obtained by $\partial^2 \bar{F}(t_1, t_2)/\partial t_1 \partial t_2$, and then for $t > 0$

$$P(t < T_1 < T_2) = \frac{\lambda_1}{\lambda} e^{-\lambda t},$$

$$P(t < T_2 < T_1) = \frac{\lambda_2}{\lambda} e^{-\lambda t},$$

where $\lambda = \lambda_1 + \lambda_2 + \lambda_{12}$. It follows that $P(T_1 = T_2 > t) = \frac{\lambda_{12}}{\lambda} e^{-\lambda t}$. So (T_1, T_2) has density f on the region $\{(t_1, t_2) : t_1 > 0, t_2 > 0, t_1 \neq t_2\}$ with respect to two-dimensional Lebesgue measure, and a density g on the line $\{(t, t) : t > 0\}$ in the sense that $P(T_1 = T_2 \leq t) = \int_0^t g(u) du$ with respect to one-dimensional Lebesgue measure. Density functions f and g are given by

$$f(t_1, t_2) = \begin{cases} \lambda_1(\lambda_2 + \lambda_{12}) e^{-\lambda_1 t_1 - (\lambda_2 + \lambda_{12}) t_2}, & 0 < t_1 < t_2 \\ \lambda_2(\lambda_1 + \lambda_{12}) e^{-(\lambda_1 + \lambda_{12}) t_1 - \lambda_2 t_2}, & 0 < t_2 < t_1, \end{cases} \tag{2.14}$$

and

$$g(t) = \lambda_{12} e^{-\lambda t}. \tag{2.15}$$

In addition, $P(T_1 = T_2) = \int_0^\infty g(t) dt = \frac{\lambda_{12}}{\lambda} \geq 0$.

The Laplace transform $\psi(s, t) = E(e^{-sT_1 - tT_2})$ is given by, from (2.14) and (2.15),

$$\psi(s, t) = \frac{(\lambda_1 + \lambda_{12})(\lambda_2 + \lambda_{12})(\lambda + s + t) + st\lambda_{12}}{(s + t + \lambda)(s + \lambda_1 + \lambda_{12})(t + \lambda_2 + \lambda_{12})}.$$

It follows that the marginal expectations, variances, and correlation coefficient can be obtained. In particular, the correlation coefficient $\rho(T_1, T_2) = \frac{\lambda_{12}}{\lambda} \geq 0$.

For the n-dimensional multivariate exponential distribution (MVE), the joint reliability function is given by

$$\bar{F}(t_1, \ldots, t_n) = \exp\left\{ -\left[\sum_{i=1}^n \lambda_i t_i + \sum_{i<j}^n \lambda_{ij} \max(t_i, t_j) + \sum_{i<j<k}^n \lambda_{ijk} \max(t_i, t_j, t_k) + \cdots + \lambda_{1 \ldots n} \max(t_1, \ldots, t_n) \right] \right\}, \tag{2.16}$$

where $t_i \geq 0$ and the $2^n - 1$ different λ-symbols are non-negative constants. By setting $t_i = 0$, we have a $(n-1)$-dimensional MVE distribution. Hence the marginal distributions of all order are MVE. For example,

$$\bar{F}(t_1, t_2, 0) = \exp\left\{-\left[\sum_{i=1}^{2}(\lambda_i + \lambda_{i3})t_i + (\lambda_{12} + \lambda_{123}) \max(t_1, t_2)\right]\right\}$$

is a BVE.

We next evaluate the reliability functions for a k-out-of-n:F system whose components lifetimes are characterized by the MVE. Recall in Sect. 2.2.1.2 that the time of failure of a k-out-of-n:F system may be represented in terms of the kth order statistic, i.e., $R_{k:n}(t) = P(T_{k:n} > t)$. The system reliability functions for 1-out-of-2:F and 2-out-of-2:F are $R_{1:2}(t) = e^{-\lambda t}$ and $R_{2:2}(t) = e^{-(\lambda_1+\lambda_{12})t} + e^{-(\lambda_2+\lambda_{12})t} - e^{-\lambda t}$, respectively, where $\lambda = \lambda_1 + \lambda_2 + \lambda_{12}$. The MVE distribution gives a formidably large task to get the system reliability functions, as there are $2^n - 1$ different $\lambda's$ in Eq. (2.16). For simplicity, we consider the case of identical component lifetime distributions. Let

$$\begin{aligned}
\lambda_i &= \beta_1, \quad i = 1, \ldots, n, \\
\lambda_{ij} &= \beta_2, \quad i, j = 1, \ldots, n,\ i \neq j, \\
&\ \vdots \\
\lambda_{1\ldots n} &= \beta_n,
\end{aligned}$$

then the joint reliability function (2.16) becomes

$$\begin{aligned}
\bar{F}(t_1, \ldots, t_n) = \exp\Bigg\{-\Bigg[\beta_1 \sum_{i=1}^{n} t_i + \beta_2 \sum_{i<j}^{n} \max(t_i, t_j) \\
+\beta_3 \sum_{i<j<k}^{n} \max(t_i, t_j, t_k) + \cdots + \beta_n \max(t_1, \ldots, t_n)\Bigg]\Bigg\}. \quad (2.17)
\end{aligned}$$

The system reliability functions for 1-out-of-n:F and n-out-of-n:F are, respectively,

$$R_{1:n}(t) = \bar{F}(t, \ldots, t) = \exp\left\{-\left[\binom{n}{1}\beta_1 + \binom{n}{2}\beta_2 + \cdots + \binom{n}{n}\beta_n\right]t\right\},$$

and

$$\begin{aligned}
R_{n:n}(t) &= P\left(\bigcup_{k=1}^{n}\{T_k > t\}\right) \\
&= \sum_{k=1}^{n} P(T_k > t) - \sum_{1 \leq i_1 \leq i_2 \leq n} P(T_{i_1} > t, T_{i_2} > t) + \cdots + (-1)^{n-1} P\left(\bigcap_{k=1}^{n}\{T_k > t\}\right).
\end{aligned}$$

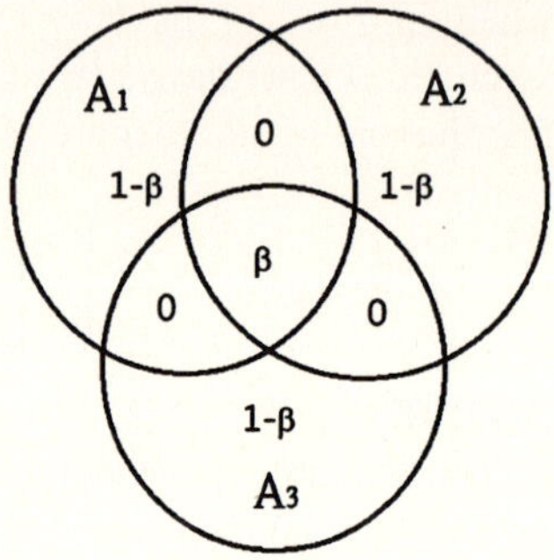

Fig. 2.7 Failure events for a system with three identical components when using the beta-factor model

From (2.17), for $k = 1, \ldots, n$, we have

$$\begin{aligned}
&P(T_1 > t, \ldots, T_k > t) \\
&\quad = P(T_1 > t, \ldots, T_k > t, T_{k+1} = 0, \ldots, T_n = 0) \\
&\quad = \exp\left\{-\left[\beta_1 kt + \beta_2\left(\binom{n}{2} - \binom{n-k}{2}\right)t + \beta_3\left(\binom{n}{3} - \binom{n-k}{3}\right)t + \cdots + \beta_n t\right]\right\} \\
&\quad = \exp\left\{-\sum_{i=0}^{n-1}\sum_{j=1}^{k}\binom{n-j}{i}\beta_{i+1}t\right\},
\end{aligned}$$

where we use the equality

$$\sum_{j=1}^{k}\binom{n-j}{r} = \binom{n}{r+1} - \binom{n-k}{r+1}, \qquad r = 1, \ldots, n-1,$$

where $\binom{x}{y} = 0$ if $x < y$. Hence

$$R_{n:n}(t) = \sum_{k=1}^{n}(-1)^{k-1}\binom{n}{k} \cdot \exp\left\{-\sum_{i=0}^{n-1}\sum_{j=1}^{k}\binom{n-j}{i}\beta_{i+1}t\right\}.$$

Reliability modeling of common cause failures was introduced in the nuclear power industry more than 30 years ago [108]. According to NEA [79] a common cause failure is defined as "a dependent failure in which two or more component fault states exist simultaneously, or within a short time interval, and are a direct result of a shared cause." Common cause failures are especially important for redundant components, and they may be classified in two main types: (i) multiple failures that occur at the same time due to a common cause, and (ii) multiple failures that occur due to a common cause, but not necessarily at the same time.

The *beta-factor model* is the most commonly used common cause failure model. Figure 2.7 gives failure events for a system with three identical components when using the beta-factor model. Extensions of the beta-factor model have been suggested, for example, the binomial failure rate model, the alpha factor model, the multiple Greek letter model, and the multiple beta-factor model [38].

As for time-varying failure rates, there are considerable studies in the topic of common cause failures for the system reliability modeling and analysis, such as models apply to systems with exponential time-to-failure distributions [7, 17, 28]; components being affected by at most a single common-cause [105, 110]; a single common-cause affects all components of a system [5]. Zhang and Horigome [125] present a method for analyzing availability & reliability of repairable systems with common cause failures among components. Xing [117] studies the reliability analysis of computer network systems by allowing for multiple common cause that can affect different subsets of system components.

2.2.2.2 Freund's Model

Freund [29] proposed a bivariate extension of the exponential distribution by allowing the failure rate of the surviving component to be affected after the failure of another component. Freund's bivariate distribution is absolutely continuous and possesses the bivariate lack of memory property [12]. Suppose components T_1 and T_2 follow independently exponential distributions with parameters α_1 and α_2, respectively, until one of components fails. When component $T_1(T_2)$ has failed, the parameter of the surviving component $T_2(T_1)$ changes to $\alpha_2'(\alpha_1')$. T_1 and T_2 are not independent, since the failure of one component changes the parameter of the surviving one. Consequently, the joint density of (T_1, T_2) is

$$f(t_1,t_2) = \begin{cases} \alpha_1\alpha_2' e^{-(\alpha_1+\alpha_2-\alpha_2')t_1-\alpha_2' t_2}, & 0 < t_1 < t_2 \\ \alpha_2\alpha_1' e^{-\alpha_1' t_1-(\alpha_1+\alpha_2-\alpha_1')t_2}, & 0 < t_2 < t_1. \end{cases} \tag{2.18}$$

For convenience, we call this distribution the FBVE.

The joint reliability function of (T_1, T_2) is, for $0 < t_1 < t_2$,

$$\begin{aligned} \bar{F}(t_1,t_2) &= \int_{t_2}^{\infty}\int_{t_1}^{t_2} \alpha_1\alpha_2' e^{-(\alpha_1+\alpha_2-\alpha_2')x-\alpha_2' y}\,dx\,dy + \bar{F}(t_2,t_2) \\ &= \frac{\alpha_1}{\alpha_1+\alpha_2-\alpha_2'} e^{-(\alpha_1+\alpha_2-\alpha_2')t_1-\alpha_2' t_2} + \frac{\alpha_2-\alpha_2'}{\alpha_1+\alpha_2-\alpha_2'} e^{-(\alpha_1+\alpha_2)t_2}. \end{aligned}$$

It is similar for $0 < t_2 < t_1$. Thus the joint reliability function of the FBVE is given by

$$\bar{F}(t_1,t_2) = \begin{cases} \kappa_1 e^{-(\alpha_1+\alpha_2-\alpha_2')t_1-\alpha_2' t_2} + (1-\kappa_1)e^{-(\alpha_1+\alpha_2)t_2}, & t_1 < t_2 \\ \kappa_2 e^{-(\alpha_1+\alpha_2-\alpha_1')t_2-\alpha_1' t_1} + (1-\kappa_2)e^{-(\alpha_1+\alpha_2)t_1}, & t_2 < t_1, \end{cases} \tag{2.19}$$

where $\kappa_1 = \frac{\alpha_1}{\alpha_1+\alpha_2-\alpha_2'}$ and $\kappa_2 = \frac{\alpha_2}{\alpha_1+\alpha_2-\alpha_1'}$. T_1 and T_2 are independent when $\alpha_1' = \alpha_1$ and $\alpha_2' = \alpha_2$.

The Laplace transform $\psi(s,t)$ is given by

$$\psi(s,t) = E(e^{-sT_1-tT_2}) = (\alpha_1+\alpha_2+s+t)^{-1}\left[\frac{\alpha_1'\alpha_2}{\alpha_1'+s}+\frac{\alpha_1\alpha_2'}{\alpha_2'+t}\right].$$

We remark that the moment generating function $E(e^{sT_1+tT_2})$ given by Freund [29] was wrong. It follows that the marginal expectations, variances, and correlation coefficient can be obtained. In particular, the correlation coefficient is

$$\rho(T_1,T_2) = \frac{\alpha_1'\alpha_2' - \alpha_1\alpha_2}{\sqrt{(\alpha_1'^2+2\alpha_1\alpha_2+\alpha_2^2)(\alpha_2'^2+2\alpha_1\alpha_2+\alpha_1^2)}}.$$

It can be shown that $-1/3 < \rho(T_1,T_2) < 1$.

The marginals are not exponentials; each is a mixture of exponentials

$$f_1(t_1) = (1-\kappa_2)(\alpha_1+\alpha_2)e^{-(\alpha_1+\alpha_2)t_1} + \kappa_2\alpha_1' e^{-\alpha_1' t_1},$$
$$f_2(t_2) = (1-\kappa_1)(\alpha_1+\alpha_2)e^{-(\alpha_1+\alpha_2)t_2} + \kappa_1\alpha_2' e^{-\alpha_2' t_2},$$

provided that $\alpha_1+\alpha_2-\alpha_1' \neq 0$ and $\alpha_1+\alpha_2-\alpha_2' \neq 0$. The marginal density $f_i(t_i)$ contains various parameter ranges $\alpha_1' < \alpha_1$, $\alpha_1' = \alpha_1$, $\alpha_1 < \alpha_1' < \alpha_1+\alpha_2$ and $\alpha_1+\alpha_2 < \alpha_1'$. If $\alpha_1+\alpha_2-\alpha_1' = 0$ and $\alpha_1+\alpha_2-\alpha_2' = 0$, then

$$f_1(t_1) = (\alpha_1+\alpha_1'\alpha_2 t_1)e^{-\alpha_1' t_1},$$
$$f_2(t_2) = (\alpha_2+\alpha_2'\alpha_1 t_2)e^{-\alpha_2' t_2}.$$

Thus $f_i(t_i)$ is a mixture of an exponential density $g_i(t_i)$ and a gamma density $g_i^*(t_i)$, i.e.,

$$f_1(t_1) = \frac{\alpha_1}{\alpha_1'}g_1(t_1) + \frac{\alpha_2}{\alpha_1'}g_1^*(t_1),$$

where $g_1(t_1) = \alpha_1' e^{-\alpha_1' t_1}$ and $g_1^*(t_1) = \alpha_1'^2 t_1 e^{-\alpha_1' t_1}$.

From (2.19), the reliability functions for 1-out-of-2:F and 2-out-of-2:F systems are, respectively, $R_{1:2}(t) = e^{-(\alpha_1+\alpha_2)t}$ and

$$R_{2:2}(t) = \begin{cases} \kappa_1 e^{-\alpha_2' t} + \kappa_2 e^{-\alpha_1' t} + (1-\kappa_1-\kappa_2)e^{-(\alpha_1+\alpha_2)t}, & \alpha_1+\alpha_2 \neq \alpha_1', \alpha_2' \\ \kappa_1 e^{-\alpha_2' t} + (1-\kappa_1+\alpha_2 t)e^{-(\alpha_1+\alpha_2)t}, & \alpha_1+\alpha_2 = \alpha_1' \neq \alpha_2' \\ \kappa_2 e^{-\alpha_1' t} + (1-\kappa_2+\alpha_1 t)e^{-(\alpha_1+\alpha_2)t}, & \alpha_1+\alpha_2 = \alpha_2' \neq \alpha_1' \\ [1+(\alpha_1+\alpha_2)t]\, e^{-(\alpha_1+\alpha_2)t}, & \alpha_1+\alpha_2 = \alpha_1' = \alpha_2', \end{cases}$$

where $\kappa_1 = \frac{\alpha_1}{\alpha_1+\alpha_2-\alpha_2'}$ and $\kappa_2 = \frac{\alpha_2}{\alpha_1+\alpha_2-\alpha_1'}$.

Example 2.6. When $\alpha_1 = \alpha_2 = \alpha$ and $\alpha_1' = \alpha_2' = 2\alpha$, the system reliability $R_{2:2}(t)$ is given by

$$R_{2:2}(t) = (1 + 2\alpha t)e^{-2\alpha t}.$$

From (2.4), the system's model failure rate function is

$$\lambda(t) = \frac{4\alpha^2 t}{1 + 2\alpha t}.$$

This system is related to *cascading failures* models, see Singpurwalla [100].

For a n-component system of this type, the joint density can be formulated in which the parameters of the surviving components change as components fail. Let the n components fail at times $T_{1:n} \leq T_{2:n} \leq \cdots \leq T_{n:n}$. For $j = 1, \ldots, n$ and $k = 1, \ldots, j-1$, $T_{j:n}$ has failure rate $\alpha_j^{(k)}$ after k of the components have failed, i.e., for $j = 1, \ldots, n$,

$$T_{j:n} : \alpha_j \quad \overset{\text{1 failed}}{\longrightarrow} \quad \alpha_j^{(1)} \quad \overset{\text{2 failed}}{\longrightarrow} \quad \alpha_j^{(2)} \quad \overset{\text{3 failed}}{\longrightarrow} \quad \cdots \quad \overset{(j-1)\text{ failed}}{\longrightarrow} \quad \alpha_j^{(j-1)}.$$

The joint density of $T_1, \ldots, T_n$ is, similar to (2.18),

$$f(t_1, \ldots, t_n) = \prod_{k=1}^{n} \left[\alpha_k^{(k-1)} \prod_{j=k}^{n} \exp\left(-\alpha_j^{(k-1)} (t_{k:n} - t_{k-1:n}) \right) \right],$$

where $0 = t_{0:n} \leq t_{1:n} \leq \cdots \leq t_{n:n}$. We call this distribution the FMVE.

The formulae become more complicated as the number of components becomes larger than two. However, if we assume these different components are the same versions with equal parameters, then the reliability functions have simpler forms. If k components failed, the conditional distribution of the surviving components will have exponential distributions with parameters α_{k+1}, for $k = 0, 1, \ldots, n-1$. That is, for $j = 1, \ldots, n$,

$$T_{j:n} : \alpha_1 \quad \overset{\text{1 failed}}{\longrightarrow} \quad \alpha_2 \quad \overset{\text{2 failed}}{\longrightarrow} \quad \alpha_3 \quad \overset{\text{3 failed}}{\longrightarrow} \quad \cdots \quad \overset{(j-1)\text{ failed}}{\longrightarrow} \quad \alpha_j.$$

The joint density of $T_1, \ldots, T_n$ becomes

$$f(t_1, \ldots, t_n) \prod_{k=1}^{n} \alpha_k \exp\left[-\alpha_k (n - k + 1)(t_{k:n} - t_{k-1:n})\right],$$

where $0 = t_{0:n} \leq t_{1:n} \leq \cdots \leq t_{n:n}$. The spacings $D_k = T_{k:n} - T_{k-1:n}$ are independent exponential distributions with mean $1/\theta_k$, where

$$\theta_k = (n-k+1)\alpha_k, \quad k = 1, \ldots, n.$$

The kth order statistics $T_{k:n} = \sum_{i=1}^{k} D_i$ can be considered as the time of failure of a k-out-of-n:F system. Thus the reliability function for the 1-out-of-n:F system is $R_{1:n}(t) = P(T_{1:n} > t) = e^{-\theta_1 t} = e^{-n\alpha_1 t}$. In order to obtain the reliability function for the n-out-of-n:F system, we need to obtain the p.d.f. of $T_{k:n}$. The Laplace transform of $T_{k:n}$ is

$$\psi(s) = E(e^{-sT_{k:n}}) = \prod_{i=1}^{k} \frac{\theta_i}{\theta_i + s}.$$

The p.d.f. of $T_{k:n}$ is obtained by expressing the Laplace transform in partial fraction form as

$$\sum_{i=1}^{k} p_i \frac{\theta_i}{\theta_i + s},$$

where

$$p_i = \prod_{j=1, j \neq i}^{n} \frac{\theta_j}{\theta_j - \theta_i}.$$

Thus the p.d.f. of $T_{k:n}$ is

$$f_{T_{k:n}}(t) = \sum_{i=1}^{k} p_i \theta_i \exp[-\theta_i t].$$

It follows that the reliability function for the n-out-of-n:F system is

$$R_{n:n}(t) = P(T_{n:n} > t) = \int_t^{\infty} f_{T_{n:n}}(v)dv = \sum_{i=1}^{n} p_i \exp[-\theta_i t],$$

that is,

$$R_{n:n}(t) = \sum_{i=1}^{n} \left[\prod_{j=1, j \neq i}^{n} \frac{(n-j+1)\alpha_j}{(n-j+1)\alpha_j - (n-i+1)\alpha_i} \right] \exp[-(n-i+1)\alpha_i t], \tag{2.20}$$

provided that $(n-j+1)\alpha_j - (n-i+1)\alpha_i \neq 0$, for $j \neq i$. The reliability function $R_{n:n}(t)$ is a mixture of exponentials, since $\sum_{i=1}^{n} p_i = 1$.

Example 2.7. Suppose that a system consists of n identical components connected in parallel. Let us suppose that failures of certain components lead to increase in the functional load applied to the components still in operation. The failure rate of each component is independent of time but depends on the number of components that have not failed. At a given instant, $(n-k+1)$ components are in operation, then the failure rate of each of them is equal to α_k. The reliability function is given by (2.20). It can be shown that the mean lifetime of the system is [32]

$$\frac{1}{n\alpha_1} + \frac{1}{(n-1)\alpha_2} + \cdots + \frac{1}{\alpha_n}.$$

Load-sharing model is originated from the study of the failure properties of composite materials using the concept of fiber bundles. The model can be either static or time-dependent. Experiments usually favor the dynamic failure fiber bundle models [50]. The load-sharing rule is an important element of the load-sharing models. It governs how the loads on the working components change after some components in the system fail. There are equal, local, monotone, and non-monotone load-sharing rules, and the equal rule is the one most applied. FBVE is a simple load-sharing model and is applicable to almost all load-sharing rule.

Studies of k-out-of-n system related to FMVE are Scheuer [94], Lin et al. [53], and Amari and Misra [4]. Amari et al. [6] study tampered failure rate load-sharing k-out-of-n:G systems. Huang and Xu [41], Liu [62], and Lu [63] study reliability functions of load-sharing redundant systems with non-exponential or arbitrary failure distributions. Shao and Lamberson [97] models a load-sharing k-out-of-n:G system by a continuous-time Markov chain and solves its differential equations using inverse Laplace transforms. We return to this model in Sect. 2.5.2.

2.3 Markov Models

Markov models are commonly used to perform reliability analysis of engineering systems and fault-tolerant systems. They are also used to handle reliability and availability analysis of repairable systems. We first give notations and several properties of stochastic processes. Next, we explore Markov chains focusing on criteria of recurrent/transient state, and long-run probabilities. We then discuss basic properties of the homogeneous Poison process, which is one of the most important stochastic processes. The nonhomogeneous Poisson process (NHPP) is an time-dependent variation of the homogeneous Poisson process. NHPP models has been applied to various fields, such as repairable systems and software reliability modeling. The discussion is then going to the continuous-time Markov chain, including the birth, the death, and the birth-death processes. It is not an easy task to solve the state equations. A number of solution techniques exist, such as analytical solution [9], Laplace-Stieltjes transforms [92], numerical integration and computer-assisted evaluation [88].

2.3.1 Basic Properties of Stochastic Processes

A *stochastic process* is a mathematical model for the occurrence, at each moment after the initial time, of a random phenomenon. Mathematically, a stochastic process is a collection of random variables $\{X(t), t \in T\}$ defined on a common probability space indexed by a suitable index set T which describes the evolution of some systems. That is, for each $t \in T$, $X(t)$ is a random variable. Generally these random variables $X(t)$ are dependent. Stochastic processes are then characterized by different dependency relationships among $X(t)$. We often interpreted t as time and call $X(t)$ the *state* of the process at time t. If T is a countable set, we call $\{X(t) : t \in T\}$ a discrete-time stochastic process. If T is a continuum, we call $\{X(t) : t \in T\}$ a continuous-time stochastic process.

Often $T = [0, \infty)$ if the system evolves in continuous time. For example, $X(t)$ might be the number of people in a queue at time t, or the accumulated claims paid by an insurance company in $[0, t]$. Alternatively, we could have $T = \{0, 1, \ldots\}$ if the system evolves in discrete time. Then $X(n)$ might represent the number of arrivals to a queue during the service interval of the nth customer.

Any *realization* of a stochastic process $\{X(t) : t \in T\}$ is called a *sample path*. The space of the possible values of each $X(t)$ is called the *state space* S. If S is a countable set, we call $\{X(t) : t \in T\}$ a discrete state stochastic process. If S is a continuum, we call $\{X(t) : t \in T\}$ a continuous state stochastic process. Several common properties of stochastic processes are given as follows.

Markov property: A Markov process is a stochastic process having the Markov property that, given the present state at time s and all past states, the future state at $t + s$ is independent of the past and depends only on the present state at s. Formally, if $\forall s, t \geq 0$ and $0 \leq u \leq s$,

$$P(X(t+s) = j | X(s) = i, X(u) = x(u), 0 \leq u < s) = P(X(t+s) = j | X(s) = i).$$

A Markov process having a countable state space is called a *Markov chain*. A discrete-time Markov process is simply called a Markov chain. Random walk is a Markov chain. The Poisson process and birth-death processes are *continuous-time Markov chains*. A Markov process for which all realization are continuous functions is called a *diffusion process*. Brownian motion is a diffusion process.

Homogeneous (or Stationary): A stochastic process is said to be homogeneous or stationary if the transition probability $p_{ij}(t) = P(X(t+s) = j | X(s) = i)$ is independent of s. Many continuous-time Markov chain we consider will have homogeneous property. Let τ_i be the amount of time that the process stays in state i. By the Markovian property, $P(\tau_i > s + t | \tau_i > s) = P(\tau_i > t)$ for all $s, t \geq 0$. Hence, the random variable τ_i is memoryless and must be exponentially distributed. That is, the amount of time a continuous-time Markov chain spends in each state, before jumping to the next state, is exponentially distributed.

Independent increments: A continuous-time stochastic process $\{X(t), t \in T\}$ is said to have *independent increments* if, for all $0 \leq t_0 < t_1 < t_2 < \cdots < t_n$, the random variables $X(t_1) - X(t_0), X(t_2) - X(t_1), \ldots, X(t_n) - X(t_{n-1})$ are jointly independent, where the difference $X(t_i) - X(t_{i-1})$ is called the *increment*. That is, for all n,

$$P(X(t_0, t_1) = k_1, \ldots, X(t_{n-1}, t_n) = k_n) = \prod_{i=1}^{n} P(X(t_{i-1}, t_i) = k_i),$$

where $X(t_2) - X(t_1) = X(t_1, t_2)$. If the index set contains a smallest index t_0, it is also assumed that $X(t_0), X(t_1) - X(t_0), X(t_2) - X(t_1), \ldots, X(t_n) - X(t_{n-1})$ are jointly independent. *Brownian motions*, *Poisson processes* and *Lévy processes* have homogeneous independent increments.

Martingale processes: A stochastic process $\{X_t : t \in T\}$ is said to be a *martingale* process if, for $t_1 < \cdots < t_n < t_{n+1}$,

1. $E(|X_t|) < \infty, \quad \forall t,$
2. $E\left(X_{t_{n+1}} | X_{t_1}, X_{t_2}, \ldots, X_{t_n}\right) = X_{t_n},$
3. For any $t \in T$ the random variable $X(t)$ is $\mathcal{F}_t$-measurable.

Taking expectation of (2) gives $E(X_{t_{n+1}}) = E(X_{t_n})$, and so $E(X_{t_n}) = E(X_{t_1})$ for all t's. For any $s \leq t$, $s, t \in T$, if $E(X(t)|\mathcal{F}_s) \geq X(s)$ P-a.s., then we have the definition of a submartingale; if $E(X(t)|\mathcal{F}_s) \leq X(s)$ P-a.s., then we have a supermartingale.

A martingale is a generalized version of a fair game. Martingale concepts and methodologies have provided a far-reaching apparatus vital to the analysis of all kinds of functionals of stochastic processes. In particular, martingale constructions serve decisively in the investigating of stochastic models of *diffusion type*.

2.3.2 Markov Chains

2.3.2.1 Transition Probabilities

A Markov chain with discrete-state space S is a collection of random variables X_n with the Markov property

$$\begin{aligned} P(X_{n+1} &= j | X_0 = i_0, X_1 = i_1, \ldots, X_{n-1} = i_{n-1}, X_n = i) \\ &= P(X_{n+1} = j | X_n = i) = p_{ij}, \end{aligned}$$

for all states $i_0, i_1, \ldots, i_{n-1}, i, j$ and all $n \geq 0$. p_{ij} is called a 1-step *transition probability*, representing the probability that a process in state i will next make a transition into state j. We write p_{ij} instead of $p_{ij}^{n,n+1}$ to represent the homogeneous transition probabilities.

We often display the 1-step transition probabilities $p_{ij} = P(X_{n+1} = j|X_n = i)$ in a matrix $\mathbf{P} = [p_{ij}]$, called the *transition probability matrix* (or simply transition matrix), such that

$$\mathbf{P} = [p_{ij}] = \begin{bmatrix} p_{00} & p_{01} & p_{02} & \cdots \\ p_{10} & p_{11} & p_{12} & \cdots \\ \vdots & \vdots & \vdots & \ldots \\ p_{i0} & p_{i1} & p_{i2} & \cdots \\ \vdots & \vdots & \vdots & \ldots \end{bmatrix},$$

where $p_{ij} \geq 0\ \forall i, j \geq 0$, and $\sum_{j=0}^{\infty} p_{ij} = 1, i = 0, 1, \ldots$.

A Markov chain $\{X_n, n = 0, 1, 2, \ldots\}$ is completely defined by its (1-step) transition probability matrix and its initial distribution (i.e., the specification of a probability distribution on the state of the process at time 0):

$$\begin{aligned} & P\left(X_0 = i_0, X_1 = i_1, \ldots, X_n = i_n\right) \\ & = P\left(X_n = i_n | X_0 = i_0, X_1 = i_1, \ldots, X_{n-1} = i_{n-1}\right) \\ & \quad \cdot P\left(X_0 = i_0, X_1 = i_1, \ldots, X_{n-1} = i_{n-1}\right) \\ & = p_{i_{n-1},i_n} \cdot p_{i_{n-2},i_{n-1}} \cdot \cdots \cdot p_{i_1,i_2} \cdot p_{i_0,i_1} \cdot P(X_0 = i_0), \end{aligned}$$

where $P(X_0 = i_0)$ is the *initial probability*.

The analysis of a Markov chain concerns mainly the calculation of the probability of the possible realizations of the process. Central in these calculations are the n-step transition probability matrices $\mathbf{P}^{(n)} = [p_{ij}^{(n)}]$ where $p_{ij}^{(n)}$ is the n-step transition probability and $p_{ij}^{(n)} = P(X_{n+m} = j|X_m = i), n \geq 0, i, j \geq 0$. The interest is to calculate $p_{ij}^{(n)}$ in terms of the transition probability p_{ij}.

2.3.2.2 Chapman-Kolmogorov Equations

The Chapman-Kolmogorov equations calculate the $(n+m)$-step transition probability $p_{ij}^{(n+m)}$ by summing over all the intermediate state k at time n and moving to state j from state k at the remaining time m: $\mathbf{P}^{(n+m)} = \mathbf{P}^{(n)} \cdot \mathbf{P}^{(m)}$. A consequence of the Chapman-Kolmogorov equations is that $\mathbf{P}^{(n)} = \mathbf{P}^n$, i.e., the n-step transition probability matrix is equal to the nth power of the 1-step transition matrix.

The distribution of position at any time, $P(X_n = j)$, is given by the initial distribution and the n-step transition probabilities: $(n \geq 1)$

$$P(X_n = j) = \sum_{i=0}^{\infty} P(X_0 = i)P(X_n = j|X_0 = i) = \sum_{i=0}^{\infty} P(X_0 = i)p_{ij}^{(n)}.$$

Example 2.8. Consider a system with three state components situated in a shock environment that generates shocks according to a Poisson process with intensity λ. The three states can be described as 0 or good, 1 or degraded, and 2 or failed. When a component is hit by a shock it will (independently of the other components) pass from state 0 to state 1 with probability $p_1 = 1 - q_1$ and from state 1 to state 2 with probability $p_2 = 1 - q_2$ ($p_1 \neq p_2$). Let P_k be the probability that a component in state 0 survives k shocks. Then

$$P_k = q_1^k + \sum_{j=0}^{k-1} q_1^j p_1 q_2^{k-1-j} = \frac{p_1 q_2^k - p_2 q_1^k}{q_2 - q_1}.$$

2.3.2.3 Stationary Distributions

Let $w = [w_0, w_1, w_2, \ldots]$ and $\sum_{i=0}^{\infty} w_i = 1$. Then w is said to be a *stationary distribution* (or equilibrium vector, steady-state vector) for the Markov chain if $w\,\mathbf{P} = w$. If w is a stationary distribution, then $\forall n \geq 1$,

$$w\mathbf{P}^n = (w\mathbf{P})\mathbf{P}^{n-1} = w\mathbf{P}^{n-1} = \cdots = w\mathbf{P} = w.$$

The chain is sometimes said to be in *equilibrium*.

Example 2.9. Consider a two-state Markov chain with the state space $S = \{0, 1\}$ and the transition probability matrix $P = \begin{bmatrix} 1-\alpha & \alpha \\ \beta & 1-\beta \end{bmatrix}$, $0 < \alpha, \beta < 1$. The stationary distribution of the Markov chain is $\left(\frac{\beta}{\alpha+\beta}, \frac{\alpha}{\alpha+\beta}\right)$.

2.3.2.4 Classification of States

- *Accessible*: If $p_{ij}^{(n)} > 0$ for some $n \geq 0$. That is, the state j is accessible from the state i if there is positive probability that in a finite number of steps a Markov chain moves from i to j.
- *Communicate*: If $\exists\, m, n \geq 0$ such that $p_{ij}^{(m)} > 0$ and $p_{ji}^{(n)} > 0$. That is, two states communicate if each is accessible from the other.
- *Irreducible*: "Communicate with" is an equivalence relation, which breaks state space S up into disjoint equivalence classes. Two states that communicate are in the same class. If there is just one class, the chain is irreducible (all states communicate with each other), otherwise it is *Reducible*. A *closed* class is one we cannot leave. A closed class containing exactly one state is called *absorbing*.
- *Transient/recurrent*: A state is transient if it is not certain we shall return to it sometime, otherwise it is recurrent. Formally,

$$P(X_n = i \text{ for some } n \geq 1 | X_0 = i) \begin{cases} = 1, & \text{state } i \text{ is recurrent} \\ < 1, & \text{state } i \text{ is transient.} \end{cases}$$

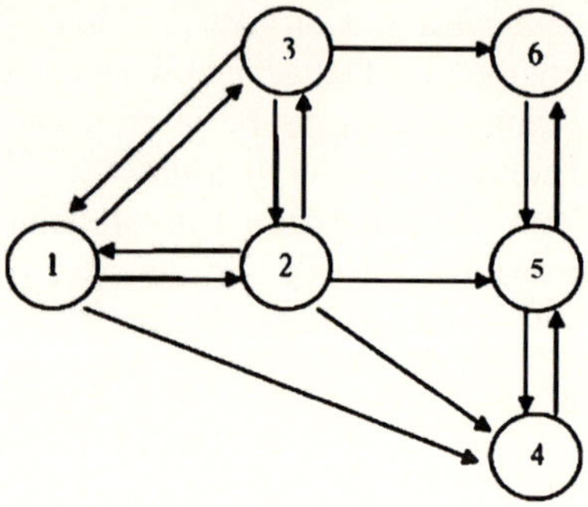

Fig. 2.8 A six-state system transitions

- *Periodicity*: Period $d(i)$ of a state i is the greatest common divisor of those time $n \geq 1$ taken to return to i, given we start there. That is, $d(i) = \gcd\{n : n \geq 1, p_{ii}^{(n)} > 0\}$. In particular, if $d(i) = 1$, then the state i is called *aperiodic*.

Example 2.10. If $S = \{1, 2, 3, 4, 5, 6\}$ and the transition probability matrix is given by

$$\mathbf{P} = \begin{bmatrix} 0 & 1/3 & 1/3 & 1/3 & 0 & 0 \\ 1/3 & 0 & 1/3 & 1/6 & 1/6 & 0 \\ 1/3 & 1/3 & 0 & 0 & 0 & 1/3 \\ 0 & 0 & 0 & 0 & 1 & 0 \\ 0 & 0 & 0 & 2/3 & 0 & 1/3 \\ 0 & 0 & 0 & 0 & 1 & 0 \end{bmatrix}.$$

Then $\{1, 2, 3\}$: transient, aperiodic; $\{4, 5, 6\}$: irreducible closed set, recurrent, period 2 (Fig. 2.8).

2.3.2.5 First Hitting Probability

The states of a Markov chain are classified independently from two viewpoints: (i) recurrent and transient states, and (ii) periodic and aperiodic states. The vast majority of Markov chains we deal with are aperiodic, and criteria for a state to be recurrent or transient is closely related to the *first hitting probability*. The first hitting probability is defined as

$$f_{ij}^{(n)} = P(T_j = n | X_0 = i), \qquad n \geq 1,$$

representing the probability that, starting from state i, the first transition into state j occurs at time n. The random variable T_j represents the time (i.e., number of steps) for the first transition into state j, i.e., $T_j = \min\{n \geq 1 : X_n = j\}$. If $\forall\, n \geq 1$, $X_n \neq j$, then define $T_j = \infty$. Define $f_{ij} = \sum_{n=1}^{\infty} f_{ij}^{(n)}$ to be the probability that the chain ever visits j, starting from i. For $i \neq j$, $f_{ij} > 0$ if and only if j is accessible from i. Of course, $f_{ij} \leq 1$. Now we have a criterion for a state to be recurrent or transient.

Definition 2.1. State j is recurrent if and only if $f_{jj} = 1$, and transient if and only if $f_{jj} < 1$.

If state j is an absorbing state, then $P(T_j = 1 | X_0 = j) = 1$. This implies that $f_{jj} = 1$. Thus an absorbing state is recurrent. If a Markov chain is in state j and j is recurrent, the probability that the process will return to that state is 1. Since the process is a Markov chain, this is equivalent to the process beginning once more from state j, and with probability 1, it will once again return to that state. Repeating this argument leads to the conclusion that state j will be entered *infinitely often*. Hence, a recurrent state has the property that the expected number of time periods that the process is in state j is infinite. If state j is transient, then the probability of reentering state j is f_{jj} (< 1) and the probability of not reentering state j is $1 - f_{jj}$. Then the expected number of time periods that the process is in state j is finite and given by $\frac{1}{1-f_{jj}}$.

2.3.2.6 A Criterion for Recurrent and Transient

Theorem 2.2. *An important formula that connect $f_{ij}^{(n)}$ and $p_{ij}^{(n)}$ is*

$$p_{ij}^{(n)} = \sum_{k=1}^{n} f_{ij}^{(k)} p_{jj}^{(n-k)}, \qquad \forall i, j \text{ and } 1 \leq n < \infty.$$

Theorem 2.3. *A criterion for identifying a state to be recurrent or transient in terms of the n-step transition probabilities $p_{ij}^{(n)}$ is: (1) State j is recurrent if and only if $\sum_{n=1}^{\infty} p_{jj}^{(n)} = \infty$, (2) State j is transient if and only if $\sum_{n=1}^{\infty} p_{jj}^{(n)} < \infty$.*

2.3.2.7 Examples of Recurrent/Transient

Example 2.11. Consider a two-state Markov chain with states $\{0, 1\}$ and the transition probability matrix $P = \begin{bmatrix} 1 & 0 \\ 1/2 & 1/2 \end{bmatrix}$. It is obvious that $\sum_{n=1}^{\infty} p_{00}^{(n)} = 1 + 1 + \cdots = \infty$ and $\sum_{n=1}^{\infty} p_{11}^{(n)} = \frac{1}{2} + \left(\frac{1}{2}\right)^2 + \cdots = 1 < \infty$, which implies that state 0 is recurrent and state 1 is transient.

Example 2.12. Consider a three-state Markov chain with states $\{1, 2, 3\}$ and the transition probability matrix

$$P = \begin{bmatrix} \alpha & 1-\alpha & 0 \\ 0 & 0 & 1 \\ 1-\beta & 0 & \beta \end{bmatrix}.$$

Since $f_{11} = \sum_{n=1}^{\infty} f_{11}^{(n)} = \alpha + \sum_{n=3}^{\infty}(1-\alpha)(1-\beta)\beta^{n-3} = \alpha + (1-\alpha)\frac{(1-\beta)}{(1-\beta)} = 1$, then $f_{11} = 1$, and consequently the state 1 is recurrent.

2.3.2.8 Ergodic

Ergodic states are important in the classification of chains, and in proving the existence of limiting probability distributions.

Definition 2.2. The *mean recurrence time* μ_{jj} denotes the expected time of transitions needed to return to state j, starting from state j:

$$\mu_{jj} = E(T_j | X_0 = j) = \begin{cases} \sum_{n=1}^{\infty} n f_{jj}^{(n)} & \text{if } j \text{ is recurrent} \\ \infty & \text{if } j \text{ is transient.} \end{cases}$$

μ_{jj} may be infinite even if j is recurrent. The recurrent state j is called *positive recurrent* if $\mu_{jj} < \infty$ and it is *null recurrent* if $\mu_{jj} = \infty$.

It can be shown that for a finite-state Markov chain, all recurrent states are positive recurrent. Note that $\mu_{ij} = E(T_{ij})$ is the *mean first passage time*, where $T_{ij} = \min\{n \geq 0 : X_n = j | X_0 = i\}$.

Definition 2.3. Positive recurrent states that are aperiodic are called ergodic states. A Markov chain is said to be ergodic if all its states are ergodic states.

Example 2.13. Consider a two-state Markov chain $S = \{1, 2\}$ with transition probability matrix $\mathbf{P} = \begin{bmatrix} 1/3 & 2/3 \\ 1/4 & 3/4 \end{bmatrix}$. Given $X_0 = 1$, the chain enters state 2 as soon as it leaves state 1. Hence

$$P(T_{12} = n) = \frac{2}{3}\left(\frac{1}{3}\right)^{n-1}, \qquad n \geq 1,$$

and

$$\mu_{12} = \sum_{n=1}^{\infty} n f_{12}^{(n)} = \sum_{n=1}^{\infty} n \frac{2}{3}\left(\frac{1}{3}\right)^{n-1} = \frac{3}{2}.$$

Likewise, first return to state 1 at the nth step occurs after $n-2$ consecutive visits to state 2, so

$$P(T_1 = n) = \begin{cases} \frac{1}{3}, & n = 1 \\ \frac{2}{3}\left(\frac{3}{4}\right)^{n-2}\frac{1}{4}, & n \geq 2. \end{cases}$$

Hence

$$\mu_{11} = \sum_{n=1}^{\infty} n f_{11}^{(n)} = \frac{1}{3} + \sum_{n=2}^{\infty} n \frac{2}{3} \left(\frac{3}{4}\right)^{n-2} \frac{1}{4} = \frac{11}{3}.$$

Example 2.14. If the transition probability matrix is given by

$$\mathbf{P} = \begin{bmatrix} \alpha & 1-\alpha & 0 \\ 0 & 0 & 1 \\ 1-\beta & 0 & \beta \end{bmatrix}.$$

The state 1 is recurrent and its mean recurrence time is given by

$$\mu_{11} = \sum_{n=1}^{\infty} n f_{11}^{(n)} = \alpha + (1-\alpha)(1-\beta) \sum_{n=3}^{\infty} n \beta^{n-3} = \frac{3 - 2\alpha - 2\beta + \alpha\beta}{1-\beta} < \infty.$$

Hence state 1 is positive recurrent. Because $p_{11}^{(n)} > 0$ for $n \geq 3$, state 1 is aperiodic. Hence state 1 is ergodic (and also states 2 and 3).

Theorem 2.4. *An irreducible chain has a stationary distribution* π *if and only if all the states are positive recurrent; in this case,* π *is the unique stationary distribution and is given by* $\pi_j = 1/\mu_{jj}$ *for each* $j \in S$*, where* μ_{jj} *is the mean recurrence time of* j.

Example 2.15. Consider the Markov chain

$$P = \begin{bmatrix} 0 & 1/3 & 2/3 \\ 2/3 & 0 & 1/3 \\ 1/3 & 2/3 & 0 \end{bmatrix}.$$

The unique stationary distribution is $\pi = (1/3, 1/3, 1/3)$. Thus, $\mu_{11} = \pi_1^{-1} = 3$, $\mu_{22} = \pi_2^{-1} = 3$, and $\mu_{33} = \pi_3^{-1} = 3$. The mean recurrence time for all states is 3.

Theorem 2.5. *If we have an irreducible aperiodic chain:*

(1) If it is transient, then $p_{ij}^{(n)} \longrightarrow 0 \; \forall i, j$;
(2) If it is positive recurrent, then $p_{ij}^{(n)} \longrightarrow \pi_j = 1/\mu_{jj}$;
(3) If it is null recurrent, then $p_{ij}^{(n)} \longrightarrow 0 \; \forall i, j$*, and the mean time to return to any state is infinite.*

2.3.2.9 Limiting Probabilities in Finite-State Markov Chains

For a finite-state Markov chain: (i) not all states are transient, and (ii) there are no null recurrent states. An irreducible finite chain must be positive recurrent. If the irreducible finite-state Markov chain is aperiodic, there exists a unique and positive stationary distribution which is the limiting distribution.

For a Markov chain, all states can be classified into several recurrent classes $C_1, \ldots, C_m$, and remaining transient states which form a transient set T. Then all sets $C_1, \ldots, C_m, T$ are disjoint. Rearranging (or relabeling) all the states so that the irreducible classes are together and transient states are last, the Markov chain can be rewritten in the form

$$P = \begin{array}{c} \\ C_1 \\ C_2 \\ \vdots \\ C_m \\ T \end{array} \begin{array}{c} \begin{array}{cccccc} C_1 & C_2 & \ldots & C_m & T \end{array} \\ \begin{pmatrix} P_1 & 0 & \ldots & 0 & 0 \\ 0 & P_2 & \ldots & 0 & 0 \\ \vdots & \vdots & \ddots & \vdots & \vdots \\ 0 & 0 & \ldots & P_m & 0 \\ R_1 & R_2 & \ldots & R_m & Q \end{pmatrix} \end{array},$$

where $P_1, \ldots, P_m$ are the transition matrices for recurrent classes $C_1, \ldots, C_m$, Q is a square matrix for describing the transitions among all transient states for T, and $R_1, \ldots, R_m$ are (not necessarily square) matrices describing the transitions from all transient states to the corresponding recurrent classes $C_1, \ldots, C_m$. Let us consider the limiting probabilities $\lim_{n\to\infty} P_{ij}^{(n)}$ for all states. Assume that all positive recurrent states are *aperiodic*, we have

$$\lim_{n\to\infty} p_{ij}^{(n)} = \frac{1}{\mu_{jj}} \qquad (i, j \in C_k; k = 1, \ldots, m),$$

$$\lim_{n\to\infty} p_{ij}^{(n)} = 0 \qquad (i \in C_k, j \in C_l; k \neq l),$$

$$\lim_{n\to\infty} p_{ij}^{(n)} = 0 \qquad (i \in C_k, j \in T),$$

$$\lim_{n\to\infty} p_{ij}^{(n)} = \frac{f_{ij}}{\mu_{jj}} \qquad (i \in T, j \in C_k),$$

$$\lim_{n\to\infty} p_{ij}^{(n)} = 0 \qquad (i, j \in T).$$

Theorem 2.6. *For a finite-state Markov chain, the eventual transition probability f_{ij} from state $i \in T$ to state $j \in C_k$ $(k = 1, \ldots, m)$ satisfies*

$$f_{ij} = \sum_{l \in C_k} P_{il} + \sum_{l \in T} P_{il} f_{lj} \qquad (i \in T, j \in C_k),$$

or in a matrix form

$$[f_{ij}] = [I - Q]^{-1} R_k \cdot \mathbf{1},$$

where I is a identity matrix and $\mathbf{1}$ is a column vector of all the components' unity.

Example 2.16. Consider a Markov chain with the state space $S = \{1, 2, 3, 4\}$ and the transition probability

$$P = \begin{bmatrix} 1 & 0 & 0 & 0 \\ 0 & 0.3 & 0.7 & 0 \\ 0 & 0.5 & 0.5 & 0 \\ 0.2 & 0 & 0.1 & 0.7 \end{bmatrix}.$$

We know that $\{1\}$ is an absorbing state, $\{2, 3\}$ are positive recurrent states, and $\{4\}$ is a transient state. For the recurrent class $\{1\}$, we have $P_1 = [1]$ and the limiting distribution $\pi_1 = 1/\mu_{11} = 1$. For the recurrent class $\{2, 3\}$, we have the submatrix $P_2 = \begin{bmatrix} 0.3 & 0.7 \\ 0.5 & 0.5 \end{bmatrix}$ and the stationary distribution is $(\pi_2, \pi_3) = (1/\mu_{22}, 1/\mu_{33}) = (5/12, 7/12)$. For the transient set $\{4\}$,

$$[I - Q]^{-1} = [10/3],$$

$$[I - Q]^{-1} R_1 \cdot \mathbf{1} = [10/3][2/10][1] = 2/3,$$

$$[I - Q]^{-1} R_2 \cdot \mathbf{1} = [10/3][0 \quad 1/10]\begin{bmatrix} 1 \\ 1 \end{bmatrix} = 1/3.$$

That is, $f_{41} = 2/3$ and $f_{42} = f_{43} = 1/3$. Thus, $P_{41}^{(\infty)} = f_{41}/\mu_{11} = 2/3$. $P_{42}^{(\infty)} = f_{42}/\mu_{22} = 5/36$, and $P_{43}^{(\infty)} = f_{43}/\mu_{33} = 7/36$. The limiting probability matrix is

$$P^{\infty} = \lim_{n \to \infty} P^n = \begin{bmatrix} 1 & 0 & 0 & 0 \\ 0 & 5/12 & 7/12 & 0 \\ 0 & 5/12 & 7/12 & 0 \\ 2/3 & 5/36 & 7/36 & 0 \end{bmatrix}.$$

2.3.3 *Poisson Processes*

The Poisson process is the simplest stochastic process that arises in many applications, especially as a model for the 'counting process' (or 'arrival processes') that counts the number of failures through time. A stochastic process $\{N(t), t \geq 0\}$ is said to be a counting process if $N(t)$ represents the total number of 'events' that have occurred up to time t.

2.3.3.1 Poisson Process

Definition 2.4. A counting process $\{N(t), t \geq 0\}$ is said to be a Poisson process with *rate* $\lambda > 0$ if

(i) $N(0) = 0$,
(ii) The process has independent increments.

(iii) The probability that k events take place for any interval t is Poisson distributed with mean λt. That is, $\forall\ s, t \geq 0, P(N(t+s) - N(s) = k) = e^{-\lambda t}\frac{(\lambda t)^k}{k!}$, $k = 0, 1, 2, \ldots$.

It follows from condition (iii) that a Poisson process has stationary increments and also that $E[N(t)] = \lambda t$, which explains why λ is called the 'rate' of the process. To determine if an arbitrary counting process is actually a Poisson process, we must show that conditions (i)–(iii) are satisfied. Conditions (i) and (ii) can usually be verified from our knowledge of the process. However, it is not at all clear how we would determine that condition (iii) is satisfied, and for this reason we need an alternative definition of a Poisson process.

Definition 2.5. A counting process $\{N(t), t \geq 0\}$ is said to be a Poisson process with rate $\lambda > 0$ if the following conditions are satisfied:

(i) $N(0) = 0$,
(ii) $\{N(t), t \geq 0\}$ has stationary (or homogeneous) independent increments,
(iii) $P(N(h) = 1) = \lambda h + o(h)$,
(iv) $P(N(h) \geq 2) = o(h)$.

From this definition, we derive the probabilities $P_n(t) = P(N(t) = n)$, $n = 0, 1, 2, \ldots$, which denotes the probability that n events occur in $[0, t]$, and $\sum_{n=0}^{\infty} P_n(t) = 1$. To find the distribution of $N(t)$, we first derive a differential equation for $P_0(t)$.

$$\begin{aligned} P_0(t+h) &= P(N(t+h) = 0) \\ &= P(N(t) = 0)P(N(t+h) - N(t) = 0|N(t) = 0) \\ &= P_0(t)P_0(h) = P_0(t)[1 - \lambda h + o(h)], \end{aligned}$$

where the final two equations follow from (ii), (iii), and (iv). Hence

$$\frac{P_0(t+h) - P_0(t)}{h} = -\lambda P_0(t) + \frac{o(h)}{h}.$$

Letting $h \to 0$ yields $P_0^{'}(t) = -\lambda P_0(t)$. Next, for $n \geq 1$,

$$\begin{aligned} P_n(t+h) &= P(N(t+h) = n) \\ &= P(N(t) = n, N(t+h) - N(t) = 0) \\ &\quad + P(N(t) = n-1, N(t+h) - N(t) = 1) \\ &\quad + P(N(t+h) = n, N(t+h) - N(t) \geq 2). \end{aligned}$$

By (iv), the last term in the above is $o(h)$; hence, by (ii), we obtain

$$\begin{aligned} P_n(t+h) &= P_n(t)P_0(h) + P_{n-1}(t)P_1(h) + o(h) \\ &= (1 - \lambda h)P_n(t) + \lambda h P_{n-1}(t) + o(h). \end{aligned}$$

Thus,

$$\frac{P_n(t+h) - P_n(t)}{h} = -\lambda P_n(t) + \lambda P_{n-1}(t) + \frac{o(h)}{h}.$$

Letting $h \to 0$ yields $P_n'(t) = -\lambda P_n(t) + \lambda P_{n-1}(t)$. Thus, we need to solve a family of differential equations:

$$P_0'(t) = -\lambda P_0(t),$$
$$P_n'(t) = -\lambda P_n(t) + \lambda P_{n-1}(t), \quad n = 1, 2, \ldots.$$

Subject to (i), $P_0(0) = 1$ and $P_n(0) = 0$ for all $n \geq 1$, solving recursively we obtain

$$P_n(t) = e^{-\lambda t} \frac{(\lambda t)^n}{n!}, \quad n = 0, 1, 2, \ldots.$$

Thus, for a fixed t, $N(t) \sim \text{Poisson}(\lambda t)$.

From the above results, Definition 2.5 implies Definition 2.4. In fact, two definitions are equivalent. It is obvious that the process has stationary increment from (iii) of Definition 2.4. Moreover, from (iii) of Definition 2.4, we have

$$P\{N(h) = 1\} = \lambda h e^{-\lambda h} = \lambda h + o(h),$$
$$P\{N(h) = k\} = \frac{(\lambda h)^k}{k!} e^{-\lambda h} = o(h), \quad k = 2, 3, \ldots,$$

where we use

$$e^{-\lambda h} = 1 - \lambda h + \frac{(\lambda h)^2}{2!} - \frac{(\lambda h)^3}{3!} + \cdots = 1 - \lambda h + o(h).$$

Thus Definition 2.4 implies Definition 2.5. A Poisson process implies all properties in Definitions 2.4 and 2.5.

2.3.3.2 Interarrival Time and Arrival Time Distributions

For a Poisson process with rate $\lambda > 0$, the interarrival times X_n $(n = 1, 2, \ldots)$ are i.i.d. exponential r.v.s with parameter λ:

$$\begin{aligned} P(X_1 > t) &= P(N(t) = 0) = e^{-\lambda t}, \\ P(X_2 > t | X_1 = s) &= P(0 \text{ event in } (s, s+t] | X_1 = s) \\ &= P(0 \text{ event in } (s, s+t]) = e^{-\lambda t}, \end{aligned}$$

i.e., the conditional distribution $P(X_2 > t | X_1 = s)$ is independent of X_1, and $X_2 \sim \text{Exp}(\lambda)$. Repeating the same argument yields the desired results.

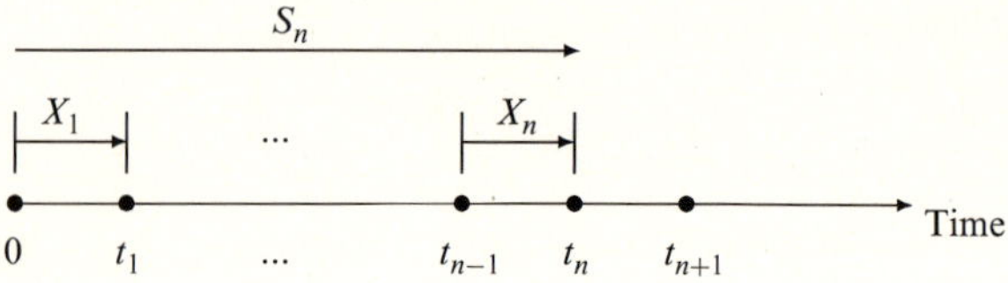

Fig. 2.9 Interarrival times and waiting times

The assumption of stationary and independent increments is equivalent to asserting that, at any point in time, the Poisson process probabilistically restarts itself. In other words, the process has no memory, and hence exponential interarrival times are to be expected. Thus, we have a third definition for the Poisson process.

Definition 2.6. A Poisson process with rate $\lambda > 0$ is a *renewal process* with the exponential interarrival distribution $F(t) = 1 - e^{-\lambda t}$.

The arrival time of the nth event, also called the waiting time until the nth event (from the origin), is $S_n = \sum_{i=1}^{n} X_i$, where $S_0 = 0$ (Fig. 2.9). It can be shown that S_n has a gamma distribution with shape parameter n and scale parameter λ. Use the independent increment assumption as follows:

$$\begin{aligned} P(t < S_n < t + dt) &= P(N(t) = n - 1) \cdot P(1 \text{ event in } (t, t + dt)) + o(dt) \\ &= \frac{e^{-\lambda t}(\lambda t)^{n-1}}{(n-1)!} \lambda dt + o(dt). \end{aligned}$$

Dividing by dt and then letting it approach 0, we have $f_{S_n}(t) = \lambda e^{-\lambda t} \frac{(\lambda t)^{n-1}}{(n-1)!}$.

2.3.3.3 Some Examples of the Poisson Process

Example 2.17 (Superposition). The pooled process $\{N_1(t) + N_2(t), t \geq 0\}$ of two independent Poisson processes $\{N_1(t), t \geq 0\}$ and $\{N_2(t), t \geq 0\}$ with respective rates λ and μ is again a Poisson process with rate $\lambda + \mu$:

$$\begin{aligned} P(N_1(t) + N_2(t) = n) &= \sum_{k=0}^{n} P(N_1(t) = k, N_2(t) = n - k) \\ &= e^{-(\lambda+\mu)t} \frac{[(\lambda + \mu)t]^n}{n!}. \end{aligned}$$

Example 2.18 (Decomposition). Let $\{N(t), t \geq 0\}$ be a Poisson process with rate λ and let the consecutive events follow the Bernoulli trials. For instance, each arrival of failures at an assembly line can be classified as a major and a minor with probability p and probability $1 - p$, respectively. The classified processes

$\{N_1(t), t \geq 0\}$ and $\{N_2(t), t \geq 0\}$ of major failures and minor failures are independent Poisson processes with rates $p\lambda$ and $(1-p)\lambda$, respectively:

$$P(N_1(t) = k, N_2(t) = n-k) = P(N_1(t) = k, N_2(t) = n-k | N(t) = n) \cdot P(N(t) = n)$$
$$= e^{-p\lambda t} \frac{(p\lambda t)^k}{k!} \cdot e^{-(1-p)\lambda t} \frac{((1-p)\lambda t)^{n-k}}{(n-k)!}.$$

Example 2.19. Let $\{N(t), t \geq 0\}$ be a Poisson process with rate λ. For $s < t$ and $k < n$

$$P(N(s) = k | N(t) = n) = \frac{P(N(s) = k, N(t) - N(s) = n-k)}{P(N(t) = n)}$$
$$= \binom{n}{k} \frac{s^k (t-s)^{n-k}}{t^n} \sim \text{Binomial}\left(n, \frac{s}{t}\right).$$

Example 2.20. Given two independent Poisson process $\{N_1(t), t \geq 0\}$ and $\{N_2(t), t \geq 0\}$ with rates λ_1 and λ_2

$$P(N_1(t) = k | N_1(t) + N_2(t) = n) = \frac{P(N_1(t) = k, N_2(t) = n-k)}{P(N_1(t) + N_2(t) = n)}$$
$$= \binom{n}{k} \left(\frac{\lambda_1}{\lambda_1 + \lambda_2}\right)^k \left(\frac{\lambda_2}{\lambda_1 + \lambda_2}\right)^{n-k} \sim \text{Binomial}\left(n, \frac{\lambda_1}{\lambda_1 + \lambda_2}\right).$$

2.3.3.4 The Order Statistic Property

Suppose that exactly one event of a Poisson process has taken place by time t. The time at which the event occurred is a uniform distribution over $[0, t]$: For $s \leq t$,

$$P(X_1 \leq s | N(t) = 1) = \frac{P(N(s) = 1, N(t) - N(s) = 0)}{P(N(t) = 1)} = \frac{s}{t}.$$

This result is generalized as the following order statistic property.

Theorem 2.7. *Suppose $\{N(t), t \geq 0\}$ is a Poisson process, and let $S_k = \sum_{i=1}^{k} X_i$ be the kth arrival time. Given that $N(t) = n$, the n arrival times $S_1, S_2, \ldots, S_n$ have the same distribution as the order statistics corresponding to n independent random variables uniformly distributed on the interval* $[0, t]$.

Example 2.21 (A Counter Model). Suppose that a device is subject to shocks that occur in accordance with a Poisson process having rate λ. The ith shock gives rise to a damage D_i. The $D_i, i \geq 1$, are assumed to be i.i.d. and also to be independent of $\{N(t), t \geq 0\}$, where $N(t)$ denotes the number of shocks in $[0, t]$. The damage due to a shock is assumed to decrease exponentially in time. That is, if a shock has an initial damage D, then a time t later its damage is $De^{-\alpha t}$. If we suppose that the damage are additive, then $D(t)$, the damage at t, can be expressed as $D(t) = \sum_{i=1}^{N(t)} D_i e^{-\alpha(t-S_i)}$, where S_i represents the arrival time of the ith shock. The question is to find $E[D(t)]$.

Conditioning on $N(t)$ yields

$$E[D(t)|N(t) = n] = E\left[\sum_{i=1}^{n} D_i e^{-\alpha(t-S_i)}|N(t) = n\right]$$
$$= E\left[D\right] e^{-\alpha t} E\left[\sum_{i=1}^{n} e^{\alpha S_i}|N(t) = n\right].$$

Letting $U_1, \ldots, U_n$ be i.i.d. $U(0, t)$ r.v.'s, then, by Theorem 2.7,

$$E\left[\sum_{i=1}^{n} e^{\alpha S_i}|N(t) = n\right] = E\left[\sum_{i=1}^{n} e^{\alpha U_{i:n}}\right] = E\left[\sum_{i=1}^{n} e^{\alpha U_i}\right] = \frac{n}{\alpha t}(e^{\alpha t} - 1).$$

Hence,

$$E[D(t)|N(t)] = \frac{N(t)}{\alpha t}(1 - e^{-\alpha t})E[D],$$

and, taking expectations,

$$E[D(t)] = \frac{\lambda E[D]}{\alpha}(1 - e^{-\alpha t}).$$

2.3.3.5 Nonhomogeneous Poisson Process

Definition 2.7. The counting process $\{N(t), t \geq 0\}$ is said to be a nonstationary or nonhomogeneous Poisson process (NHPP) with *intensity function* $\lambda(t)$, $t \geq 0$ if:

(i) $N(0) = 0$,
(ii) $\{N(t), t \geq 0\}$ has independent increments,
(iii) $P(N(t + h) - N(t) = 1) = \lambda(t)h + o(h)$,
(iv) $P(N(t + h) - N(t) \geq 2) = o(h)$.

The NHPP has independent increments but does not have stationary property. Thus, there is the possibility that events may be more likely to occur at certain times than at other times. It can be shown that $N(t + s) - N(t) \sim \text{Poisson}(m(t + s) - m(t))$, where $m(t) = \int_0^t \lambda(u)du$. Thus, $N(t)$ is Poisson distributed with mean $m(t)$, and for this reason $m(t)$ is called the *mean value function* of the process. We use Definition 2.7 to find the probability $P_n(s) = P(N(t+s) - N(t) = n)$. For fix t,

$$\begin{aligned} P_0(s+h) &= P(N(t+s+h) - N(t) = 0) \\ &= P(N(t+s) - N(t) = 0) \cdot P(N(t+s+h) - N(t+s) = 0) \\ &= P_0(s)[1 - \lambda(t+s)h + o(h)]. \end{aligned}$$

Hence

$$\frac{P_0(s+h) - P_0(s)}{h} = -\lambda(t+s)P_0(s) + \frac{o(h)}{h}.$$

Letting $h \to 0$ yields

$$P_0'(s) = -\lambda(t+s)P_0(s). \tag{2.21}$$

By integration and setting $P_0(0) = 1$, we have $P_0(s) = e^{-[m(t+s)-m(t)]}$. Similarly, by the same procedure used in the Poisson process, we have

$$P(N(t+s) - N(t) = n) = e^{-[m(t+s)-m(t)]}\frac{[m(t+s) - m(t)]^n}{n!}, \quad n = 0, 1, 2, \ldots.$$

In particular,

$$P(N(t) = n) = e^{-m(t)}\frac{[m(t)]^n}{n!}, \quad n = 0, 1, 2, \ldots.$$

The entire course of a NHPP is determined by the distribution of first failure time:

$$P(X_{n+1} > x | S_1 = s_1, \ldots, S_n = s_n) = P(X_1 > s_n + x | X_1 > s_n), \; n = 0, 1, 2, \ldots,$$

where X's are interarrival times and S's are arrival times. In addition, setting $t = 0$ in (2.21), the intensity function of a NHPP satisfies $\frac{d}{dt}EN(t) = \frac{d}{dt}[-\ln \overline{F}(t)]$, the failure rate of first failure time. For example, if the first failure time has a Weibull distribution $F(t) = 1 - \exp(-\alpha t^{\beta})$ with a shape parameter β and a scale parameter α, then the mean value function of the corresponding NHPP is $m(t) = \alpha t^{\beta}$. Such a NHPP is called a *Weibull process*, because first failure time has the Weibull distribution.

A NHPP model also corresponds to what is called *minimal repairs*, meaning that the system after repair is only as good as it was immediately before the failure. This means that the intensity function $\lambda(t)$ of the failure process immediately after the failure is the same as it was immediately before the failure, and hence is exactly as it would be if no failure had ever occurred. Thus we must have $\lambda(t) = h(t)$, where $h(t)$ is the failure rate of the first failure time. That is, the intensity function $\lambda(t)$ of the failure process is completely defined by the failure rate $h(t)$, and the mean value function of a NHPP is $m(t) = \int_0^t h(w)dw$. Formally,

$$P(X_{n+1} > x|S_n = t) = P(X_1 > t + x|X_1 > t) = \frac{\overline{F}(t+x)}{\overline{F}(t)} = e^{-[m(t+x)-m(t)]}. \tag{2.22}$$

Thus, the conditional distribution of time to failure, given $S_n = t$, is $F_i(x|S_n = t) = 1 - e^{-[m(t+x)-m(t)]}$. The conditional expectation is given by

$$\int_0^\infty [1 - F_i(x|S_n = t)]dx = \int_0^\infty \exp\left[-\int_t^{t+x} \lambda(w)dw\right]dx.$$

For example, if $\lambda(t) = t$ then the conditional expectation is

$$\int_0^\infty \exp\left[-\int_t^{t+x} wdw\right]dx = \int_0^\infty \exp\left[-\frac{1}{2}\left((t+x)^2 - t^2\right)\right]$$
$$dx = (2\pi)^{1/2} e^{t^2/2}[1 - \Phi(t)],$$

where Φ is the standard normal distribution function.

Parzen [81] gives the unconditional distributions of the nth failure time S_n:

$$P(S_n > t) = P(N(t) \le n-1) = \sum_{j=0}^{n-1} e^{-m(t)} \frac{(m(t))^j}{j!} = \int_{m(t)}^\infty \frac{x^{n-1}}{\Gamma(n)} e^{-x} dx. \tag{2.23}$$

The p.d.f. of S_n is

$$f_{S_n}(t) = \lambda(t) e^{-m(t)} \frac{[m(t)]^{n-1}}{(n-1)!}. \tag{2.24}$$

From (2.22) and (2.24),

$$P(X_{n+1} > x) = \int_0^\infty \lambda(s) e^{-m(x+s)} \frac{[m(s)]^{n-1}}{(n-1)!} ds. \tag{2.25}$$

By comparing (2.25) with (2.22) we see explicitly that the interarrival times are not independent, as is stated in Cox and Lewis [20].

From (2.23), the expectation of S_n is

$$ES_n = \int_0^\infty P(S_n > t)dt = \int_0^\infty m^{-1}(x) \frac{x^{n-1}}{\Gamma(n)} e^{-x} dx,$$

where m^{-1} is the inverse function of m. For example, the Weibull process where $m(t) = \alpha t^{\beta}, \alpha > 0$, then $m^{-1}(x) = \left(\frac{x}{\alpha}\right)^{1/\beta}$ and

$$ES_n = \int_0^{\infty} \left(\frac{x}{\alpha}\right)^{1/\beta} \frac{x^{n-1}}{\Gamma(n)} e^{-x} dx = \frac{\Gamma(n + \frac{1}{\beta})}{\alpha^{1/\beta} \cdot \Gamma(n)}.$$

2.3.3.6 Some Properties of NHPP

1. Like homogeneous Poisson process, NHPPs are closed under superposition, that is, the sum of a number of NHPPs is also a NHPP. Generally, we may mix the failure time data from different failure processes assumed to be NHPP and obtain an overall NHPP with a mean value function which is the sum of the mean value functions of the underlying NHPP models.
2. If events occur at $0 < t_1 < t_2 < \cdots < t_n < t$, the likelihood function of a NHPP is

$$f(t_1, \ldots, t_n) = \left\{\prod_{i=1}^{n} \lambda(t_i)\right\} \exp\left\{-\int_0^t \lambda(u) du\right\}.$$

Assume that at $[s_{i-1}, s_i)$, the actual observed cumulative number of detected faults is n_i. The likelihood function is

$$L(n_1, \ldots, n_k) = \prod_{i=1}^{n} e^{-[m(s_i)-m(s_{i-1})]} \frac{[m(s_i) - m(s_{i-1})]^{n_i}}{n_i!}.$$

3. Any NHPP can be transformed to a homogeneous Poisson process through an appropriate time-transformation, and vice versa. Specifically, let $m(t)$ be a continuous nondecreasing function of t. Then $S_1, S_2, \ldots$ are arrival times in a NHPP $\{N(t), t \geq 0\}$ with $EN(t) = m(t)$ if and only if $m(S_1), m(S_2), \ldots$ are the arrival times in a homogeneous Poisson process with rate one. Lindqvist et al. [55] study the *trend renewal process*, by allowing above rate one homogeneous Poisson process to be any renewal process.
4. In software reliability modeling, we are interested in the cumulative number of failures $X(t)$ experienced up to time $t \geq 0$. We assume that the number of initial faults in a software is a finite unknown constant N, and $\{X(t), t \geq 0\}$ is a pure linear birth process with birth rate ϕ. If we write $Y(t) = N - X(t)$, then $\{Y(t), t \geq 0\}$ is a linear death process representing the number of remaining faults or undiscovered faults up to time t. If N has a Poisson distribution with parameter θ, then $X(t)$ and $Y(t)$ are independent and the unconditional distribution of $X(t)$ is a NHPP with the mean value function $m(t) = \theta(1-e^{-\phi t})$.
5. (M/G/∞ queue) Suppose that customers arrive at a service station in accordance with a Poisson process with rate λ. Upon arrival the customer is immediately

served by one of an infinite number of possible servers, and the service times are assumed to be independent with a common distribution G. We want to find the distribution of the number of customers that have completed service by time t and the distribution of the number of customers that are being served at time t. Consider an entering customer a type-I customer if it completed its service by time t and a type-II customer if it does not complete its service by time t. If the customer enters at time s, $s \leq t$, then it will be a type-I customer if its service time is less than $t-s$, and since the service time distribution is G, the probability of this will be $G(t-s)$. Similarly, a customer entering at time s will be a type-II customer with probability $1-G(t-s)$. Let $N_i(t)$ represents the number of type-i events that occur by time t $(i = 1, 2)$, then $N_1(t)$ and $N_2(t)$ are independent Poisson random variables having respective mean $E[N_1(t)] = \lambda \int_0^t G(y)dy$ and $E[N_2(t)] = \lambda \int_0^t [1-G(y)]dy$, respectively. Details of this example, and further examples of this type, we refer the readers to Ross [93].

2.3.4 *Birth and Death Processes*

2.3.4.1 Chapman-Kolmogorov Equations

Definition 2.8. A stochastic process $\{X(t), t \geq 0\}$ is a continuous-time Markov chain if it takes on values in the set S (state space) of nonnegative integers and satisfies the Markov property:

$$\begin{aligned} P(X(t_{n+1}) &= x_{n+1}|X(t_1) = x_1, X(t_2) = x_2, \ldots, X(t_n) = x_n) \\ &= P(X(t_{n+1}) = x_{n+1}|X(t_n) = x_n), \end{aligned}$$

for $0 \leq t_1 < t_2 < \cdots < t_n < t_{n+1}$ and $x_1, x_2, \ldots, x_n, x_{n+1} \in S$.

Definition 2.9. The continuous-time Markov chain is called time-homogeneous or stationary if the transition probability $p_{ij}(s,t) = P(X(t+s) = j|X(s) = i)$ is independent of s, for all i, j, s, t, and we write

$$p_{ij}(t) = P(X(t+s) = j|X(s) = i) = P(X(t) = j|X(0) = i),$$

where $p_{ij}(t)$ is the probability that a Markov chain, presently in state i, will be in state j after an additional time t.

By the *Markov property* and the *homogeneity*, the amount of time a continuous-time Markov chain spends in each state is exponentially distributed.

If we put $p_{ij}(t)$ into the ith row, the jth column of a matrix, then the resulting matrix $\mathbf{P}(t) = (p_{ij}(t))$ is called the transition matrix of $\{X(t), t \geq 0\}$. The family $\{\mathbf{P}(t), t \geq 0\}$ is a stochastic semigroup; that is, it satisfies:

(i) $\mathbf{P}(t)$ is stochastic ($0 \leq p_{ij}(t) \leq 1$ and row sums equal to 1),

(ii) $\mathbf{P}(0) = \mathbf{I}$ (the identity matrix), (2.26)

(iii) The Chapman-Kolmogorov equation: $\mathbf{P}(t+s) = \mathbf{P}(t)\mathbf{P}(s)$.

2.3.4.2 Infinitesimal Generator

A matrix $\mathbf{Q}$, called the *infinitesimal generator* (or generator, Q-matrix, transition rate matrix), plays a role similar to that of a 1-step transition matrix in discrete-time Markov chains. Later in (2.30), we will see that there is a one-to-one correspondence between the infinitesimal generator $\mathbf{Q}$ and the transition matrix $\mathbf{P}(t)$.

Assume that the probability of 2 or more transitions in the interval $(t, t+h)$ is $o(h)$. We are interested in the behavior of $p_{ij}(h)$ for small h. Following (2.26), it turns out that $p_{ij}(h)$ is approximately linear in h when h is small. That is, there exist constants $\{q_{ij}; i, j \in S\}$ such that

$$p_{ij}(h) \approx \begin{cases} q_{ij}h & \text{if } i \neq j \\ 1 + q_{ii}h & \text{if } i = j. \end{cases}$$

It can be shown that each transition probability $p_{ij}(t)$ is uniformly continuous in $t \geq 0$, and $p_{ij}(t)$ is differentiable with respect to $t \geq 0$ [45]. Let us define

$$\mathbf{Q} \equiv \mathbf{P}'(0+) \qquad \text{or} \qquad \mathbf{Q} = \lim_{h \to 0+} \frac{\mathbf{P}(h) - \mathbf{I}}{h}.$$

Since $\mathbf{P}(0) = \mathbf{I}$, we have

$$q_{ij} = \begin{cases} \lim_{h \to 0+} \frac{p_{ij}(h)}{h} = p'_{ij}(0) & i \neq j \\ \lim_{h \to 0+} \frac{p_{ii}(h)-1}{h} = -\sum_{j \neq i} p'_{ij}(0) & i = j, \end{cases}$$

with $q_{ij} \geq 0$ for $i \neq j$ and $q_{ii} = -\sum_{j \neq i} q_{ij} \leq 0$. We call q_{ij} the *transition rate* from state i to state j. State i is an *absorbing* state if $q_{ii} = 0$.

$(-q_{ii})$ is the expected number of times that the process leaves state i per unit of time spent in state i. It is also called the *mean sojourn rate* of state i. Thus $(-q_{ii})$ is the reciprocal of the expected *holding time* that the process spends in state i per visit to state i. Similarly, q_{ij} is the transition rate from state i to state j in the sense that q_{ij} is the expected number of times that the process leaves from state i to state j per unit of time spent in state i. Thus $(-q_{ii}) = \sum_{j \neq i} q_{ij}$.

The information of the generator is a combination of the *embedded Markov chain* and the mean sojourn rates. Not only can the generator be obtained from the embedded Markov chain and mean sojourn rates, but the reverse is also true. For the generator matrix, the absolute value of the diagonal elements gives the mean sojourn rates. The transition matrix for the embedded Markov chain is then obtained by dividing the off-diagonal element by the absolute value of that row's diagonal element. The diagonal elements of the transition matrix for the embedded Markov chain are zero (except an absorbing state). $q_{ij}/(-q_{ii})$, for example, is the conditional probability of a transition from state i to state j given that a transition from state i has taken place. If the holding time in each state follows an arbitrary distribution other than exponential, the involved process is a *semi-Markov process*.

Example 2.22. Suppose a Markov process with state space $\{a, b, c\}$ has a generator

$$\mathbf{Q} = \begin{bmatrix} -2 & 2 & 0 \\ 2 & -4 & 2 \\ 1 & 4 & -5 \end{bmatrix}.$$

The mean sojourn rates are $\lambda_a = 2$, $\lambda_b = 4$, and $\lambda_c = 5$, and the transition matrix for the embedded Markov chain is

$$\mathbf{P} = \begin{bmatrix} 0 & 1 & 0 \\ 0.5 & 0 & 0.5 \\ 0.2 & 0.8 & 0 \end{bmatrix}.$$

2.3.4.3 Steady-State Probabilities

The steady-state probabilities are obtained directly from the generator $\mathbf{Q}$. Let $\boldsymbol{\pi}$ be a vector of steady-state probability matrix, then $\boldsymbol{\pi}$ is the solution to $\boldsymbol{\pi}\mathbf{Q} = \mathbf{0}$, with row sum of $\boldsymbol{\pi}$ is 1. This equation is similar to $\boldsymbol{\pi}\mathbf{P} = \boldsymbol{\pi}$. Since Markov matrices have row sum of 1, the right-hand side of $\boldsymbol{\pi}\mathbf{P} = \boldsymbol{\pi}$ is *one times* $\boldsymbol{\pi}$. Since the generator has row sums of 0, the right-hand side of $\boldsymbol{\pi}\mathbf{Q} = \mathbf{0}$ is *zero times* $\boldsymbol{\pi}$.

Example 2.23. A 3-state Markov process $S = \{0, 1, 2\}$ with generator $\mathbf{Q} = \begin{bmatrix} -2 & 2 & 0 \\ 2 & -3 & 1 \\ 0 & 2 & -2 \end{bmatrix}$ has the steady-state distribution $(\pi_0, \pi_1, \pi_2) = (2/5, 2/5, 1/5)$.

2.3.4.4 Eigenvalue Approach

The eigenvalue approach is useful for finding the transition probabilities of a finite states Markov process. In principle, if we can calculate all the eigenvalues and their associated column eigenvectors for the generator $\mathbf{Q}$, then we can calculate $\mathbf{P}(t)$ analytically. Assume that $\mathbf{w}_1, \ldots, \mathbf{w}_n$ are eigenvalues of $\mathbf{Q}$ and $\mathbf{u}_1, \ldots, \mathbf{u}_n$ are their associated right eigenvectors. We also assume that all the eigenvalues are different. We can write $\mathbf{Q} = \mathbf{UWU}^{-1}$, where $\mathbf{U} = [\mathbf{u}_1, \ldots, \mathbf{u}_n]$, $\mathbf{U}^{-1}$ is the inverse matrix of $\mathbf{U}$, and $\mathbf{W}$ is the orthogonal matrix with eigenvalues on the main diagonal. Then $\mathbf{Q}^n = \mathbf{UW}^n\mathbf{U}^{-1}$. It follows thats

$$\mathbf{P}(t) = \mathbf{I} + \sum_{i=1}^{\infty} \frac{(\mathbf{Q}t)^i}{i!} = \mathbf{I} + \sum_{i=1}^{\infty} \frac{\mathbf{U}(\mathbf{W}t)^i\mathbf{U}^{-1}}{i!} = \mathbf{U}e^{\mathbf{W}t}\mathbf{U}^{-1}. \tag{2.27}$$

Example 2.24. Consider an electronic device that is either "on standby" or "in use." The on-standby period is exponentially distributed with a mean of 20 s, and the in-use period is exponentially distributed with a mean of 12 s. Because the exponential assumption is satisfied, a Markov process $\{Y(t); t \geq 0\}$ with state space $E = \{0, 1\}$ can be used to model the electronic device as it alternates between being on standby and in use. State 0 denotes on standby, and state 1 denotes in use. The generator is given by $\mathbf{Q} = \begin{bmatrix} -3 & 3 \\ 5 & -5 \end{bmatrix}$, where the time unit is minute. The long-run probability of being in use is $\frac{3}{8}$. Now, we are interested in the time-dependent probabilities. Two eigenvalues are $\mathbf{w}_1 = 0$ and $\mathbf{w}_2 = -8$, and the corresponding eigenvectors are $(1 \;\; 1)'$ and $(3 \;\; -5)'$, respectively. Therefore, from (2.27), we have

$$\mathbf{P}(t) = \mathbf{U}e^{\mathbf{W}t}\mathbf{U}^{-1} = \begin{bmatrix} 1 & 3 \\ 1 & -5 \end{bmatrix}\begin{bmatrix} 1 & 0 \\ 0 & e^{-8t} \end{bmatrix}\begin{bmatrix} 0.625 & 0.375 \\ 0.125 & -0.125 \end{bmatrix}$$

$$= \begin{bmatrix} 0.625 + 0.375e^{-8t} & 0.375 - 0.375e^{-8t} \\ 0.625 - 0.625e^{-8t} & 0.375 + 0.625e^{-8t} \end{bmatrix}, \quad t \geq 0.$$

The probability that the electronic device is in use at time t given that it started in use at time 0 is

$$P(Y(t) = 1 | Y(0) = 1) = 0.375 + 0.625e^{-8t}.$$

Example 2.25. Consider an *alternating renewal process* whose infinitesimal generator is

$$\mathbf{Q} = \begin{bmatrix} -\lambda & \lambda \\ \mu & -\mu \end{bmatrix}.$$

Two eigenvalues are $\mathbf{w}_1 = 0$ and $\mathbf{w}_2 = -(\lambda + \mu)$, and the corresponding eigenvectors are $(1 \;\; 1)'$ and $(\lambda \;\; -\mu)'$, respectively. Thus, from (2.27), we have

$$\mathbf{P}(t) = \begin{bmatrix} \dfrac{\mu}{\lambda+\mu} + \dfrac{\lambda}{\lambda+\mu}e^{-(\lambda+\mu)t} & \dfrac{\lambda}{\lambda+\mu} - \dfrac{\lambda}{\lambda+\mu}e^{-(\lambda+\mu)t} \\ \dfrac{\mu}{\lambda+\mu} - \dfrac{\mu}{\lambda+\mu}e^{-(\lambda+\mu)t} & \dfrac{\lambda}{\lambda+\mu} + \dfrac{\mu}{\lambda+\mu}e^{-(\lambda+\mu)t} \end{bmatrix}, \quad t \geq 0.$$

2.3.4.5 Kolmogorov Backward/Forward Equations

When the eigenvalue approach is not applicable, we need an alternative method to find transition probabilities $p_{ij}(t)$. Applying the Chapman-Kolmogorov equations $\mathbf{P}(t + s) = \mathbf{P}(t)\mathbf{P}(s)$, then

$$\frac{\mathbf{P}(t+h) - \mathbf{P}(t)}{h} = \frac{\mathbf{P}(t)\mathbf{P}(h) - \mathbf{P}(t)}{h} = \mathbf{P}(t)\frac{\mathbf{P}(h) - \mathbf{I}}{h}, \quad h > 0.$$

This implies

$$\mathbf{P}'(t) = \mathbf{P}(t)\mathbf{Q} \quad \text{or} \quad p'_{ij}(t) = \sum_k p_{ik}(t)q_{kj}. \tag{2.28}$$

This is Kolmogorov *forward* differential equation. Similarly, using $\mathbf{P}(t+s) = \mathbf{P}(s)\mathbf{P}(t)$, we have

$$\frac{\mathbf{P}(t+h) - \mathbf{P}(t)}{h} = \frac{\mathbf{P}(h)\mathbf{P}(t) - \mathbf{P}(t)}{h} = \frac{\mathbf{P}(h) - \mathbf{I}}{h}\mathbf{P}(t), \quad h > 0.$$

This implies

$$\mathbf{P}'(t) = \mathbf{Q}\mathbf{P}(t) \quad \text{or} \quad p'_{ij}(t) = \sum_k q_{ik}p_{kj}(t). \tag{2.29}$$

This is Kolmogorov *backward* differential equation. Subject to the boundary condition $\mathbf{P}(0) = \mathbf{I}$, the unique solution for both forward and backward equations is given by

$$\mathbf{P}(t) = e^{\mathbf{Q}t} = \sum_{i=0}^{\infty} \frac{(\mathbf{Q}t)^i}{i!}, \quad t \geq 0. \tag{2.30}$$

2.3.4.6 Birth Process

Poisson Process: A Poisson process discussed in Sect. 2.3.3.1 is a pure birth process. The generator of a Poisson process with state $S = \{0, 1, 2, \ldots\}$ is

$$\mathbf{Q} = \begin{bmatrix} -\lambda & \lambda & 0 & 0 & \ldots \\ 0 & -\lambda & \lambda & 0 & \ldots \\ 0 & 0 & -\lambda & \lambda & \ldots \\ \vdots & & & \ddots & \end{bmatrix}.$$

From the Kolmogorov forward equations (2.28), we have

$$p'_{ij}(t) = p_{ij}(t)q_{jj} + p_{i,j-1}(t)q_{j-1,j} = -\lambda p_{ij}(t) + \lambda p_{i,j-1}(t).$$

Pure Birth Process: The pure (or simple) birth process is a natural and the simplest generalization of the Poisson process. It permits the chance of an event occurring at a given instant of time to depend upon the number of events which have already occurred.

Definition 2.10. A pure birth process $\{X(t), t \geq 0\}$ is a continuous-time Markov chain with positive parameters $\{\lambda_k, k = 0, 1, 2, \ldots\}$ called *birth rates* satisfying the following statements.

1. $P(X(t+h) - X(t) = 1 | X(t) = k) = \lambda_k h + o_{1,k}(h)$,
2. $P(X(t+h) - X(t) = 0 | X(t) = k) = 1 - \lambda_k h + o_{2,k}(h)$,
3. $P(X(t+h) - X(t) \geq 2 | X(t) = k) = o_{3,k}(h)$,
4. $X(0) = 0$.

The infinitesimal generator of a pure birth process $\{X(t), t \geq 0\}$ is

$$\mathbf{Q} = \begin{pmatrix} -\lambda_0 & \lambda_0 & 0 & 0 & 0 \ldots \\ 0 & -\lambda_1 & \lambda_1 & 0 & 0 \ldots \\ 0 & 0 & -\lambda_2 & \lambda_2 & 0 \ldots \\ \vdots & \vdots & \vdots & \vdots & \vdots \end{pmatrix}.$$

To find the distribution of $X(t)$, we let $P_n(t) = P(X(t) = n)$. The probability $P_n(t)$ is the transition probability $P(X(t) = n | X(0) = 0)$. Using the Kolmogorov forward equations (2.28), we have

$$P_0'(t) = -\lambda_0 P_0(t),$$
$$P_n'(t) = -\lambda_n P_n(t) + \lambda_{n-1} P_{n-1}(t), \quad n = 1, 2, \ldots,$$

with the initial conditions $P_0(0) = 1$ and $P_n(0) = 0$, $n = 1, 2, \ldots$. Solving recursively, we have

$$P_0(t) = e^{-\lambda_0 t},$$

$$P_1(t) = \lambda_0 \left[\frac{e^{-\lambda_0 t}}{\lambda_1 - \lambda_0} + \frac{e^{-\lambda_1 t}}{\lambda_0 - \lambda_1} \right], \quad (\lambda_0 \neq \lambda_1),$$

$$P_2(t) = \lambda_0 \lambda_1 \left[\frac{e^{-\lambda_0 t}}{(\lambda_1 - \lambda_0)(\lambda_2 - \lambda_0)} + \frac{e^{-\lambda_1 t}}{(\lambda_0 - \lambda_1)(\lambda_2 - \lambda_1)} + \frac{e^{-\lambda_2 t}}{(\lambda_0 - \lambda_2)(\lambda_1 - \lambda_2)} \right],$$

$$\vdots$$

$$P_n(t) = \lambda_0 \lambda_1 \ldots \lambda_{n-1} \left[\psi_{0,n} e^{-\lambda_0 t} + \psi_{1,n} e^{-\lambda_1 t} + \cdots + \psi_{n,n} e^{-\lambda_n t} \right],$$

where

$$\psi_{k,n} = [(\lambda_0 - \lambda_k) \ldots (\lambda_{k-1} - \lambda_k)(\lambda_{k+1} - \lambda_k) \ldots (\lambda_n - \lambda_k)]^{-1}$$

$(\lambda_j \neq \lambda_k$ unless $j = k)$.

To find a general transition probability $P_{ij}(t)$, we again use the Kolmogorov forward equations (2.28) and yield the differential equations

$$\begin{aligned} P_{ii}'(t) &= -\lambda_i P_{ii}(t), \\ P_{ij}'(t) &= P_{i,j-1}(t)q_{j-1,j} + P_{i,j}(t)q_{jj} \\ &= \lambda_{j-1} P_{i,j-1}(t) - \lambda_j P_{ij}(t), \quad t \geq 0 \quad (j > i; \ P_{ij}(t) = 0 \text{ if } j < i), \end{aligned}$$

with $P_{ii}(0) = 1$ and $P_{ij}(0) = 0$, $j \neq i$. Solving recursively we get

$$\begin{aligned} P_{ii}(t) &= e^{-\lambda_i t}, \\ P_{ij}(t) &= \lambda_{j-1} \int_0^t e^{-\lambda_j (t-s)} P_{i,j-1}(s) ds, \quad j > i. \end{aligned} \tag{2.31}$$

Yule Process: The Yule process is a well-known example of a pure birth process in biology. It characterizes the reproduction of living organisms, in which under certain conditions (sufficient food, no mortality, no migration, etc.) the probability of a birth at a given instant is proportional to the population size at that time. That is, the Yule process is a linear birth process with birth rate $\lambda_k = k\lambda$ $(k = 0, 1, 2, \ldots)$.

We first start with a single individual at time 0, i.e., $X(0) = 1$. The infinitesimal generator with $S = \{0, 1, 2, \ldots\}$ is

$$\mathbf{Q} = \begin{pmatrix} 0 & 0 & 0 & 0 & 0 & \ldots \\ 0 & -\lambda & \lambda & 0 & 0 & \ldots \\ 0 & 0 & -2\lambda & 2\lambda & 0 & \ldots \\ \vdots & \vdots & \vdots & \vdots & \vdots & \end{pmatrix}.$$

Let $P_n(t) = P(X(t) = n)$, given $X(0) = 1$. The Kolmogorov forward equations (2.28) is

$$\begin{aligned} P_0'(t) &= 0, \\ P_n'(t) &= -n\lambda P_n(t) + (n-1)\lambda P_{n-1}(t), \quad n = 1, 2, \ldots, \end{aligned}$$

with the initial conditions $P_1(0) = 1$ and $P_n(0) = 0$ $(n = 2, 3, \ldots)$. Solving recursively we get $P_0(t) = 0$ and

$$P_n(t) = e^{-\lambda t}(1 - e^{-\lambda t})^{n-1}, \quad n = 1, 2, \ldots. \tag{2.32}$$

Thus, starting with a single individual at time 0, $X(t)$ is geometric distributed with mean $e^{\lambda t}$ and variance $(1 - e^{-\lambda t})e^{2\lambda t}$.

Example 2.26. Applying the binomial theorem into (2.32), we write

$$P_n(t) = e^{-\lambda t}(1-e^{-\lambda t})^{n-1} = e^{-\lambda t} \cdot \sum_{k=0}^{n-1} \binom{n-1}{k} (-1)^k e^{-\lambda k t}.$$

Let $\tilde{P}_n(s) = \int_0^\infty P_n(t)e^{-st}\,dt$ be the Laplace transform of $P_n(t)$ for the linear birth process with $X(0) = 1$. Then

$$\tilde{P}_n(s) = \int_0^\infty e^{-\lambda t} \cdot \sum_{k=0}^{n-1} \binom{n-1}{k} (-1)^k e^{-\lambda k t} e^{-st}\,dt = \sum_{k=0}^{n-1} (-1)^k \binom{n-1}{k} \frac{1}{s+(k+1)\lambda}.$$

Next, if the population starts with γ individuals, its size at time t will be the sum of γ i.i.d. geometric r.v.'s, and thus have a negative binomial distribution. Let $P_n^\gamma(t) = P(X(t) = n|X(0) = \gamma)$, then

$$P_n^\gamma(t) = \binom{n-1}{\gamma-1} \left(e^{-\lambda t}\right)^\gamma \left(1-e^{-\lambda t}\right)^{n-\gamma}, \quad n = \gamma, \gamma+1, \ldots.$$

To obtain a general transition probability $P_{ij}(t)$, we substitute $\lambda_j = j\lambda$ into (2.31) and solve them recursively to get the desired results.

The following is an example of the Yule process as a epidemic model, which may be used to model faults propagation in network or software reliability.

Example 2.27. Consider a population of m individuals that at time 0 consists of one "infected" and $m-1$ "susceptibles". Once infected an individual remains in that state forever and suppose that in any time interval h any given infected person will cause, with probability $\alpha h + o(h)$, any given susceptible to become infected. Let $X(t)$= # of infected individuals at time t, the $\{X(t), t \geq 0\}$ is a pure birth process with

$$\lambda_n = \begin{cases} (m-n)n\alpha, & n = 1, \ldots, m-1 \\ 0, & \text{otherwise.} \end{cases}$$

Let T be the time until the total population is infected, then $T = \sum_{i=1}^{m-1} T_i$, where T_i is the time from i infectives to $i+1$ infectives ($i = 1, \ldots, m-1$). Because $T_i \sim \text{Exp}((m-i)i\alpha)$, the mean and variance of T are, respectively,

$$E[T] = \sum_{i=1}^{m-1} \frac{1}{(m-i)i\alpha} = \frac{1}{m\alpha} \sum_{i=1}^{m-1} \left(\frac{1}{m-i} + \frac{1}{i} \right) \approx \frac{2\log(m-1)}{m\alpha},$$

$$\text{Var}(T) = \frac{1}{\alpha^2} \sum_{i=1}^{m-1} \left(\frac{1}{(m-i)i} \right)^2.$$

2.3.4.7 Death Process

Pure Death Process:

Definition 2.11. A pure (or simple) death process $\{X(t), t \geq 0\}$ is a continuous-time Markov chain with parameters $\{\mu_k, k = 1, 2, \ldots, n\}$ called *death rates* satisfying the following properties:

(i) $P(X(t+h) - X(t) = -1|X(t) = k) = \mu_k h + o_{1,k}(h)$,
(ii) $P(X(t+h) - X(t) = 0|X(t) = k) = 1 - \mu_k h + o_{2,k}(h)$,
(iii) $P(X(t+h) - X(t) \leq -2|X(t) = k) = o_{3,k}(h),\ k \geq 0$,
(iv) $X(0) = n$.

It is natural that in the pure birth process, the events take place infinitely often for an infinite interval of time. However, in the pure death process, at most n events take place for any interval of time, since no event take place once the process reaches state 0, an absorbing state.

Let $P_k(t) = P(X(t) = k|X(0) = n)$, $k = 0, 1, \ldots, n$, be the transition probability with the initial condition $X(0) = n$. The Kolmogorov forward equations (2.28) for the pure death process are

$$P_n'(t) = -\mu_n P_n(t),$$
$$P_k'(t) = -\mu_k P_k(t) + \mu_{k+1} P_{k+1}(t), \quad k = 1, 2, \ldots, n-1,$$
$$P_0'(t) = \mu_1 P_1(t),$$

subject to the initial conditions $P_n(0) = 1$ and $P_k(0) = 0$ $(k = 1, 2, \ldots, n-1)$. Solving recursively we obtain

$$P_n(t) = e^{-\mu_n t},$$

$$P_{n-1}(t) = \mu_n \left[\frac{e^{-\mu_{n-1} t}}{\mu_n - \mu_{n-1}} + \frac{e^{-\mu_n t}}{\mu_{n-1} - \mu_n} \right],$$

$$P_{n-2}(t) = \mu_n \mu_{n-1} \left[\frac{e^{-\mu_n t}}{(\mu_{n-1} - \mu_n)(\mu_{n-2} - \mu_n)} + \frac{e^{-\mu_{n-1} t}}{(\mu_n - \mu_{n-1})(\mu_{n-2} - \mu_{n-1})} \right.$$
$$\left. + \frac{e^{-\mu_{n-2} t}}{(\mu_n - \mu_{n-2})(\mu_{n-1} - \mu_{n-2})} \right],$$

$$\vdots$$

$$P_0(t) = \mu_n \mu_{n-1} \ldots \mu_1 \left[\psi_{n,0} e^{-\mu_n t} + \psi_{n-1,0} e^{-\mu_{n-1} t} + \cdots + \psi_{0,0} e^{-\mu_0 t} \right],$$

where

$$\psi_{k,0} = [(\mu_n - \mu_k) \ldots (\mu_{k+1} - \mu_k)(\mu_{k-1} - \mu_k) \ldots (\mu_0 - \mu_k)]^{-1}$$

$(\mu_j \neq \mu_k$ unless $j = k)$.

Linear Death Process: If $\mu_k = k\mu$ ($k = 1, 2, \ldots, n$), i.e., the death rate is proportional to the number of members alive in a population. With the initial conditions $P_n(0) = 1$ and $P_k(0) = 0$ ($k = 1, 2, \ldots, n-1$), we have

$$P_k(t) = \binom{n}{k}(e^{-\mu t})^k(1 - e^{-\mu t})^{n-k}, \quad k = 0, 1, 2, \ldots, n.$$

That is, $X(t) \sim \text{Binomial}(n, e^{-\mu t})$. The mean and variance are, respectively,

$$E[X(t)] = ne^{-\mu t} \quad \text{and} \quad \text{Var}[X(t)] = ne^{-\mu t}(1 - e^{-\mu t}).$$

2.3.4.8 Birth and Death Processes

Birth and Death Process

Definition 2.12. A birth and death process $\{X(t), t \geq 0\}$ is a continuous-time Markov chain on the states $\{0, 1, 2, \ldots\}$ with stationary transition probability $P_{ij}(t) = P(X(t+s) = j | X(s) = i)$ satisfying

1. $P_{i,i+1}(h) = \lambda_i h + o_{1,i}(h)$,
2. $P_{i,i-1}(h) = \mu_i h + o_{2,i}(h)$,
3. $P_{ii}(h) = 1 - (\lambda_i + \mu_i)h + o_{3,i}(h)$,
4. $P_{ij}(0) = \sigma_{ij}$, (i.e., $P_{ii}(0) = 1$, $P_{ij}(0) = 0$ if $j \neq i$),
5. $\mu_0 = 0, \lambda_0 > 0, \lambda_i, \mu_i > 0, i = 1, 2, \ldots,$

where λ_i are called the birth rates and μ_i are called the death rates.

The infinitesimal generator of a birth and death process $\{X(t), t \geq 0\}$ is

$$\mathbf{Q} = \begin{pmatrix} -\lambda_0 & \lambda_0 & 0 & 0 & 0 \ldots \\ \mu_1 & -(\lambda_1 + \mu_1) & \lambda_1 & 0 & 0 \ldots \\ 0 & \mu_2 & -(\lambda_2 + \mu_2) & \lambda_2 & 0 \ldots \\ 0 & 0 & \mu_3 & -(\lambda_3 + \mu_3) & \lambda_3 \ldots \\ \vdots & \vdots & \vdots & \vdots & \vdots \end{pmatrix}.$$

The amount of time the process in state k before making a transition into a different state is $\text{Exp}(\lambda_k + \mu_k)$. From the Kolmogorov forward equations (2.28), we have

$$\begin{aligned} P'_{ij}(t) &= P_{i,j-1}(t)q_{j-1,j} + P_{ij}(t)q_{jj} + P_{i,j+1}(t)q_{j+1,j} \\ &= \lambda_{j-1}P_{i,j-1}(t) - (\lambda_j + \mu_j)P_{ij}(t) + \mu_{j+1}P_{i,j+1}(t), \quad j \geq 1, \end{aligned}$$

$$P'_{i0}(t) = -\lambda_0 P_{i0}(t) + \mu_1 P_{i1}(t), \quad (\mu_0 = 0, \lambda_{-1} = 0).$$

From the Kolmogorov backward equations: $P'_{ij}(t) = \sum_{k=0}^{\infty} q_{ik} P_{kj}(t)$, we have

$$\begin{aligned} P'_{ij}(t) &= q_{i,i-1} P_{i-1,j}(t) + q_{ii} P_{ij}(t) + q_{i,i+1} P_{i+1,j}(t) \\ &= \mu_i P_{i-1,j}(t) - (\lambda_i + \mu_i) P_{ij}(t) + \lambda_i P_{i+1,j}(t), \quad i \geq 1, \\ P'_{0j}(t) &= -\lambda_0 P_{0j}(t) + \lambda_0 P_{1j}(t). \end{aligned}$$

For both cases, the initial conditions are $P_{ij}(0) = \delta_{ij}$. The transition probabilities $P_{ij}(t)$ may in principle be calculated from a knowledge of the birth and death rates, although in practice these functions rarely have nice forms. It is an easier matter to determine the asymptotic behavior of the process as $t \to \infty$. There is a theoretical interest involving the transition probabilities $P_{ij}(t)$ called Karlin-McGregor representation, see Doom [109].

A Linear Growth Model with Immigration: A birth-death process is called a linear growth process with immigration if

$$\begin{aligned} \mu_n &= n\mu, \quad n \geq 1, \\ \lambda_n &= n\lambda + \theta, \quad n \geq 0. \end{aligned}$$

Such processes occur naturally in the study of biological reproduction and population growth. Each individual is assumed to give birth at an exponential rate λ; in addition, there is an exponential rate of increase θ of the population due to an external source such as immigration. Hence, $\lambda_n = n\lambda + \theta$. Deaths are assumed to occur at an exponential rate μ for each member of the population, and hence $\mu_n = n\mu$. Substitute $\lambda_n = n\lambda + \theta$ and $\mu_n = n\mu$ into the Kolmogorov forward equations (2.28):

$$\begin{aligned} P'_{i0}(t) &= -\theta P_{i0}(t) + \mu P_{i1}(t), \quad (\lambda_0 = 0), \\ P'_{ij}(t) &= (\lambda(j-1) + \theta) P_{i,j-1}(t) - ((\lambda + \mu) j + \theta) P_{ij}(t) \\ &\quad + \mu(j+1) P_{i,j+1}(t), \quad j \geq 1. \end{aligned}$$

If $X(0) = i$ and we multiply the jth equation by j and sum, it can be shown that the expected value $M(t) = E[X(t)] = \sum_{j=1}^{\infty} j P_{ij}(t)$ satisfies the differential equation

$$M'(t) = \theta + (\lambda - \mu) M(t),$$

with the initial condition $M(0) = i$.

2.4 Software Reliability Models

2.4.1 Introduction

Musa et al. [76] pointed out that three of the most significant software product characteristics are *quality*, *cost*, and *time of delivery*. It seems unlikely that anyone can create a software product simultaneously satisfying high quality, rapid delivery, and low cost, so trade-offs are required among the characteristics. Quantitative measures exist for cost and schedule, but the quantification of quality is more difficult. The quality of a software product can be represented by *software quality attributes* which are multidimensional properties and even hierarchical structures. Before a software product is released for service, it is developed through software development life cycle (SDLC): requirements, design, implementation, testing, and validation. Determination and measurement of user desirable software quality attributes is a subject of *software engineering* which concerns the study of the SDLC in various aspects.

Software reliability is defined as the probability of failure free software operation for a specified period of time in a specified environment [76]. Software reliability is an attribute of software quality and it has proved to be the most readily quantifiable of the attributes of software quality. Many software reliability models are applied to the testing or debugging phase of software development. There is an implicit assumption in most works on software reliability that the failure data obtained during the test and development phase represents its behaviour in the use environment [56].

Early software reliability models were deterministic models which are also called *software metrics* or *complexity metrics*. Software metrics are defined by analyzing the structure of the program or the flowgraph of the program. These models usually empirically measure the software quality attributes. They are used in the early phases of the SDLC to predict the number of faults, or are used in the maintenance phase for assessing and controlling the quality of a software product [37, 46, 68, 91]. Nowadays software reliability models refer to probabilistic ones and they have become the mainstream of software reliability study. The subject of software reliability has rapidly developed and spawned some original probability models of its own [22].

A *fault* is defined as a defect in a system that may cause an *error* during its operation. If an error affects the service to be provided by a system, a *failure* occurs. Because of the characteristics of software products, the operational environments of software, and the variety of development processes, the definition of software failure differs from application to application and should be defined clearly in the specifications. For example, a response time of 30 s could be a serious failure for an air traffic control system, but acceptable for an airline reservation system.

There is no physical process that determines when a program should fail, rather there occur in the program certain faults of logic or syntax that might cause the program to crash or produce results contrary to the specifications. Upon execution of a program, an input state is translated into an output state. Hence, a program can be regarded as a function mapping the input space (the set of all input states) to the output space (the set of all output states). Different users have different operational profiles which will determine the probabilities of selection of different inputs during execution. Software failures may not be 'generated' stochastically but may be 'detected' in a random way [42]. It justifies the use of stochastic models.

A large number of statistical software reliability models have been proposed over the past decade for evaluating the reliability of computer software. For reviews of the existing models, see [2, 33, 52, 64, 76, 83, 102, 104]. One of the main approaches of those models is to obtain results from assumptions about the stochastic behavior of how failures occur during software's running time. In this approach, the methods deal with the failure process by characterizing the time pattern of failure occurrences.

It is of great interest to classify or unify the large amount of software reliability models. Model classification and unification are useful in research in a sense that similar models are related to each other, and new and useful models can be built based on this. Section 2.4.2 reviews classification and unification of software reliability models. The most cited software reliability models are the Jelinski-Moranda model [42], henceforth the JM model, and the Goel-Okumoto model [34], henceforth the GO model. Essentially, the JM model corresponds to a (bounded) pure birth process, while the GO model corresponds to an NHPP process. In Sect. 2.4.3, we review the assumptions of the JM model. Sections 2.4.4–2.4.6 introduce several types of SRGMs by relations between the transition rate functions and the conditional failure rates, including i.i.d. order statistic models, time-between-failure models, and NHPP models. In Sect. 2.4.7, a more general setting of self-exciting processes is discussed.

2.4.2 Taxonomy of Software Reliability Models

Software reliability models can be broadly divided into the so-called white-box and black-box models. The white-box (or architecture-based) models incorporates information on the structure of the software in the models, while the black-box models disregard the internal structure of the system and represent solely its external failure behavior. A good review of architecture-based models is given by Goševa-Popstojanova and Trivedi [36]. The black-box approach is the most prevalent approach, and they can be further classified as *static* and *dynamic*. Static models include input-domain related models, complexity models (deterministic), and error seeding models (probabilistic). Error seeding models are also called capture-recapture models, and hypergeometric distribution is the basic tool [14]. The dynamic models consider the failure process as a stochastic process, and the reliability of the software changes at different stages of testing and debugging. Classifications of black-box software reliability models have been discussed by Goel [33], Gokhale et al. [35], Musa et al. [76], Pham [83], and Singpurwalla and Wilson [102]. Recently, Sharma et al. [98] have provide a figure which thoroughly classifies software reliability model according to SDLC. Models are classified into six categories: early prediction models, software reliability growth models (SRGMS), input domain based models, architecture based models, hybrid black box models, and hybrid white box models.

By unification we mean a mechanism which can generate new models and accommodate some existing models. Several such unifications had been proposed. For example, Langberg and Singpurwalla [51] unified some earlier models by adopting Bayesian method, Miller [70] discussed general order statistic models, Chen and Singpurwalla [18] unified software reliability models by self-exciting processes, Pham [86] generalized NHPP by extending the differential equations of the GO model, Wang [113] generalized existing software reliability models by considering both properties of mixture and self-exciting, and Huang et al. [40] presented how several existing software reliability growth models based on NHPPs can be derived according to the concept of weighted arithmetic, weighted geometric, or weighted harmonic means.

For many SRGMs, it is usually assumed that a fault is removed immediately on detection and correction, and no new faults are introduced. The so-called perfect debugging. In addition, the removal time is always assumed to be negligible. However, it is more practical to consider effect of debugging and timing of debugging as the imperfect removals happen in practice.

Effect of debugging:

- Perfect removal: detected faults are removed successfully,
- Imperfect removal: detected faults are not removed successfully and no new fault is introduced,
- Fault introduction: new faults may be introduced by incorrect debugging.

Timing of debugging:

- Immediate: all detected faults instantaneously repaired at time of first manifestation,
- Time-delayed: a fault may manifest itself repeatedly before being debugged.

The term *imperfect debugging* is used for both imperfect removal and fault introduction. In terms of the number of remaining faults, under certain assumptions (see Sect. 2.4.3), the imperfect removal process can be described by a simple death process [13, 33], and the imperfect debugging process can be modeled by a birth-death process [49].

Intuitively, we may assume a probability p for a fault to be removed successfully, and consider that there are two types of failures. The first type of failures are caused by inherent faults and the second type of failures are caused by introduced faults. A simple way for describing introduced faults is to assume that new faults are introduced by a homogeneous Poisson process with some parameter θ. For examples, Tokuno and Yamada [107] assume that failures caused by inherent faults follow Moranda's geometric model [72], while Yamada [118] assumes that inherent

Table 2.1 Assumptions from the JM model

	Initial fault content:
(A1)	The number of initial faults is a finite unknown constant, say N
	Independent of faults:
(A2)	Faults are assumed to be statistically independent
	Constant failure rates:
(A3)	Each fault has the same constant detection rate, say ϕ
	Fault removal (debugging) process:
(A4)	A fault is removed immediately on detection and correction, and no new faults are introduced. The removal time is negligible

faults follows NHPP models. Zeephongsekul et al. [124], based on NHPPs, consider that primary failures generate secondary-faults under imperfect debugging. Pham et al. [86] propose generalized NHPP models for modeling imperfect debugging.

An intrinsic assumption in most models is that a fault is removed immediately on detection. Consequently, the failure detected process and the fault corrected process are regarded as the same. An alternative approach, using queueing theorem, studies the detection process and the correction process separately, see Yang [122], Dohi et al. [23], and Huang and Huang [39].

We will not pursue further about imperfect debugging and time-delay correction models. However, in Sect. 2.5.3, a bivariate counting process is applied to a 2-version programming system to illustrating the effect of imperfect debugging.

2.4.3 *The JM Model*

The JM model [42] is the first software reliability model to be widely known and used. Assumptions of the JM model are important but seems not realistic. Many models are proposed hereafter to challenge the JM model. However, most of them are propagated from the JM model by extending one or more assumptions. Assumptions from the JM model are listed in Table 2.1. Assumption (A4) is known as *perfect debugging*. A program will exhibit the phenomenon of reliability growth after fault detection and correction, and thus perfect debugging models are usually called software reliability growth models (SRGMs). Assumption (A3) is a basic property of the exponential distribution. Hence the time to failure for each fault has the exponential distribution with parameter ϕ. From assumption (A2), given N and ϕ, the T_i are independent exponential distributions with parameters $(N-i+1)\phi$. The p.d.f. of T_i is given by $f_i(\tau|\phi, N) = (N-i+1)\phi \cdot e^{-(N-i+1)\phi\tau}$, $\tau > 0$. Let $h_i(\tau)$ be the conditional failure rate of T_i, then $h_i(\tau) = (N-i+1)\phi$. Thus the failure rates of the program at any time are proportional to the number of remaining faults, and the sequence of failure rates shows a decreasing trend. The MTTF of T_i is $\{(N-i+1)\phi\}^{-1}$, an increasing function of i.

Table 2.2 Five types of software reliability models

	$S_{i-1} \le t < S_i$	
(a)	Specify $r_i(t) = (N-i+1)\psi(t)$	Then $h_i(\tau) = (N-i+1)\psi(\tau + S_{i-1})$
(b)	Specify $r_i(t) = (N-i+1)\psi_i(t)$	Then $h_i(\tau) = (N-i+1)\psi_i(\tau + S_{i-1})$
(c)	Specify $r_i(t) = \lambda(t)$	Then $h_i(\tau) = \lambda(\tau + S_{i-1})$
	$0 \le \tau < S_i - S_{i-1}$	
(d)	Specify $h_i(\tau) = (N-i+1)\psi(\tau)$	Then $r_i(t) = (N-i+1)\psi(t - S_{i-1})$
(e)	Specify $h_i(\tau) = \psi_i(\tau)$	Then $r_i(t) = \psi_i(t - S_{i-1})$

Let $\{X(t), t \ge 0\}$ be the software failure process where $X(t)$ is the cumulative number of failures experienced up to time $t \ge 0$. For the JM model, the failure process $\{X(t), t \ge 0\}$ is a linear birth continuous-time Markov chain with failure state space $\{0, 1, \ldots, N\}$. This failure process satisfies two properties: *homogeneity* and *conditional orderliness*. The property of conditional orderliness prevents the process having multiple simultaneous failures.

For a conditional orderly failure process $\{X(t), t \ge 0\}$ with state space $\{0, 1, \ldots\}$, the *transition rate* $r_i(t)$ is the probability of encountering a new fault in the next instant of time dt, given that there are $(i-1)$ failures in $[0, t)$, i.e., for $i = 1, \ldots,$

$$r_i(t) = \lim_{\Delta t \to 0+} \frac{1}{\Delta t} P(X(t+\Delta t) - X(t) = 1 | X(t) = i-1), \quad S_{i-1} \le t < S_i. \tag{2.33}$$

Recall that the conditional *failure rate* of T_i is defined by

$$h_i(\tau) = \lim_{\Delta t \to 0+} \frac{1}{\Delta t} P(\tau < T_i \le \tau + \Delta t | T_i > \tau), \quad \tau \ge 0. \tag{2.34}$$

$h_i(\tau)$ is assumed to exist for all the T_i. Relations between $r_i(t)$ and $h_i(\tau)$ can be established in the following theorem.

Theorem 2.8. *Let* $0 \le S_1 \le S_2 \le \ldots$ *be the indexed failure times and let* $T_i = S_i - S_{i-1}, i = 1, \ldots$. *The relations between the transition rate* $r_i(t)$ *given in Eq. (2.33) and the conditional failure rate* $h_i(\tau)$ *given in Eq. (2.34) are, for* $i = 1, \ldots,$

$$r_i(t) = h_i(t - S_{i-1}), \quad S_{i-1} \le t < S_i, \tag{2.35}$$

$$h_i(\tau) = r_i(\tau + S_{i-1}), \quad 0 \le \tau < S_i - S_{i-1}. \tag{2.36}$$

Thus, if we specify $r_i(t)$ then we have $h_i(\tau)$, and vice versa. Five types of software reliability models are given in Table 2.2. The JM model is a special example of all cases with $r_i(t) = h_i(\tau) = (N-i+1)\phi$.

Type (a) is the class of *i.i.d. order statistic models*, type (c) is the class of NHPP models, types (d) and (e) are the class of *time-between-failure models* (or *concatenated failure rates models*), and type (b) seems not be considered by the existing models. In type (b) we retain the explicit factor $(N-i+1)$ rather than subsuming it into ψ_i, because it is more intelligible to do so. Note that types (a), (b) and (d) have finite fault content while types (c) and (e) may have infinite fault content.

Theorem 2.9. *Let $0 \le S_1 \le S_2 \le \dots$ be the indexed failure times and let $T_i = S_i - S_{i-1}, i = 1, \dots$. The joint density function of $T_1, \dots, T_k$ is given by*

$$f(t_1, \dots, t_k) = \prod_{j=1}^{k} \left\{ h_j(t_j) \exp\left[-\int_0^{t_j} h_j(\tau) d\tau \right] \right\},$$

and the joint density function of $S_1, \dots, S_k$ is

$$f(s_1, \dots, s_k) = \prod_{j=1}^{k} \left\{ h_j(s_j - s_{j-1}) \exp\left[-\int_0^{s_j - s_{j-1}} h_j(\tau) d\tau \right] \right\},$$

where the conditional failure rates $h_i(\tau)$ are defined by Eq. (2.34). When the $h_j(\tau)$ are given as types (a), (b) and (d) of Table 2.2, k must be less than or equal to the initial number of faults.

Theorem 2.9 is useful to obtain the likelihood functions of parameters. We can then find estimates of parameters by applying the usual estimation techniques, such as the maximum likelihood approach or the least squares approach.

Example 2.28. The JM model [42]: $h_j(\tau) = (N - j + 1)\phi$

$$f(t_1, \dots, t_k) = \frac{N!}{(N-k)!} \phi^k \exp\left[-\phi \sum_{j=1}^{k} (N - j + 1) t_j \right], \quad k \le N,$$

or

$$f(s_1, \dots, s_k) = \frac{N!}{(N-k)!} \phi^k \exp\left[-\phi \left(\sum_{j=1}^{k} s_j + (N-k) s_k \right) \right], \quad k \le N.$$

Example 2.29. The Littlewood model [57]: $h_j(\tau) = (N - j + 1)\alpha / (\beta + \tau + S_{j-1})$

For $k \le N$,

$$f(t_1, \dots, t_k) = \frac{N!}{(N-k)!} \alpha^k \beta^{N\alpha} \left[\prod_{j=1}^{k-1} \left(\frac{1}{\beta + t_1 + \dots + t_j} \right)^{\alpha+1} \right] \times \left(\frac{1}{\beta + t_1 + \dots + t_k} \right)^{(N-k+1)\alpha+1},$$

or

$$f(s_1,\ldots,s_k) = \frac{N!}{(N-k)!}\alpha^k \beta^{N\alpha}\left[\prod_{j=1}^{k-1}\left(\frac{1}{\beta+s_j}\right)^{\alpha+1}\right]\left(\frac{1}{\beta+s_k}\right)^{(N-k+1)\alpha+1}.$$

Example 2.30. The Schick-Wolverton model [95]: $h_j(\tau) = (N-j+1)\phi\tau$

For $k \le N$,

$$f(t_1,\ldots,t_k) = \frac{N!}{(N-k)!}\phi^k\left[\prod_{j=1}^{k} t_j\right]\exp\left[-\frac{\phi}{2}\sum_{j=1}^{k}(N-j+1)t_j^2\right],$$

or

$$f(t_1,\ldots,t_k) = \frac{N!}{(N-k)!}\phi^k\left[\prod_{j=1}^{k}(s_j-s_{j-1})\right]\exp\left[-\frac{\phi}{2}\sum_{j=1}^{k}(N-j+1)(s_j-s_{j-1})^2\right].$$

Example 2.31. The Littlewood-Verrall model [60]: $h_j(\tau) = \alpha/(\tau+\psi_j)$

$$f(t_1,\ldots,t_k) = \prod_{j=1}^{k}\left[\frac{\alpha}{\psi_j}\left(\frac{\psi_j}{t_j+\psi_j}\right)^{\alpha+1}\right],$$

or

$$f(t_1,\ldots,t_k) = \prod_{j=1}^{k}\left[\frac{\alpha}{\psi_j}\left(\frac{\psi_j}{s_j-s_{j-1}+\psi_j}\right)^{\alpha+1}\right].$$

2.4.4 I.I.D. Order Statistic Models

Let assumptions (A1), (A2) and (A4) in Table 2.1 hold. Assumption (A3) states that every fault has the same deterministic effect on the overall failure rate. However, faults may have different failure rates. To generalize this, we consider two approaches. First, assumption (A3) can be interpreted by a Poisson shock model. A fault causes a failure to occur according to a Poisson process with rate ϕ. We may assume that shocks follow a Poisson process with rate $\psi(t)$ representing that faults have different failure rates. Second, recall that failure rates of T_i, time between failures, in the JM model are $h_i(\tau) = (N-i+1)\phi$. Each failure rate of T_i is a summation of failure rates of remaining faults that are equally ϕ. It has been argued that the detection rates $\{h_i\}$ depend on the size of faults subsets which may not be of the same size [56]. Faults with larger size tend to be removed first. Thus we may assume that the sequence $\{h_i\}$ is a stochastic process.

Table 2.3 Type (a):$r_i(t) = (N - i + 1)\psi(t)$; i.i.d. order statistics models

(M1)	Exponential	$\psi(t) = \phi$	JM model [42]
		$F(t) = 1 - \exp(-\phi t)$	$(\phi > 0)$
(M2)	Pareto	$\psi(t) = \alpha/(\beta + t)$	Littlewood [57]
		$F(t) = 1 - [\beta/(\beta + t)]^\alpha$	$(\alpha, \beta > 0)$
(M3)	Gamma(2,ϕ)	$\psi(t) = \phi^2 t/(1 + \phi t)$	
		$F(t) = 1 - (1 + \phi t)\exp(-\phi t)$	$(\phi > 0)$
(M4)	Logistic	$\psi(t) = \phi/[1 + c\ \exp(-\phi t)]$	
		$F(t) = [1 - \exp(-\phi t)]/[1 + c\ \exp(-\phi t)]$	$(\phi, c > 0)$
(M5)	Weibull	$\psi(t) = \alpha\beta t^{\beta-1}$	Wagoner [112]
		$F(t) = 1 - \exp(-\alpha t^\beta)$	$(\alpha > 0, \beta > 0)$
(M6)	Rayleigh	$\psi(t) = \phi t$	
		$F(t) = 1 - \exp(-\phi t^2/2)$	$(\phi > 0)$
(M7)	Parabola	$\psi(t) = \phi(-at^2 + bt + c)$	
		$F(t) = 1 - \exp[\phi(-at^3/3 + bt^2/2 + ct)]$	$(a, b, c > 0)$

In the first approach, we assume that shocks follow a Poisson process with rate $\psi(t)$. This in turn generalizes the exponential distribution by a distribution F such that

$$F(t) = 1 - \exp\left\{-\int_0^t \psi(s)ds\right\}. \tag{2.37}$$

One way to obtain a reasonable common distribution F is using the doubly stochastic process. Assume that time to failures caused by faults are independent exponential distributions with rates $\phi_1, \phi_2, \ldots, \phi_N$. If the rates ϕ have a mixing distribution G, then we have i.i.d. mixture distributions given by

$$F(t) = \int_0^\infty (1 - e^{-\phi t})dG(\phi) = 1 - \int_0^\infty e^{-\phi t}dG(\phi). \tag{2.38}$$

So that the unconditional failure rates are

$$\psi(t) = \frac{f(t)}{\overline{F}(t)} = \frac{\int_0^\infty \phi e^{-\phi t}dG(\phi)}{\int_0^\infty e^{-\phi t}dG(\phi)}.$$

Thus the models become the *i.i.d. order statistic models* [70]. From Eqs. (2.37) and (2.38), the Laplace transform of the mixing distribution G is given by

$$G^*(t) = \int_0^\infty e^{-\phi t}dG(\phi) = \exp\left\{-\int_0^t \psi(s)ds\right\}.$$

Let $X(t)$ be the cumulative number of failures up to time t, then $X(t)$ has a binomial distribution similar to that in the JM model. The mean value function of $X(t)$ is $m(t) = NF(t)$. Models in this class, listed in Table 2.3, correspond to type (a) of Table 2.2.

Table 2.4 Type (e): $h_i(\tau) = \psi_i(\tau)$; time-between-failure-models

(M8)	Pareto	$h_i(\tau) = \alpha/(\tau + \psi_i)$	Littlewood-Verrall [60]
		$F_i(\tau) = 1 - [\psi_i/(\tau + \psi_i)]^\alpha$	$(\alpha > 0, \psi \uparrow i)$
(M9)	Geometric	$h_i(\tau) = Dk^{i-1}$	Moranda [72]
		$F_i(\tau) = 1 - \exp[-Dk^{i-1}\tau]$	$(D > 0, 0 < k < 1)$
(M10)	Power	$h_i(\tau) = \phi(N - i + 1)^\gamma$	Xie [116]
		$F_i(\tau) = 1 - \exp[-\phi(N - i + 1)^\gamma \tau]$	$(\phi > 0, \gamma > 1)$

Model (M2) is proposed by Littlewood [57]. He assumed the mixing distribution G to be a gamma distribution with the scale parameter β and the shape parameter α, namely, $dG(\phi) = \frac{\beta^\alpha}{\Gamma(\alpha)}\phi^{\alpha-1}e^{-\beta\phi}$, $\alpha, \beta > 0$. Models (M3) and (M4) can be viewed as variations of Yamada et al. [120, 121] in Table 2.6. $\psi(t)$ in model (M5) is decreasing for $0 < \beta < 1$, constant for $\beta = 1$, and increasing for $\beta > 1$. In model (M6), the distribution function F is known as the Rayleigh distribution, a Weibull distribution with shape parameter 2. In model (M7), $\psi(t)$ is a parabola which increases and reaches a maximum, then declines. Both models (M6) and (M7) are variations of Schick and Wolverton [95, 96] in Table 2.5.

2.4.5 *Time Between Failures Models*

In the second approach of extension from assumption (A3), partially due to the difficulty in defining what a program fault is, the failure rates of the T_i, times between successive failures, are given by $h_i(\tau) = \Lambda_i$, $i = 1, 2, \ldots$, which form a stochastic process. To describe reliability growth, we may choose $\{\Lambda_i\}$ such that the Λ_i are stochastically decreasing with i, i.e., $P(\Lambda_{i+1} < \lambda) \geq P(\Lambda_i < \lambda), \forall \lambda \geq 0$, or equivalently such that the T_i are stochastically increasing with i, i.e., $P(T_{i+1} > t) \geq P(T_i > t)$, $\forall t \geq 0$. If the Λ_i have a mixing distribution G_i, then the mixture distribution is

$$F_i(\tau) = 1 - \int_0^\infty e^{-\lambda_i \tau} dG_i(\lambda_i).$$

The posterior failure rate is

$$h_i(\tau) := \psi_i(\tau) = \frac{\int_0^\infty \lambda_i e^{\lambda_i \tau} dG_i(\lambda_i)}{\int_0^\infty e^{-\lambda_i \tau} dG_i(\lambda_i)}.$$

In some cases, one can simply specify a particular function $\psi_i(\tau)$. This class of models, listed in Table 2.4, corresponds to type (e) of Table 2.2.

The above approach, however, does not depend on the number of remaining faults $(N - i + 1)$. An alternative way is to assume that $h_i(\tau) = (N - i + 1)\psi(\tau)$. This can

Table 2.5 Type (d): $h_i(\tau) = (N - i + 1)\psi(\tau)$; time-between-failure-models

(M11)	Rayleigh	$h_i(\tau) = (N - i + 1)\phi\tau$ $F_i(\tau) = 1 - \exp\left(-\int_0^\tau h_i(s)ds\right)$	Schick-Wolverton [95] $(\phi > 0)$
(M12)	Parabola	$h_i(\tau) = (N - i + 1)\phi(-a\tau^2 + b\tau + c)$ $F_i(\tau) = 1 - \exp\left(-\int_0^\tau h_i(s)ds\right)$	Schick-Wolverton [96] $(a, b, c > 0)$

be explained as that faults are i.i.d. within each T_i with common failure rate $\psi(\tau)$. The distribution functions of T_i are

$$F_i(\tau) = 1 - \exp\left(-\int_0^\tau h_i(s)ds\right).$$

This class of models, listed in Table 2.5, corresponds to type (d) of Table 2.2.

In model (M8), Littlewood and Verrall [60, 61] assumed that the Λ_i are independent gamma random variables with the shape parameter α and the scale parameter ψ_i, namely, $g(\lambda_i|\alpha, \psi_i) = \frac{\psi_i^\alpha}{\Gamma(\alpha)}\lambda_i^{\alpha-1}\exp(-\psi_i\lambda_i)$. Model (M9) is the JM geometric de-eutrophication model suggested by Moranda [72]. The failure rates are constant for each T_i, but they decrease geometrically in i after each failure. D is the program failure rate during the first interval and k is a constant such that the rates form a converging geometric series. When a failure occurs and a fault is removed, the drop in the failure rates is $D(1-k)k^{i-1}$. Since $h_i(t) \to 0$ when $i \to \infty$, thus the fault content is infinite. Model (M9) has been generalized to the so-called *proportional models*, see Moranda [73], Gaudoin and Soler [30], and Gaudoin et al. [31].

Model (M10) is a power-type function of the number of remaining faults where $0^\gamma = 0$ and $\gamma > 1$ such that $h_i(t)$ is a convex function denoting that early stage of debugging has higher detection probability than later. Models (M11) and (M12) in Table 2.5 are different from models (M6) and (M7) in Table 2.3, although with the same name of distributions. For example, in model (M11) the failure rate drops to 0 immediately after the $(i-1)$th failure, while in model (M6) the failure rate drops to $(N-i+1)\phi\tau$ where τ is the time to the $(i-1)$th failure. In both models (M11) and (M6), the failure rate increases linearly with slope $(N-i+1)$ until the ith failure.

2.4.6 *NHPP Models*

Assume that assumptions (A2) and (A4) in Table 2.1 hold, and that the number of initial faults N in assumption (A1) has a Poisson prior distribution with finite mean $\theta > 0$. One way of constructing an NHPP is to assume that N is Poisson distributed in the i.i.d. order statistics models discussed in Sect. 2.4.4. Let F be the common distribution of faults, usually continuous with positive support. For each fault, given that we haven't found it by time s, the conditional probability that we find it in time

Table 2.6 Type (c): $r_i(t) = \lambda(t)$; NHPP models with $m(\infty) < \infty$

(M13)	Exponential	$\lambda(t) = \theta\phi e^{-\phi t}$ $m(t) = \theta(1 - e^{-\phi t})$	GO model [34] $(\phi > 0)$
(M14)	Pareto	$\lambda(t) = \theta\alpha\beta^{\alpha}/(\beta + t)^{\alpha+1}$ $m(t) = \theta[1 - (\beta/(\beta + t))^{\alpha}]$	Miller [70] $(\alpha, \beta > 0)$
(M15)	Gamma(2,ϕ)	$\lambda(t) = \theta\phi^2 te^{-\phi t}$ $m(t) = \theta\left[1 - (1 + \phi t)e^{-\phi t}\right]$	Yamada et al. [120] $(\phi > 0)$
(M16)	Logistic	$\lambda(t) = \theta(1 + c)\phi e^{-\phi t}(1 + ce^{-\phi t})^2$ $m(t) = \theta(1 - e^{-\phi t})/(1 + ce^{-\phi t})$	Yamada et al. [121] $(\phi, c > 0)$
(M17)	Weibull	$\lambda(t) = \theta\alpha\beta t^{\beta-1}\exp(-\alpha t^{\beta})$ $m(t) = \theta[1 - \exp(-\alpha t^{\beta})]$	Goel [33] $(\alpha > 0, 0 < \beta < 1)$
(M18)	Logistic	$\lambda(t) = \theta\phi\alpha e^{-\phi t}/(1 + \alpha e^{-\phi t})^2$ $m(t) = \theta/(1 + \alpha e^{-\phi t})$	Yamada-Osaki [119] $(\alpha, \phi > 0)$

interval $(s, t]$ is $P(s < \tau \leq t|\tau > s) = 1 - \bar{F}(t)/\bar{F}(s)$. Then for a fixed number n of faults and $n \geq u$, we have, for $k = 0, 1, \ldots, n - u$,

$$P(X(t) - X(s) = k|N = n, X(s) = u) = \binom{n-u}{k}\left(1 - \frac{\bar{F}(t)}{\bar{F}(s)}\right)^k \left(\frac{\bar{F}(t)}{\bar{F}(s)}\right)^{n-u-k},$$

which is the probability that we find k of $n - u$ independent faults in time interval $(s, t]$. Thus, for $0 < s < t$,

$$X(t) - X(s)|N, X(s) \sim \text{Binomial}(N - X(s), 1 - \bar{F}(t)/\bar{F}(s)).$$

In particular, $X(t)|N \sim \text{Binomial}(N, 1 - \bar{F}(t))$. If N has a Poisson distribution with parameter θ, then it can be shown that $\{X(t), t \geq 0\}$ is a NHPP with the mean value function $m(t) = \theta F(t)$.

In Table 2.6, we list models in which the expected number of initial faults are finite, $m(\infty) = \lim_{t\to\infty} m(t) = \theta < \infty$. Model (M14), mentioned by Miller [70], is the NHPP version of Littlewood model (M2). Models (M15) and (M16) are discussed by Yamada et al. [120, 121], and are referred to as the delayed S-shaped and the inflection S-shaped SRGMs, respectively. Both models have increasing failure rates $\psi(t)$. Model (M17) is proposed by Goel [33] to describe the situation that the intensity function first increases and then decreases. Model (M18) is the logistic growth curve model suggested by Yamada and Osaki [119]. However, we have $F(0) = 1/(1 + \alpha) > 0$. If we modify $F(t)$ by

$$F'(t) = \left(F(t) - \frac{1}{1+\alpha}\right)\frac{1+\alpha}{\alpha},$$

such that $F'(0) = 0$ and $F'(\infty) = 1$, then we have the same model as (M16).

Table 2.7 Type (c): $r_i(t) = \lambda(t)$; NHPP models with $m(\infty) \to \infty$

(M19)	Poisson ($N \to \infty$ in (M1))	$\lambda(t) = \lambda$ $m(t) = \lambda t$	Musa-Ackerman [74] ($\lambda > 0$)
(M20)	Power law process ($N \to \infty$ in (M5))	$\lambda(t) = \alpha\beta t^{\beta-1}$ $m(t) = \alpha t^{\beta}$	Crow [21] ($\alpha, \beta > 0$)
(M21)	Logarithmic Poisson ($N \to \infty$ in (M2))	$\lambda(t) = \lambda_0/(\lambda_0\phi t + 1)$ $m(t) = \ln(\lambda_0\phi t + 1)/\phi$	Musa-Okumoto [75] ($\phi, \lambda_0 > 0$)

Another class of NHPP models can be generated directly from various mean value functions $m(t)$ or intensity functions $\lambda(t)$. Some of them are originally used in hardware system reliability, and are referred to as *reliability growth models*. A growth function (the intensity function or the mean value function) is used to represent the growth of system reliability after improvement. For the models listed in Table 2.7, the expected number of initial faults are infinite, i.e., $m(\infty) = \lim_{t\to\infty} m(t) \to \infty$.

Model (M19) is the homogeneous Poisson process appeared in Musa and Ackerman [74]. Model (20) is the Duane model or the Duane process. In light of Duane's discovery [24], Crow [21] proposes the model (M20) where the intensity function $\lambda(t) = \alpha\beta t^{\beta-1}$ is the failure rate of a Weibull distribution and the time to the first failure is Weibull distributed. Hence this model is commonly referred to as the Weibull process. As Ascher and Feingold [8] point out, the name of the Weibull process is improper. We refer this model as the *power law process* according to them. The logarithm of the mean value function is $\ln m(t) = \ln\alpha + \beta\ln t$ which is consistent with Duane's empirical observation that the cumulative number of failures versus cumulative operation hours is approximately linear on the log-log plot. It is clear that $0 < \beta < 1$ corresponds to the decreasing intensity function. In this case, the intensity function $\lambda(t) \to \infty$ as $t \to 0$ is an unrealistic property for real systems. See Littlewood [58] for a modified Duane model. Model (M14), with inverse power law intensity function $\lambda(t) = \theta\alpha\beta^{\lambda}(\beta + t)^{-\alpha-1}$, can be viewed as a modified model of (M20).

Model (M21) is Musa and Okumoto's logarithmic Poisson model [75]. They postulate that the intensity function $\lambda(t)$ is a exponentially decreasing function of the mean value function $m(t)$ given by $\lambda(t) = \lambda_0\exp(-\phi m(t))$, where λ_0 represents the initial value of $\lambda(t)$, and ϕ measures the relative decrease in the failure intensity that follows every failure. Since $\lambda(t) = dm(t)/dt$ and $m(0) = 0$, we have $m(t) = \ln(\lambda_0\phi t + 1)/\phi$ and $\lambda(t) = \lambda_0/(\lambda_0\phi t + 1)$. By the fact that $m(t)$ is an NHPP and a logarithmic function of t, model (M21) is then called the logarithmic Poisson model.

In fact, it can be shown that Model (M21) is a limit case of the Littlewood model (M2), the Poisson model (M19) is a limit case of the JM model (M1), and that the power law process (M20) is a limit case of the Wagoner model (M5).

2.4.7 *Self-exciting Point Processes*

2.4.7.1 Self-exciting Point Process

Let $\{X(t), t \geq 0\}$ be a counting process, which characterizes failure occurrences of a computer software, defined on a *filtered probability space* $(\Omega, \mathcal{F}, (\mathcal{F}_t)_{t\geq 0}, \mathcal{P})$ with $\{X(t), t \geq 0\}$ adapted to the filtration $(\mathcal{F}_t)_{t\geq 0}$ where $\mathcal{F}_t := \sigma\{X(s), 0 \leq s \leq t\}$. A *filtration* $\{\mathcal{F}_t, t \geq 0\}$ is a family of sub-σ-fields of $\mathcal{F}$ such that $\mathcal{F}_s \subseteq \mathcal{F}_t$ for all $s \leq t$ in $[0, \infty)$. A probability space endowed with such a filtration is called a filtered probability space. We assume that the filtration is *standard*. That is, the filtration satisfies *right-continuous*: $\mathcal{F}_t = \mathcal{F}_{t^+} := \bigcap_{s>t} \mathcal{F}_s$, $\forall t \geq 0$, and *complete*: $\mathcal{F}_0$ contains all of the $\mathcal{P}$-null sets in $\mathcal{F}$. A counting process $\{X(t), t \geq 0\}$ is said to be *adapted* to the *filtration* $(\mathcal{F}_t)_{t\geq 0}$ if $X(t)$ is $\mathcal{F}_t$-measurable for each t. Thus $X(t)$ is known when $\mathcal{F}_t$ is known at time t, i.e., $\mathcal{F}_t$ represents the history or information available up to time t.

A counting process is said to be self-exciting if it depends on the entire or some fractions of its history $(\mathcal{F}_t)_{t\geq 0}$ that affects or excites the intensity function $\Lambda(t|\, \mathcal{F}_t)$ of the process defined as

$$\Lambda(t|\, \mathcal{F}_t) = \lim_{\Delta t \to 0+} \frac{1}{\Delta t} E[X(t + \Delta t) - X(t)|\, \mathcal{F}_t].$$

Assume that the process does not have failures which occur simultaneously. In other words, the process satisfies the conditional orderliness property [19]: for any $\mathcal{G}_t \subseteq \mathcal{F}_t$ and $t \geq 0$, as $\delta \downarrow 0$,

$$P(X(t + \delta) - X(t) \geq 2|\, \mathcal{G}_t) = o(\delta) P(X(t + \delta) - X(t) = 1|\, \mathcal{G}_t).$$

When such is the case, the intensity function becomes

$$\Lambda(t|\, \mathcal{F}_t) = \lim_{\Delta t \to 0+} \frac{1}{\Delta t} P(X(t + \Delta t) - X(t) = 1|\, \mathcal{F}_t), \quad S_{X(t)} \leq t < S_{X(t)+1}, \tag{2.39}$$

where S_i is the time of occurrence of the ith failure with $S_0 \equiv 0$. Let $T_i = S_i - S_{i-1}$ be the interarrival time, for $i = 1, 2, \ldots$.

A self-exciting point process takes account of the points generated by the failure times so that, for example, $\mathcal{F}_t = \{X(t), S_1, \ldots, S_{X(t)}\}$. A point process is said to have no-memory if $\Lambda(t|\, \mathcal{F}_t) = \Lambda(t)$; it has 0-memory if $\Lambda(t|\, \mathcal{F}_t) = \Lambda(t|X(t))$; and it has m-memory if $\Lambda(t|\, \mathcal{F}_t) = \Lambda(t|X(t), S_{X(t)}, \ldots, S_{X(t)-m+1})$, for $m = 1, 2, \ldots$. It has been shown that if $\{X(t), t \geq 0\}$ is m_2-memory then it is m_1-memory, for $0 \leq m_1 \leq m_2 < \infty$, and a counting process is 0-memory if and only if it is a Markov process. In addition, a self-exciting point process is 1-memory with intensity function

$$\Lambda(t|X(t), S_{X(t)}) = f(X(t), t - S_{X(t)}),$$

for some function f if and only if the times between failures $T_1, T_2, \ldots$ are independent [103]. Note that the intensity function $\Lambda(t|\,\mathcal{F}_t)$ is continuous from the left, and we assume for $t \geq 0$ that $E[\Lambda(t|\,\mathcal{F}_t)] < \infty$.

Recall that the expectation of $X(t)$ is called the mean value function, and we denote it by $m(t) = E[X(t)]$. The derivative of the mean value function is called the *rate of occurrence of failures* (ROCOF) representing approximate probability that a failure occurs in $(t, t+dt]$, and we denote it by $v(t) = \frac{d}{dt}E[X(t)]$. If the counting process $\{X(t), t \geq 0\}$ does not have failures which occur simultaneously, the ROCOF is equivalent to

$$v(t) = \lim_{\Delta t \to 0+} \frac{1}{\Delta t} P(X(t + \Delta t) - X(t) = 1). \tag{2.40}$$

$\Lambda(t|\,\mathcal{F}_t)$ is the probability of encountering a failure within the next instant of time dt, given the history available up to the present time t. For Poisson processes, due to the property of independent increments, the ROCOF (2.40) and the intensity function (2.39) are the same, but in general these two functions are different. When a system has only one r.v. T and we define $X(t) := \mathbb{I}\{T \leq t\}$, then the intensity function (2.39) is equivalent to the failure rate function (2.3). Thus, failure rate function is a special case of intensity function.

2.4.7.2 Classification of SRGMs

Another function which is used in reliability modeling (to include software reliability), is the concatenated failure rate function of the failure process. Given $\mathcal{F}_t$, it is defined by Chen and Singpurwalla [18],

$$H(\tau|\,\mathcal{F}_t) = \lim_{\Delta t \to 0+} \frac{1}{\Delta t} P(\tau + S_{X(t)} < S_{X(t)+1} \leq S_{X(t)} + \tau + \Delta t|\,\mathcal{F}_t), \quad \tau \geq 0. \tag{2.41}$$

The intensity function (2.39) and the concatenated failure rate function (2.41) can be viewed as a wilder sense of the transition rate function (2.33) and the conditional failure rate function (2.34).

Let us consider that a system consists of only one component whose lifetime is T, and define $X(t) = \mathbb{I}\{T \leq t\}$; thus, for a 1-component system $\Lambda(t|\,\mathcal{F}_t)$ is equivalent to $H(\tau|\,\mathcal{F}_t)$. As might be expected, there is a close relation between $\Lambda(t|\,\mathcal{F}_t)$ and $H(\tau|\,\mathcal{F}_t)$, similar to Theorem 2.8, as is shown in the following:

$$\Lambda(t|\,\mathcal{F}_t) = H(t - S_{X(t)}|\,\mathcal{F}_t), \quad S_{X(t)} \leq t < S_{X(t)+1},$$
$$H(\tau|\,\mathcal{F}_t) = \Lambda(\tau + S_{X(t)}|\,\mathcal{F}_t), \quad 0 \leq \tau < S_{X(t)+1} - S_{X(t)}.$$

Table 2.8 Relations between $\Lambda(t|\,\mathcal{F}_t)$ and $H(\tau|\,\mathcal{F}_t)$

(a)	$\Lambda(t\|\,\mathcal{F}_t) = (N - X(t))\psi(t)$	$\Longrightarrow$	$H(\tau\|\,\mathcal{F}_t) = (N - X(t))\psi(\tau + S_{X(t)})$
(b)	$\Lambda(t\|\,\mathcal{F}_t) = (N - X(t))\psi_{X(t)+1}(t)$	$\Longrightarrow$	$H(\tau\|\,\mathcal{F}_t) = (N - X(t))\psi_{X(t)+1}(\tau + S_{X(t)})$
(c)	$\Lambda(t\|\,\mathcal{F}_t) = \lambda(t)$	$\Longrightarrow$	$H(\tau\|\,\mathcal{F}_t) = \lambda(\tau + S_{X(t)})$
(d)	$H(\tau\|\,\mathcal{F}_t) = (N - X(t))\psi(\tau)$	$\Longrightarrow$	$\Lambda(t\|\,\mathcal{F}_t) = (N - X(t))\psi(t - S_{X(t)})$
(e)	$H(\tau\|\,\mathcal{F}_t) = \psi_{X(t)+1}(\tau)$	$\Longrightarrow$	$\Lambda(t\|\,\mathcal{F}_t) = \psi_{X(t)+1}(t - S_{X(t)})$

Accordingly, this relation enables us to indicate a method to classify software reliability models. The implication is that if we can specify $\Lambda(t|\,\mathcal{F}_t)$ then we have $H(\tau|\,\mathcal{F}_t)$, and vice versa. Here we give five cases as examples in Table 2.8, similar to Table 2.2, where $\lambda(\cdot)$ and $\psi(\cdot)$ are continuous functions.

Obviously, $\Lambda(t|\,\mathcal{F}_t)$ is no-memory in case (c), 0-memory in cases (a) and (b), and 1-memory in cases (d) and (e). Except for case (b), the given cases correspond to some categories of software reliability models in the context of Chen and Singpurwalla [18]. Thus, case (a) corresponds to the class of i.i.d. order statistic models, case (c) corresponds the class of NHPP models, and cases (d) and (e) correspond to the concatenated failure rate models. For $m \geq 2$ memory models, we refer the readers to Singpurwalla and Soyer [101] and Al-Mutairi et al. [3].

Observe that the relation between the intensity function (2.39) and the concatenated failure rate function (2.41) is analogous to the relation of the transition rate function (2.33) and the conditional failure rate function (2.34). The difference is on the filtration $\mathcal{F}_t = \sigma\{X(s) : 0 \leq s \leq t\}$. Based on these relations, a software reliability model can be represented simply by the intensity function according to various filtrations we might choose. Thus, a more general class of models can be constructed from a more wilder filtration.

2.4.7.3 A Self-exciting and Mixture Model

Apart from the self-exciting, another property for the model in our discussion is mixture which is assumed that individual faults come with i.i.d. random failure rates λ and that failure times have intensity function $\phi(t|\lambda, \mathcal{F}_t)$. Let $X_i(t)$ be the indicator function of failure due to fault i up to time $t \geq 0$. Therefore, the cumulative number of software failures up to time $t \geq 0$ is given by $X(t) = \sum_{i=1}^{N_0} X_i(t)$, where N_0 is the unknown initial number of faults in the software.

Now we can discuss an intensity function which have both properties of self-exciting and mixture. Assumptions are given in the following.

(i) The X_i are mutually independent, given N_0, $\{\lambda_i\}_{i\geq 0}$, and $\mathcal{F}_t$,

(ii) $$\phi(t|\lambda_i,\, \mathcal{F}_t) = \lim_{\Delta t \to 0+} \frac{1}{\Delta t} P(X_i(t + \Delta t) - X_i(t) = 1|N_0, \{\lambda_i\}_{i\geq 0},\, \mathcal{F}_t), \tag{2.42}$$

(iii) The failure process satisfies the conditional orderliness property, where λ_i is the random failure rate of faults i. The intensity function of the software can be determined and

$$\Lambda(t|N_0, \{\lambda_i\}_{i\geq 0}, \mathcal{F}_t) = \lim_{\Delta t\to 0+} \frac{1}{\Delta t} P(X(t+\Delta t) - X(t) = 1|N_0, \{\lambda_i\}_{i\geq 0}, \mathcal{F}_t). \tag{2.43}$$

The following results are provided by Wang [113].

Theorem 2.10. *Under the conditions of (2.43), if the λ_i have a mixing distribution function G, then the intensity function of $\{X(t), t \geq 0\}$, given N_0 and $\mathcal{F}_t = \sigma\{X(s), 0 \leq s \leq t\}$, is*

$$\Lambda(t|N_0, \mathcal{F}_t) = (N_0 - X(t)) \cdot \nu(\lambda, \mathcal{F}_t, G) \tag{2.44}$$

where

$$\nu(\lambda, \mathcal{F}_t, G) = \frac{\int_0^\infty \phi(t|\lambda,\ \mathcal{F}_t) \exp[-\Phi(t|\lambda,\ \mathcal{F}_t)] dG(\lambda)}{\int_0^\infty \exp[-\Phi(t|\lambda,\ \mathcal{F}_t)] dG(\lambda)},$$

and

$$\Phi(t|\lambda) = \int_0^t \phi(s|\lambda,\ \mathcal{F}_s) ds.$$

Example 2.32 (Littlewood [57]). Assume that the mixing distribution G is gamma distributed with the scale parameter β and the shape parameter α. If $\phi(t|\lambda,\ \mathcal{F}_t) \equiv \lambda$, then the intensity function is

$$\Lambda(t|N_0) = (N_0 - X(t)) \frac{\alpha}{\beta + t}.$$

This is an example corresponds to case (a) of Table 2.8.

Example 2.33 (Goel and Okumoto [34]; Miller [70]). Assume that $N_0 \sim Poisson$ (θ) and $\phi(t|\lambda,\ \mathcal{F}_t) \equiv \lambda$. If the λ_i degenerate to a constant λ_0, then the intensity function is

$$\Lambda(t) = \theta\lambda_0 e^{-\lambda_0 t}.$$

Furthermore, if the mixing distribution G is gamma distributed with the scale parameter β and the shape parameter α then the intensity function is

$$\Lambda(t) = \frac{\theta\alpha\beta^\alpha}{(\beta + t)^{\alpha+1}}.$$

These are examples correspond to case (c) of Table 2.8.

Example 2.34 (Schick and Wolverton [95]). Assume that the λ_i degenerate to a constant λ_0 and $\phi(t|\lambda_0, \mathcal{F}_t) = \lambda_0(t - S_{X(t)})$, then the intensity function is

$$\Lambda(t|N_0, \mathcal{F}_t) = (N_0 - X(t))\lambda_0(t - S_{X(t)}).$$

The concatenated failure rate function is

$$H(\tau|N_0, \mathcal{F}_t) = (N_0 - X(t))\lambda_0\tau, \quad \tau \geq 0.$$

This is an example corresponds to case (d) of Table 2.8.

2.5 Examples

2.5.1 A Model for Load-Sharing System with Common-Cause Systems

In Sect. 2.2.2, we have discussed the MVE and FMVE models. The idea here is to combine both models together. Consider a system consisting of n components whose lifetimes have the occurrence of common-cause failures and the increase of failure rates for the surviving components. Such phenomena happen especially in redundant systems [76]. A series of Poisson shocks can be used to describe such failures, where each shock may destroy components separately or simultaneously. Accordingly, failures of the system can be considered to be n possible transitions of a Markov chain. One (or more) components fails first, then one (or more) of the surviving components fails, and so on. Initially there is $2^n - 1$ independent Poisson processes that govern the occurrence of the shocks. After the system receives a shock, which destroys one (or more) of the n components, the surviving components are subject to shocks governed by another cluster of independent Poisson processes which may have different parameters (failure rates) from the previous ones due to an increased load. At this stage the number of processes will depend on the number of surviving components. The system may continue to function, depending on the structure of the system (e.g., a k-out-of-n:F system), and is waiting for receiving another shock.

Let $X = (X_t)_{t\geq 0}$ be the failure process of the system where X_t denotes the number of failed components up to time t. The state space E consists of 2^n states and we partition it into $n + 1$ classes as $E = \{E_0, E_1, \ldots, E_n\}$, where the subscript of E_i denotes number i of failed components and the elements of E_i indicate $\binom{n}{i}$ combinations of failed components. We denote these classes by the notations: $E_0 = \{0\}$, $E_1 = \{1_1, \ldots, 1_n\}, \ldots, E_n = \{n\}$, where the permutation of subscript in each class is in lexicographical order and the cardinality of E_i is $|E_i| = \binom{n}{i}$. Class E_n is the absorbing class containing only one state, and other classes constitute a partition of the set of transient states. One of the simplest example is a birth process which passes through all $n + 1$ classes by $E_0 \to E_1 \to \cdots \to E_n$. In our discussions, after visiting class E_i, the next visited class may be any class E_j with $i < j \leq n$.

Let Q be an infinitesimal generator of the failure process X, and the submatrix from class E_i to class E_j be denoted by $Q_{i,j}$ which is defined by

$$Q_{i,j} = \lim_{h\downarrow 0+} \frac{1}{h} P\{X_{t+h} \in E_j | X_t \in E_i\}, \quad \text{if } i < j,$$

$$Q_{i,j} = \lim_{h\downarrow 0+} \frac{1}{h} (P\{X_{t+h} \in E_j | X_t \in E_i\} - I), \quad \text{if } i = j,$$

where I is an identity matrix. The submatrix $Q_{i,j}$ is a $\binom{n}{i} \times \binom{n}{j}$ constant matrix which is a zero matrix if $i > j$. The entries of $Q_{i,j}$ are denoted by ${}^k\lambda_{i,j}^l$, where $k = 1, \ldots, \binom{n}{i}$ and $l = 1, \ldots, \binom{n}{j}$, and ${}^k\lambda_{i,i}^k = -\sum_{i<j} \sum_{l\neq k} {}^k\lambda_{i,j}^l < 0$.

The permutation of subscripts of the $Q_{i,j}$'s are again arranged in lexicographical order. Such arrangement guarantees that Q is a upper triangular matrix. For each state s, we let $P_s(t) = P\{X_t = s | X_0 = 0\}$ as usual. We assemble the functions $P_s(t)$ for $s \in E_j$ into a row vector with $\binom{n}{j}$ entries in lexicographical order and denote it by $P_{E_j}(t)$.

$$Q = \begin{array}{c} \\ E_0 \\ E_1 \\ E_2 \\ \vdots \\ E_{n-1} \\ E_n \end{array} \begin{array}{c} \begin{array}{ccccccc} E_0 & E_1 & E_2 & \cdots & E_{n-2} & E_{n-1} & E_n \end{array} \\ \begin{pmatrix} Q_{0,0} & Q_{0,1} & Q_{0,2} & \cdots & Q_{0,n-2} & Q_{0,n-1} & Q_{0,n} \\ 0 & Q_{1,1} & Q_{1,2} & \cdots & Q_{1,n-2} & Q_{1,n-1} & Q_{1,n} \\ 0 & 0 & Q_{2,2} & \cdots & Q_{2,n-2} & Q_{2,n-1} & Q_{2,n} \\ \vdots & \vdots & \vdots & \ddots & \vdots & \vdots & \vdots \\ 0 & 0 & 0 & \cdots & 0 & Q_{n-1,n-1} & Q_{n-1,n} \\ 0 & 0 & 0 & \cdots & 0 & 0 & 0 \end{pmatrix} \end{array}.$$

In particular, $Q_{0,0} = {}^0\lambda_{1,1}^0$, $Q_{n,n} = 0$, and for $j = 1, \ldots, n-1$,

$$Q_{j,j} = \begin{pmatrix} {}^1\lambda_{j,j}^1 & 0 & \cdots & 0 & 0 \\ 0 & {}^2\lambda_{j,j}^2 & \cdots & 0 & 0 \\ \vdots & \vdots & \ddots & \vdots & \vdots \\ 0 & 0 & \cdots & {}^{|E_j|-1}\lambda_{j,j}^{|E_j|-1} & 0 \\ 0 & 0 & \cdots & 0 & {}^{|E_j|}\lambda_{j,j}^{|E_j|} \end{pmatrix}$$

is a $|E_j| \times |E_j|$ diagonal matrix where $|E_j| = \binom{n}{j}$ the cardinality of the E_j class.

The transient state probabilities $P_{E_j}(t)$ can be shown to be

$$P_{E_0}(t) = \exp(Q_{0,0}t),$$

$$P_{E_j}(t) = \sum_{k=0}^{j-1} \sum_{0=i_0<i_1<\cdots<i_k<j} W_{0,i_1} \circ W_{i_1,i_2} \circ \cdots \circ W_{i_k,j}(t) \exp(Q_{j,j}t), \quad (2.45)$$

for $j = 1, \ldots, n-1$, where

$$W_{i,j}(t) = \int_0^t \exp[Q_{i,i}(\tau)]Q_{i,j} \exp[-Q_{j,j}(\tau)]d\tau,$$

and the function $A \circ B(t)$ is a convolution defined by $A \circ B(t) := \int_0^t A(\tau)dB(\tau)$.

The proof of (2.45) and related reliability properties are presented by Wang [114]. To illustrate the usefulness of the infinitesimal generator representation, we investigate the following 2- and 3-component systems.

For a 2-component system, state 0 denotes no failure, and State 1_1, 1_2, 2 denote the failure of components 1, 2, and both components, respectively. The infinitesimal generator of the given model is

$$\mathbf{Q} = \begin{array}{c} \\ 0 \\ 1_1 \\ 1_2 \\ 2 \end{array} \begin{array}{c} \begin{array}{cccc} 0 & 1_1 & 1_2 & 2 \end{array} \\ \begin{pmatrix} -(\lambda_1+\lambda_2+\lambda_{12}) & \lambda_1 & \lambda_2 & \lambda_{12} \\ 0 & -(\lambda_2'+\lambda_{12}) & 0 & \lambda_2'+\lambda_{12} \\ 0 & 0 & -(\lambda_1'+\lambda_{12}) & \lambda_1'+\lambda_{12} \\ 0 & 0 & 0 & 0 \end{pmatrix} \end{array}. \quad (2.46)$$

Let $P_n(t) = P(X_t = n|X_0 = 0)$ be the transient state probabilities, given the initial condition $X_0 = 0$. From the Kolmogorov forward equation (2.28) with initial condition $P_0(0) = 1$, we have the solution

$$\begin{aligned} P_0(t) &= e^{-\lambda t}, \\ P_{1_1}(t) &= \begin{cases} \frac{\lambda_1}{\lambda_1+\lambda_2-\lambda_2'}\left[e^{-(\lambda_2'+\lambda_{12})t} - e^{-\lambda t}\right], & \lambda_1+\lambda_2 \neq \lambda_2' \\ \lambda_1 t e^{-\lambda t}, & \lambda_1+\lambda_2 = \lambda_2', \end{cases} \\ P_{1_2}(t) &= \begin{cases} \frac{\lambda_2}{\lambda_1+\lambda_2-\lambda_1'}\left[e^{-(\lambda_1'+\lambda_{12})t} - e^{-\lambda t}\right], & \lambda_1+\lambda_2 \neq \lambda_1' \\ \lambda_2 t e^{-\lambda t}, & \lambda_1+\lambda_2 = \lambda_1', \end{cases} \end{aligned}$$

where $\lambda = \lambda_1 + \lambda_2 + \lambda_{12}$.

$P_0(t)$ is the probability of no component failing up to time t, and it is equivalent to the probability that the system is considered as a series structure so that $R_{1:2}(t) = P_0(t) = e^{-\lambda t}$. The system reliability for the parallel structure is the probability that at most one component fails in this 2-component system. It follows that

$$\begin{aligned} R_{2:2}(t) &= P_0(t) + P_{1_1}(t) + P_{1_2}(t) \\ &= \begin{cases} \kappa_1 e^{-(\lambda_2'+\lambda_{12})t} + \kappa_2 e^{-(\lambda_1'+\lambda_{12})t} + (1-\kappa_1-\kappa_2)e^{-\lambda t}, & \lambda_1+\lambda_2 \neq \lambda_1', \lambda_2' \\ \kappa_1 e^{-(\lambda_2'+\lambda_{12})t} + \lambda_2 t e^{-(\lambda_1'+\lambda_{12})t} + (1-\kappa_1)e^{-\lambda t}, & \lambda_1+\lambda_2 = \lambda_1' \neq \lambda_2' \\ \kappa_2 e^{-(\lambda_1'+\lambda_{12})t} + \lambda_1 t e^{-(\lambda_2'+\lambda_{12})t} + (1-\kappa_2)e^{-\lambda t}, & \lambda_1+\lambda_2 = \lambda_2' \neq \lambda_1' \\ [1+(\lambda_1+\lambda_2)t]\, e^{-\lambda t}, & \lambda_1+\lambda_2 = \lambda_1' = \lambda_2', \end{cases} \end{aligned}$$

where $\kappa_1 = \frac{\lambda_1}{\lambda_1+\lambda_2-\lambda_2'}$ and $\kappa_2 = \frac{\lambda_2}{\lambda_1+\lambda_2-\lambda_1'}$. When $\lambda_{12} = 0$, i.e., no common-cause occurs, this model reduces to the FBVE model. If there is no change of failure rate on first failure, i.e., $\lambda_1' = \lambda_1$ and $\lambda_2' = \lambda_2$, the model reduces to the BVE model.

Let $T_1(T_2)$ be the time at which component 1(2) fails. If the generator is given as (2.46), then the joint density function of T_1 and T_2 can be shown to be

$$f(t_1,t_2) = \begin{cases} \lambda_1(\lambda_2' + \lambda_{12})e^{-(\lambda_1+\lambda_2-\lambda_2')t_1-(\lambda_2'+\lambda_{12})t_2}, & 0 < t_1 < t_2 \\ \lambda_2(\lambda_1' + \lambda_{12})e^{-(\lambda_1'+\lambda_{12})t_1-(\lambda_1+\lambda_2-\lambda_1')t_2}, & 0 < t_2 < t_1, \end{cases} \tag{2.47}$$

for the absolutely continuous part, and

$$g(t) = \lambda_{12}e^{-\lambda t}, \tag{2.48}$$

for the singular part. Equations (2.47) and (2.48) lead to the joint reliability function

$$\bar{F}(t_1,t_2) = \begin{cases} \kappa_1 e^{-(\lambda_1+\lambda_2-\lambda_2')t_1-(\lambda_2'+\lambda_{12})t_2} + (1-\kappa_1)e^{-\lambda t_2}, & t_1 < t_2 \\ \kappa_2 e^{-(\lambda_1+\lambda_2-\lambda_1')t_2-(\lambda_1'+\lambda_{12})t_1} + (1-\kappa_2)e^{-\lambda t_1}, & t_2 < t_1, \end{cases} \tag{2.49}$$

where $\lambda = \lambda_1 + \lambda_2 + \lambda_{12}, \kappa_1 = \frac{\lambda_1}{\lambda_1+\lambda_2-\lambda_2'}$ and $\kappa_2 = \frac{\lambda_2}{\lambda_1+\lambda_2-\lambda_1'}$. For convenience, we call the distribution of (2.49) the bivariate exponential shared-load distribution (BVESL).

For a 3-component system, state 0 denotes no failure, and states 1_1, 1_2 and 1_3 denote respectively the failure of component 1, 2 and 3, states 2_{12}, 2_{13} and 2_{23} denote respectively the failure of two components 1 and 2, 1 and 3, and 2 and 3, and state 3 denotes the failure of all three components. The infinitesimal generator of the given model is

$$\mathbf{Q} = \begin{array}{c} \\ 0 \\ 1_1 \\ 1_2 \\ 2_{12} \\ 1_3 \\ 2_{13} \\ 2_{23} \\ 3 \end{array} \begin{array}{c} \begin{array}{cccccccc} 0 & 1_1 & 1_2 & 2_{12} & 1_3 & 2_{13} & 2_{23} & 3 \end{array} \\ \begin{pmatrix} -\alpha_0 & \lambda_1 & \lambda_2 & \lambda_{12} & \lambda_3 & \lambda_{13} & \lambda_{23} & \lambda_{123} \\ 0 & -\alpha_1 & 0 & \lambda_2' + \lambda_{12} & 0 & \lambda_3' + \lambda_{13} & 0 & \lambda_{23} + \lambda_{123} \\ 0 & 0 & -\alpha_2 & \lambda_1' + \lambda_{12} & 0 & 0 & \lambda_3' + \lambda_{23} & \lambda_{13} + \lambda_{123} \\ 0 & 0 & 0 & -\alpha_{12} & 0 & 0 & 0 & \alpha_{12} \\ 0 & 0 & 0 & 0 & -\alpha_3 & \lambda_1' + \lambda_{13} & \lambda_2' + \lambda_{23} & \lambda_{12} + \lambda_{123} \\ 0 & 0 & 0 & 0 & 0 & -\alpha_{13} & 0 & \alpha_{13} \\ 0 & 0 & 0 & 0 & 0 & 0 & -\alpha_{23} & \alpha_{23} \\ 0 & 0 & 0 & 0 & 0 & 0 & 0 & 0 \end{pmatrix} \end{array},$$

where

$$\alpha_0 = \lambda_1 + \lambda_2 + \lambda_3 + \lambda_{12} + \lambda_{13} + \lambda_{23} + \lambda_{123},$$

$$\alpha_1 = \lambda_2' + \lambda_3' + \lambda_{12} + \lambda_{13} + \lambda_{23} + \lambda_{123},$$

$$\alpha_2 = \lambda_1' + \lambda_3' + \lambda_{12} + \lambda_{23} + \lambda_{13} + \lambda_{123},$$

$$\alpha_3 = \lambda_1' + \lambda_2' + \lambda_{13} + \lambda_{23} + \lambda_{12} + \lambda_{123},$$

$$\alpha_{12} = \lambda_3'' + \lambda_{13} + \lambda_{23} + \lambda_{123},$$
$$\alpha_{13} = \lambda_2'' + \lambda_{12} + \lambda_{23} + \lambda_{123},$$
$$\alpha_{23} = \lambda_1'' + \lambda_{12} + \lambda_{13} + \lambda_{123}.$$

Note that the upper $(2^2 \times 2^2)$ submatrix denotes the failure process of components 1 and 2. In order to obtain the reliability functions of the system, we need to find the transient state probabilities $P_i(t) = P(X_t = i|X_0 = 0)$, $i = 0, 1_1, 1_2, 1_3, 2_{12}, 2_{13}, 2_{23}$. It follows that the reliability functions are $R_{1:3}(t) = P_0(t)$, $R_{2:3}(t) = P_0(t) + \sum_{i=1}^{3} P_{1_i}(t)$, and $R_{3:3}(t) = P_0(t) + \sum_{i=1}^{3} P_{1_i}(t) + \sum_{i<j} P_{2_{ij}}(t)$.

Example 2.35. Assume that the failure rates of a 3-component system are

$\lambda_1 = \lambda_2 = \lambda_3 = 0.01$ failure/h, $\lambda_{12} = \lambda_{13} = \lambda_{23} = 0.002$ failure/h, $\lambda_{123} = 0.001$ failure/h
$\lambda_1' = \lambda_2' = \lambda_3' = 0.012$ failure/h, $\lambda_1'' = \lambda_2'' = \lambda_3'' = 0.015$ failure/h.

The infinitesimal generator of the system is

$$\mathbf{Q} = \begin{array}{c} \\ 0 \\ 1_1 \\ 1_2 \\ 2_{12} \\ 1_3 \\ 2_{13} \\ 2_{23} \\ 3 \end{array} \begin{array}{c} \begin{array}{cccccccc} 0 & 1_1 & 1_2 & 2_{12} & 1_3 & 2_{13} & 2_{23} & 3 \end{array} \\ \left(\begin{array}{cccccccc}
-0.037 & 0.01 & 0.01 & 0.002 & 0.01 & 0.002 & 0.002 & 0.001 \\
0.0 & -0.031 & 0.0 & 0.014 & 0.0 & 0.014 & 0.0 & 0.003 \\
0.0 & 0.0 & -00.031 & 0.014 & 0.0 & 0.0 & 0.014 & 0.003 \\
0.0 & 0.0 & 0.0 & -0.020 & 0.0 & 0.0 & 0.0 & 0.020 \\
0.0 & 0.0 & 0.0 & 0.0 & -0.031 & 0.014 & 0.014 & 0.003 \\
0.0 & 0.0 & 0.0 & 0.0 & 0.0 & -0.020 & 0.0 & 0.020 \\
0.0 & 0.0 & 0.0 & 0.0 & 0.0 & 0.0 & -0.020 & 0.020 \\
0.0 & 0.0 & 0.0 & 0.0 & 0.0 & 0.0 & 0.0 & 0.0
\end{array}\right) \end{array}.$$

The transient state probabilities are given by

$$P_{00}(t) = e^{-0.037t},$$
$$P_{01_1}(t) = p_{01_2}(t) = p_{01_3}(t) = -1.6667e^{-0.037t} + 1.6667e^{-0.031t},$$
$$P_{02_{12}}(t) = p_{02_{13}}(t) = p_{02_{23}}(t) = 2.6275e^{-0.037t} - 4.2424e^{-0.031t} + 1.615e^{-0.02t}.$$

It follows that the system reliability functions are given by

$$R_{1:3}(t) = e^{-0.037t},$$
$$R_{2:3}(t) = 5e^{-0.031t} - 4e^{-0.037t},$$
$$R_{3:3}(t) = 4.8449e^{-0.02t} - 7.7272e^{-0.031t} + 3.8824e^{-0.037t}.$$

By integrating system reliability functions over the time interval $[0, \infty]$, we obtain the MTTFs for i-out-of-3:F systems, $i = 1, 2, 3$:

$$\text{MTTF}_{1:3} = 27.027 \text{ h}, \qquad \text{MTTF}_{2:3} = 53.182 \text{ h}, \qquad \text{MTTF}_{3:3} = 97.91 \text{ h}.$$

2.5.2 A Shared-Load k-out-of-n:G Repairable System

Shao and Lamberson [97] proposed an approach to compute and analyze a shared-load k-out-of-n:G repairable system. The assumptions of the model are given as follows:

Assumptions:

1. The failure rate of all functioning components in the system is the same and constant in each success state. When a component fails, the system state changes and the failure rate for every functioning component changes. For i components functioning, every functioning component has the constant failure rate, λ_i, $i = k, \ldots, n$. And $\lambda_n < \lambda_{n-1} < \cdots < \lambda_k$.
2. A failed component must be detected and disconnected by a controller. The probability of successful detection and switching of the controller is α, for each failure event. If the controller cannot detect and disconnect a failed unit or the controller itself has failed, the system fails. The controller failure rate is a constant, λ_c. The controller is never repaired or replaced during a mission.
3. At most r components can be in repair at one time; the constant repair rate is μ. The repair rate for j components failed is: $\mu_j = \mu \cdot \min\{j, r\}$.
4. A repaired component is as good as new and is immediately re-connected to the system.
5. The switch-over time and the time for re-energizing a repaired component are negligible.

Based on the above assumptions, a continuous-time Markov chain can be used to create a set of differential equations. The state space for the system is given below.

State Space:

- State j ($j = 0, 1, \ldots, n-k$) : j components have failed and have been disconnected from the network, the remaining $(n-j)$ components and the controller are functioning.
- State $n-k+1$: The system fails because only $(k-1)$ components are functioning. However, the system can return to the working state $(n-k)$ at a repair rate μ_{n-k+1}.
- State f: The system fails because the controller cannot detect and disconnect a failed component (Fig. 2.10).

The system state space $E = \{0, 1, \ldots, j, \ldots, n-k, n-k+1, f\}$, and $E = S \cup F$, where the failed state space F consists of failed states $n-k+1$ and f, while the success state space S includes the remaining states.

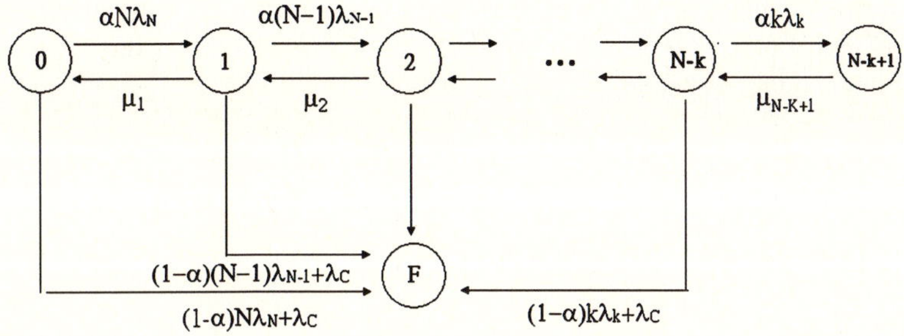

Fig. 2.10 Markov transition diagram for a shared-load k-out-of-n:G system

Generator: The infinitesimal generator of the given model is

$$
Q = \begin{array}{c} \\ \end{array}
\begin{pmatrix}
\overset{0}{-n\lambda_n - \lambda_c} & \overset{1}{\alpha n \lambda_n} & \overset{\cdots}{} & \overset{n-k}{} & \overset{n-k+1}{} & \overset{f}{(1-\alpha)n\lambda_n + \lambda_c} \\
\mu_1 & -\mu_1 - (n-1)\lambda_{n-1} - \lambda_c & & & & (1-\alpha)(n-1)\lambda_{n-1} + \lambda_c \\
 & \mu_2 & \ddots & & & \vdots \\
 & & \ddots & & & \vdots \\
 & & & \alpha(k+1)\lambda_{k+1} & & (1-\alpha)(k+1)\lambda_{k+1} + \lambda_c \\
 & & & -\mu_{n-k} - k\lambda_k - \lambda_c & \alpha k \lambda_k & (1-\alpha)k\lambda_k + \lambda_c \\
 & & & \mu_{n-k+1} & -\mu_{n-k+1} & \\
 & & & & & 0
\end{pmatrix}.
$$

Differential Equations: The set of differential equations based on the generator is

$$
\begin{aligned}
P_0'(t) &= [-n\lambda_n - \lambda_c]P_0(t) + \mu_1 P_1(t), \\
P_j'(t) &= [\alpha(n-j+1)\lambda_{n-j+1}]P_{j-1}(t) + [-\mu_j - (n-j)\lambda_{n-j} - \lambda_c]P_j(t) \\
&\quad + [\mu_{j+1}]P_{j+1}(t), \quad j = 1, 2, \ldots, n-k, \\
P_{n-k+1}'(t) &= [\alpha k \lambda_k]P_{n-k}(t) + [-\mu_{n-k+1}]P_{n-k+1}(t), \\
P_f'(t) &= \sum_{j=0}^{n-k}[(1-\alpha)(n-j)\lambda_{n-j} + \lambda_c]P_j(t).
\end{aligned}
$$

The initial conditions for this set of linear differential equations are: $P_0(0) = 1$ and $P_i(0) = 0$ for $i \in E - \{0\}$.

System Availability & Reliability: The system availability $A(t)$ is the probability that the system is surviving at time t:

$$
A(t) = \sum_{i \in S} P_i(t) = \sum_{j=0}^{n-k} P_j(t).
$$

The $P_j(t)$ can be determined by integrating the set of linear differential equations numerically. Availability can be considered as the generalized reliability when the system is repairable or component-replaceable.

System reliability is the availability of the same system without repair and replacement. Let $\mu_j = 0$, for $j = 1, 2, \ldots, n-k+1$, then the set of linear differential equations becomes

$$\begin{aligned} P_0'(t) &= -(\beta_n + \lambda_c) P_0(t), \\ P_j'(t) &= (\alpha\beta_{n-j+1}) P_{j-1}(t) - (\beta_{n-j} + \lambda_c) P_j(t), \quad j = 1, 2, \ldots, n-k, \\ P_{n-k+1}'(t) &= (\alpha\beta_k) P_{n-k}(t), \end{aligned} \tag{2.50}$$

where $\beta_j = j\lambda_j$, $j = k, k+1, \ldots, n$. Taking the Laplace transform of (2.50), we have the following expressions:

$$\begin{aligned} sP_0^*(s) - 1 &= -(\beta_n + \lambda_c) P_0^*(s), \\ sP_1^*(s) &= \alpha\beta_n P_0^*(s) - (\beta_{n-1} + \lambda_c) P_1^*(s), \\ sP_j^*(s) &= (\alpha\beta_{n-j+1}) P_{j-1}^*(s) - (\beta_{n-j} + \lambda_c) P_j^*(s), \quad j = 2, \ldots, n-k. \end{aligned}$$

Thus,

$$P_j^*(s) = \frac{\alpha\beta_{n-j+1}}{s + \beta_{n-j} + \lambda_c} P_{j-1}^*(s) = \frac{\prod_{i=1}^{j} \alpha\beta_{n-i+1}}{\prod_{i=0}^{j}(s + \beta_{n-i} + \lambda_c)}, \quad j = 1, 2, \ldots, n-k.$$

Next, we use an inverse Laplace transform to solve the above equations. The approach is to expand the product fraction into a partial fraction form:

$$\frac{1}{\prod_{i=0}^{j}(s + c_i)} = \sum_{i=0}^{j} \left[\prod_{k=0, k\neq i}^{j} \frac{1}{(c_k - c_i)} \right] \frac{1}{s + c_i}, \quad j = 1, 2, \ldots, n-k.$$

Thus, the result is $P_0(t) = e^{-(\beta_n + \lambda_c)t}$ and

$$P_j(t) = \prod_{i=1}^{j} \alpha\beta_{n-i+1} \cdot \sum_{i=0}^{j} \frac{e^{-(\beta_{n-i} + \lambda_c)t}}{\prod_{k=0, k\neq i}^{j} (\beta_{n-k} - \beta_{n-i})}, \quad j = 1, 2, \ldots, n-k. \tag{2.51}$$

The reliability function of the shared-load system is $R(t) = \sum_{j=0}^{n-k} P_j(t)$.

Special cases: (1) If the absorbing state is removed from the chain by considering perfect fault coverage ($\alpha = 1$ and $\lambda_c = 0$), the result is a birth-death process. (2) If we further assumed that the components are nonrepairable and $\lambda_i = \lambda$ for $i = 0, 1, \ldots, n$, then the shared-load system reduces to a regular k-out-of-n system.

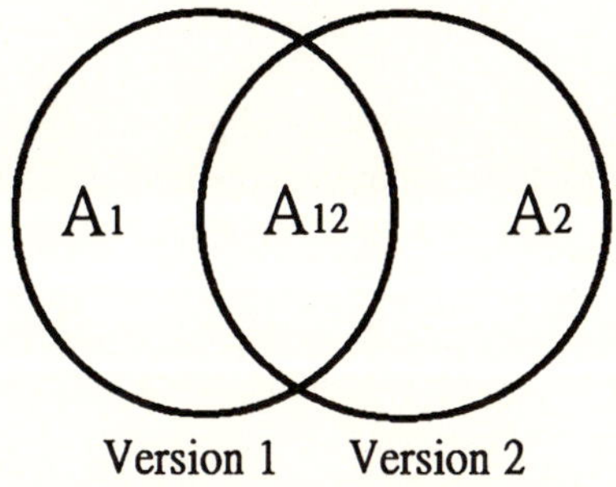

Fig. 2.11 Software faults in the 2-version software system

2.5.3 *A Dependent Model for Fault-Tolerant Software Systems During Debugging*

N-version programming (NVP) is one of the most important software fault-tolerance techniques. Numerous researchers have studied the effect of NVP for the improvement of software reliability on fault correlation theoretically [25, 27, 53, 59, 80, 87, 106] as well as empirically [15, 16, 26, 44, 47, 65, 66]. However, only a few of them consider this issue during the testing and debugging part of the software development life cycle. During testing and debugging, faults may not be successfully removed. Imperfect debugging may result in unsuccessful removal, and the introduction of new faults. In this example we review a model proposed by Wang [115], who developed a bivariate counting process for a 2VP by assuming positive dependency among versions. Considering imperfect debugging, this bivariate process characterizes dynamic changes of fault contents for each version during testing and debugging. Further explanations for the assumptions and proofs of the given model, see Wang [115].

2.5.3.1 Model Assumptions

The fault content of a 2VP system is divided into three components, according to their independence. That is, the system is composed of three mutually exclusive components, as depicted in Fig. 2.11. Here A_1 and A_2 represent independent faults which do not cause failing of the other version, and A_{12} represents related faults that affect both versions.

Assumptions:

1. Both software versions can fail on the sample input, which can be caused by either the related faults (A_{12}) or the independent faults (A_1 and A_2).
2. Related and independent faults are mutually exclusive.
3. When a failure occurs at a version, a debugging effort occurs immediately. The fault content of that version is instantaneously reduced by one, increased by one, or remains the same.
4. An independent fault can be successfully removed with high probability.

5. Debugging of a related fault may result in successful removal (reduce size of A_{12} by one), introduction of a new related fault (increase size of A_{12} by one), or unsuccessful removal (size of A_{12} has no change).
6. In terms of unsuccessful debugging of a related fault in A_{12}, it can become an independent fault in either A_1 or A_2.
7. The version-failure detection rate at any time is proportional to its current fault content (the number of faults remaining in the version at that time).
8. The fault removal rate of version 1 (2) is proportional to the current fault content of version 1 (2).
9. The fault introduction rate of version 1 is proportional to the current fault content of version 2, and vice versa. This mainly reflects the common assumption of positive correlated failures.
10. The removal rate per fault and introduction rate per fault are constant.
11. The simultaneous occurrences of multiple failures in a version are not allowed (i.e., the 'orderly property').

2.5.3.2 The Proposed Model

According to the assumptions, the model developed here is a bivariate counting process. Let $N_i(t)$ be the fault content (total number of faults) of version i at time t. The transition probabilities in $(N_1(t), N_2(t))$ during the time interval $(t, t + dt)$ are

$$P(N_1(t + dt) - N_1(t) = i, N_2(t + dt) - N_2(t) = j \,|\, N_1(t) = x, N_2(t) = y)$$
$$= f_{ij}(x, y)dt,$$

where i and j are not both zero and $f_{ij}(x, y)$ are suitable non-negative functions of x and y. Assume that

$$\begin{aligned} f_{-1,0}(x, y) &= \alpha_1 x, \quad f_{0,-1}(x, y) = \alpha_2 y, \\ f_{1,0}(x, y) &= \beta_1 y, \quad f_{0,1}(x, y) = \beta_2 x, \end{aligned} \tag{2.52}$$

where $\alpha_1, \alpha_2, \beta_1$ and β_2 are unknown constants, although it is possible to let them be functions of time. Let the joint state probability of $N_1(t)$ and $N_2(t)$ at time t be given by $P_{m,n}(t) = P(N_1(t) = m, N_2(t) = n)$. The fundamental equations governing this bivariate counting process are the following Kolmogorov forward equations [9]

$$\begin{aligned} P'_{m,n}(t) &= \alpha_1(m + 1)P_{m+1,n} + \alpha_2(n + 1)P_{m,n+1} + \beta_1 n P_{m-1,n} \\ &\quad + \beta_2 m P_{m,n-1} - [(\alpha_1 + \beta_2)m + (\alpha_2 + \beta_1)n]P_{m,n}, \quad \text{for } m, n \geq 0. \end{aligned}$$

Let $M(\theta, \phi, t)$ be the moment generating function and it is given by

$$M(\theta, \phi, t) = \sum_{m=0}^{\infty}\sum_{n=0}^{\infty} e^{m\theta + n\phi} P_{m,n}(t).$$

It can be shown that [9]

$$\frac{\partial M(\theta,\phi,t)}{\partial t} = \sum_{i,j} \left(e^{i\theta+j\phi} - 1\right) f_{ij}\left(\frac{\partial}{\partial\theta}, \frac{\partial}{\partial\phi}\right) M(\theta,\phi,t), \tag{2.53}$$

where i and j are not both zero. From (2.52) and (2.53), the partial differential equation of the moment generating function is given by

$$\frac{\partial M}{\partial t} = \left[\alpha_1(e^{-\theta} - 1) + \beta_2(e^{\phi} - 1)\right]\frac{\partial M}{\partial\theta} + \left[\alpha_2(e^{-\phi} - 1) + \beta_1(e^{\theta} - 1)\right]\frac{\partial M}{\partial\phi},$$

with initial condition $M(\theta,\phi,0) = e^{b_1\theta+b_2\phi}$. It is sometimes easier to work with the cumulant generating function, $K(\theta,\phi,t)$, which is defined by $K = \log M$. The partial differential equation of the cumulant generating function is

$$\frac{\partial K}{\partial t} = \left[\alpha_1(e^{-\theta} - 1) + \beta_2(e^{\phi} - 1)\right]\frac{\partial K}{\partial\theta} + \left[\alpha_2(e^{-\phi} - 1) + \beta_1(e^{\theta} - 1)\right]\frac{\partial K}{\partial\phi}, \tag{2.54}$$

with initial condition $K(\theta,\phi,0) = b_1\theta + b_2\phi$. Express the cumulant generating function into a Taylor expansion

$$K(\theta,\phi,t) = \sum_{i,j\geq 0} K_{ij}(t)\frac{\theta^i}{i!}\frac{\phi^j}{j!},$$

where $K_{ij}(t)$ are cumulants and $K_{00} \equiv 0$. It can be shown that the first and second order cumulants correspond to the expected number, variances, and covariance of fault contents as [9]

$$\begin{aligned}
&K_{10}(t) = E[N_1(t)], && K_{01}(t) = E[N_2(t)],\\
&K_{20}(t) = \mathrm{Var}[N_1(t)], && K_{02}(t) = \mathrm{Var}[N_2(t)],\\
&K_{11}(t) = \mathrm{Cov}[N_1(t), N_2(t)].
\end{aligned}$$

To obtain the desired cumulants, we use the following procedure. Taking the first and second order cumulants and equating coefficients of $\theta, \phi, \theta^2, \theta\phi$, and ϕ^2 on both sides of (2.54) give the following differential equations

$$\begin{aligned}
\frac{dK_{10}}{dt} &= -\alpha_1 K_{10} + \beta_1 K_{01},\\
\frac{dK_{01}}{dt} &= \beta_2 K_{10} - \alpha_2 K_{01},\\
\frac{dK_{20}}{dt} &= -2\alpha_1 K_{20} + \alpha_1 K_{10} + \beta_1 K_{01} + 2\beta_1 K_{11},
\end{aligned}$$

$$\frac{dK_{11}}{dt} = \beta_2 K_{20} - (\alpha_1 + \alpha_2)K_{11} + \beta_1 K_{02},$$

$$\frac{dK_{02}}{dt} = \beta_2 K_{10} + \alpha_2 K_{01} + 2\beta_2 K_{11} - 2\alpha_2 K_{02}. \tag{2.55}$$

These equations can be solved by the general ordinary differential equation method, using the initial conditions $K_{10}(0) = b_1$ and $K_{01}(0) = b_2$, and all other cumulants being zero. Solving $dK_{10}/dt = -\alpha_1 K_{10} + \beta_1 K_{01}$ and $dK_{01}/dt = \beta_2 K_{10} - \alpha_2 K_{01}$, we have

$$K_{10}(t) = A_1 e^{\lambda_1 t} + A_2 e^{\lambda_2 t}, \tag{2.56}$$

$$K_{01}(t) = B_1 e^{\lambda_1 t} + B_2 e^{\lambda_2 t}, \tag{2.57}$$

where A_1, A_2, B_1 and B_2 are constants and

$$\lambda_1 = \frac{-(\alpha_1 + \alpha_2) + \sqrt{(\alpha_1 - \alpha_2)^2 + 4\beta_1\beta_2}}{2}, \tag{2.58}$$

$$\lambda_2 = \frac{-(\alpha_1 + \alpha_2) - \sqrt{(\alpha_1 - \alpha_2)^2 + 4\beta_1\beta_2}}{2}. \tag{2.59}$$

Let $C = \lambda_1 - \lambda_2 = \sqrt{(\alpha_1 - \alpha_2)^2 + 4\beta_1\beta_2}$, then coefficients of A_1, A_2, B_1 and B_2 are

$$A_1 = \frac{b_1(C - \alpha_1 + \alpha_2) + 2\beta_1 b_2}{2C}, \tag{2.60}$$

$$A_2 = \frac{b_1(C + \alpha_1 - \alpha_2) - 2\beta_1 b_2}{2C}, \tag{2.61}$$

$$B_1 = \frac{\alpha_1 - \alpha_2 + C}{2\beta_1} A_1, \tag{2.62}$$

$$B_2 = \frac{\alpha_1 - \alpha_2 - C}{2\beta_1} A_2. \tag{2.63}$$

Parameters $\alpha_1, \alpha_2, \beta_1$ and β_2 can be expressed as, by substituting (2.56) and (2.57) into (2.55),

$$\alpha_1 = \frac{A_1 B_2 \lambda_1 - A_2 B_1 \lambda_2}{A_2 B_1 - A_1 B_2}, \tag{2.64}$$

$$\alpha_2 = \frac{A_2 B_1 \lambda_1 - A_1 B_2 \lambda_2}{A_1 B_2 - A_2 B_1}, \tag{2.65}$$

$$\beta_1 = \frac{A_1 A_2 C}{A_2 B_1 - A_1 B_2}, \tag{2.66}$$

$$\beta_2 = \frac{B_1 B_2 C}{A_1 B_2 - A_2 B_1}. \tag{2.67}$$

Thus, if we have numerical estimates for values of A_1, A_2, B_1, B_2 and λ_1, λ_2, then $\alpha_1, \alpha_2, \beta_1$ and β_2 are given by (2.64)–(2.67), respectively.

Since β_1 and β_2 are all positive, the roots λ_1 and λ_2 given in (2.58) and (2.59) are always real. If $\alpha_1\alpha_2 < \beta_1\beta_2$, then one of the λ_i is positive and the other is negative. If $\alpha_1\alpha_2 > \beta_1\beta_2$, both λ_1 and λ_2 are negative. This suggests that the result of eventually extinguishing the faults is certain, if $\alpha_1\alpha_2 > \beta_1\beta_2$.

Special case 1. Let $\alpha_1 = \alpha_2 \equiv \alpha$ and $\beta_1 = \beta_2 \equiv \beta$. This case describes the situation that both versions affect each other in a 'symmetric' fashion. It can be shown that the expected number of faults are $K_{10}(t) = \frac{b_1+b_2}{2}e^{-(\alpha-\beta)t} + \frac{b_1-b_2}{2}e^{-(\alpha+\beta)t}$ and $K_{01}(t) = \frac{b_1+b_2}{2}e^{-(\alpha-\beta)t} - \frac{b_1-b_2}{2}e^{-(\alpha+\beta)t}$.

Special case 2. Let $\beta_1 = 0$ and $\alpha_1 \neq \alpha_2$. In this case, only version 2 is affected by version 1. The expected number of faults are $K_{10}(t) = b_1 e^{-\alpha_1 t}$ and $K_{01}(t) = b_2 e^{-\alpha_2 t} + \frac{b_1\beta_2}{\alpha_1-\alpha_2}[e^{-\alpha_2 t} - e^{-\alpha_1 t}]$. Further, if $\alpha_1 = \alpha_2 \equiv \alpha$, the expected number of faults are $K_{10}(t) = b_1 e^{-\alpha t}$ and $K_{01}(t) = b_2 e^{-\alpha t} + b_1\beta_2 t e^{-\alpha t}$.

Special case 3. Let $\beta_1 = 0$ and $\beta_2 = 0$. This case describes that the failure behavior of both versions are s-independent. As such the expected number of faults are $K_{10}(t) = b_1 e^{-\alpha_1 t}$ and $K_{01}(t) = b_2 e^{-\alpha_2 t}$. In fact, this is the JM model [42] for two independent versions.

2.5.3.3 Expectation of Failures

The bivariate counting process by itself cannot fully describe the debugging process, for example it cannot provide the total number of failures up to time t. We need to consider both failure content and fault content. A failure occurs when the actual output deviates from the expected output as a result of executing the faulty program. For each version, when a failure has occurred it can reduce one fault, increase one fault, or give no change of the fault content. The associated fault content and failure content can also be studied by a multivariate processes. Let

$N_i(t) \equiv$ the fault content of version i at time t, $i = 1, 2$,

$M_i(t) \equiv$ the number of observed failures of version i up to time t, $i = 1, 2$.

Set

$$P(N_1(t+dt) - N_1(t) = i, N_2(t+dt) - N_2(t) = j, M_1(t+dt) - M_1(t) = k,$$
$$M_2(t+dt) - M_2(t) = l \mid N_1(t) = x, N_2(t) = y) = f_{i,j,k,l}(x, y)dt.$$

Then we assume that

$$f_{-1,0,1,0}(x,y) = \alpha_1 x,\ f_{1,0,1,0}(x,y) = \beta_1 y,$$
$$f_{0,-1,0,1}(x,) = \alpha_2 y,\quad f_{0,1,0,1}(x,y) = \beta_2 x.$$

Follow the similar procedure discussed in Sect. 2.5.3.2, we get the partial differential equation of the cumulant generating function $K(\theta_1, \theta_2, \theta_3, \theta_4)$

$$\frac{\partial K}{\partial t} = \left[\alpha_1(e^{-\theta_1+\theta_3} - 1) + \beta_2(e^{\theta_2+\theta_4} - 1)\right]\frac{\partial K}{\partial \theta_1}$$
$$+\left[\beta_1(e^{\theta_1+\theta_3} - 1) + \alpha_2(e^{-\theta_2+\theta_4} - 1)\right]\frac{\partial K}{\partial \theta_2}.$$

In addition to (2.56) and (2.57), the expected numbers of failures are

$$E[M_1(t)] = \int_0^t \left[(\alpha_1 A_1 + \beta_1 B_1)e^{\lambda_1 \tau} + (\alpha_1 A_2 + \beta_1 B_2)e^{\lambda_2 \tau}\right] d\tau,$$
$$E[M_2(t)] = \int_0^t \left[(\beta_2 A_1 + \alpha_2 B_1)e^{\lambda_1 \tau} + (\beta_2 A_2 + \alpha_2 B_2)e^{\lambda_2 \tau}\right] d\tau,$$

where A_1, A_2, B_1 and B_2 are expressed in (2.60)–(2.63), respectively. Moreover,

$$\lim_{t\to\infty} E[M_1(t)] = \begin{cases} \frac{\alpha_1 A_1+\beta_1 B_1}{(-\lambda_1)} + \frac{\alpha_1 A_2+\beta_1 B_2}{(-\lambda_2)} & \text{if } \lambda_1 < 0 \text{ and } \lambda_2 < 0 \\ \infty & \text{otherwise,} \end{cases} \quad (2.68)$$

and

$$\lim_{t\to\infty} E[M_2(t)] = \begin{cases} \frac{\beta_2 A_1+\alpha_2 B_1}{(-\lambda_1)} + \frac{\beta_2 A_2+\alpha_2 B_2}{(-\lambda_2)} & \text{if } \lambda_1 < 0 \text{ and } \lambda_2 < 0 \\ \infty & \text{otherwise.} \end{cases} \quad (2.69)$$

2.5.3.4 Reliability

Suppose that n_1 faults of version 1 and n_2 faults of version 2 still remain in the program at time t_e. The overall failure rate at that time would be $n_1(\alpha_1(t_e) + \beta_2(t_e)) + n_2(\alpha_2(t_e) + \beta_1(t_e))$. If the system fails when one of its versions fail, then the reliability of the entire system as a function of time t takes the form

$$R_{Dep}(t|t_e) = [R_1(t|t_e)]^{n_1}\,[R_2(t|t_e)]^{n_2},$$

if the system fails only when both its versions fail, the reliability of the entire system is

$$R_{Dep}(t|t_e) = 1 - \left(1 - [R_1(t|t_e)]^{n_1}\right)\left(1 - [R_2(t|t_e)]^{n_2}\right), \quad (2.70)$$

where $R_1(t|t_e) = \exp\{-(\alpha_1 + \beta_2)t\}$ and $R_2(t|t_e) = \exp\{-(\alpha_2 + \beta_1)t\}$. The function $R_1(t|t_e)$ and $R_2(t|t_e)$ can be viewed as reliabilities attributable to each of the individual faults of versions 1 and 2, respectively.

2.5.3.5 Estimation of Parameters

In fact, the actual fault contents for each version cannot be observed. However, the number of failures can be observed. Thus, we use the number of observed failures to estimate the unknown parameters, $\alpha_1, \alpha_2, \beta_1$ and β_2. We can minimize the objective function

$$S = \sum_{i=1}^{m} \left\{ (\widehat{M_1}(t_i) - EM_1(t_i))^2 + (\widehat{M_2}(t_i) - EM_2(t_i))^2 \right\},$$

i.e., the sum of two least squares functions, where $\widehat{M_1}(t_i)$ and $\widehat{M_2}(t_i)$ represent the number of observed failures of versions 1 and 2 up to time t_i, respectively. However, the estimates for $\alpha_1, \alpha_2, \beta_1$ and β_2 may not be very reliable since a small change in λ_1 and λ_2 may magnify the effect in these estimates.

2.5.3.6 Example

We illustrate the proposed model by using the observed failure data of the water reservoir control (WRC) system [106]. Water is supplied via a source pipe controlled by a source valve and removed via a drain pipe controlled by a drain valve. The WRC system achieves fault-tolerance and high reliability by using NVP software control logic with $N = 2$. This 2VP system application assumes that the reliability of the voter is 1, and that the voter can identify exactly which version(s) is failed when a failure occurs [106]. Following these assumptions, the 2VP system fails only when both its software versions fail at the same input data. This WRC system data set is listed in Table 2.9.

Apply the method in Sect. 2.5.3.5 to the WRC data set, the LSEs of $\alpha_1, \alpha_2, \beta_1$ and β_2 are

$$\hat{\alpha}_1 = 0.00568, \quad \hat{\beta}_1 = 0.000399,$$

$$\hat{\alpha}_2 = 0.00735, \quad \hat{\beta}_2 = 0.000874.$$

Based on estimated values of α_i and β_i, related measurements can be obtained. Since $\hat{\alpha}_1\hat{\alpha}_2 > \hat{\beta}_1\hat{\beta}_2$, the perfect debugging (elimination of all faults) is certain. Substituting $\hat{\alpha}_i$ and $\hat{\beta}_i$ into (2.56) and (2.57) and setting $t = 0$, we have the initial fault contents for both versions

$$\hat{b}_1 = 29.594 \quad \text{and} \quad \hat{b}_2 = 25.004.$$

Table 2.9 Failure-time normalized-data of WRC 2VP system

	Failure time			Failure time	
Failure no.	Version 1	Version 2	Failure no.	Version 1	Version 2
1	1.2	3.6	14	39.2	34.8
2	2.8	8.4	15	40.0	36.4
3	8.4	12.8	16	44.0	36.8
4	10.0	14.4	17	44.8	38.0
5	16.4	17.2	18	54.0	39.2
6	20.0	18.0	19	56.0	41.6
7	24.4	20.0	20	62.4	42.0
8	28.0	23.2	21	80.0	46.4
9	29.2	25.2	22	92.0	59.6
10	31.2	28.0	23	99.6	62.4
11	34.0	28.4	24		98.8
12	36.0	30.8	25		99.6
13	36.8	31.2	26		100.0

This shows that the initial number of faults in version 1 is bigger than that in version 2. The expected numbers of failures eventually encountered for both versions are obtained by substituting estimated values of parameters into (2.68) and (2.69). The numbers are 32.826 and 34.608 for version 1 and version 2, respectively. This shows that we expect that more failures occur in version 2 than that in version 1. Initially the number of faults in version 1 is bigger than version 2. However, we expect that version 1 has less failures eventually encountered than version 2. This shows that version 1 may have a better debugging performance than version 2. Part of the reason may be attributed to dependency between versions. From (2.56) and (2.57), we have the expected numbers of remaining faults for versions at time t

$$E[N_1(t)] = (-2.162)e^{-0.00754t} + (31.755)e^{-0.00549t},$$
$$E[N_2(t)] = (10.083)e^{-0.00754t} + (14.921)e^{-0.00549t}.$$

If these two versions are assumed to be independent of each other, then the expected numbers of remaining faults at time t are given by $E_I[N_1(t)] = (31.697)e^{-0.00564t}$ and $E_I[N_2(t)] = (33.600)e^{-0.00629t}$. From (2.70), the reliability for the 2VP system when $t_e = 250$ is

$$R_{Dep}(t|250) = 1 - (1 - e^{-0.0506t})(1 - e^{-0.04117t}).$$

References

1. O.O. Aalen, Ø. Borgan, H.K. Gjessing, *Survival and Event History Analysis: A Process Point of View* (Springer, New York, 2008)
2. A.A. Abdel-Ghaly, P.Y. Chan, B. Littlewood, Evaluation of competing software reliability predictions. IEEE Trans. Softw. Eng. **12**, 950–967 (1986)

3. D. Al-Mutairi, Y. Chen, N.D. Singpurwalla, An adaptative concatenated failure rate model for software reliability. J. Am. Stat. Assoc. **93**, 1150–1163 (1998)
4. S.V. Amari, R.B. Misra, Closed-form expressions for distribution of sum of exponential random variables. IEEE Trans. Reliab. **46**(4), 519–522 (1997)
5. S.V. Amari, J.B. Dugan, R.B. Misra, Optimal reliability of systems subject to imperfect fault coverage. IEEE Trans. Reliab. **48**(3), 275–284 (1999)
6. S.V. Amari, K.B. Misra, H. Pham, Tampered failure rate load-sharing systems: status and perspectives, in *Handbook on Performability Engineering*, ed. by K.B. Misra (Springer, London, 2008)
7. P.M. Anderson, S.K. Agarwal, An improved model for protective-system reliability. IEEE Trans. Reliab. **41**(3), 422–426 (1992)
8. H. Ascher, H. Feingold, *Repairable Systems Reliability: Modeling, Inference, Misconceptions and Their Causes* (Marcel Dekker, New York, 1984)
9. N.T.J. Bailey, *The Elements of Stochastic Processes* (Wiley, New York, 1964)
10. N. Balakrishnan, C.D. Lai, *Continuous Bivariate Distributions* (Springer, New York, 2009)
11. R.E. Barlow, F. Prochan, *Statistical Theory of Reliability and Life Testing: Probability Models* (To Begin With, Silver Spring, 1981)
12. H.W. Block, A.P. Basu, A continuous bivariate exponential extension. J. Am. Stat. Assoc. **69**, 1031–1037 (1974)
13. P.J. Boland, N.N. Chuív, Optimal times for software release when repair is imperfect. Stat. Probab. Lett. **77**, 1176–1184 (2007)
14. L.C. Briand, K.E. Emam, B.G. Freimut, A comprehensive evaluation of capture-recapture models for estimating software defect content. IEEE Trans. Softw. Eng. **26**(6), 518–540 (2000)
15. S.S. Brilliant, J.C. Knight, N.G. Leveson, Analysis of faults in an N-version software experiment. IEEE Trans. Softw. Eng. **16**(2), 238–247 (1990)
16. X. Cai, M.R. Lyu, M.A. Vouk, An experimental evaluation on reliability features of N-version programming, in *Proceeding of the 16th International Symposium on Software Reliability Engineering*, Chicago, 2005
17. K.C. Chae, G.M. Clark, System reliability in the presence of common-cause failures. IEEE Trans. Reliab. **35**(1), 32–35 (1986)
18. Y. Chen, N.D. Singpurwalla, Unification of software reliability models by self-exciting point processes. Adv. Appl. Probab. **29**, 337–352 (1997)
19. D.R. Cox, V. Isham, *Point Processes* (Chapman & Hall, London, 1980)
20. D.R. Cox, P.A.W. Lewis, *The Statistical Analysis of Series of Events* (Chapman & Hall, London, 1966)
21. L.H. Crow, Reliability analysis for complex repairable systems, in *Reliability and Biometry: Statistical Analysis of Lifelength*, ed. by F. Proschan, R.J. Serfling (SIAM, Philadelphia, 1974)
22. M.J. Crowder, A.C. Kimber, R.L. Smith, T.J. Sweeting, *Statistical Analysis of Reliability Data* (Chapman & Hall, London, 1991)
23. T. Dohi, S. Osaki, K.S. Trivedi, An infinite server queueing approach for describing software reliability growth: unified modeling and estimation framework, in *Proceedings of the 11th Asia-Pacific Software Engineering Conference*, Busan, 2004
24. J.T. Duane, Learning curve approach to reliabilitymonitoring. IEEE Trans. Aerosp. **2**, 563–566 (1964)
25. D.E. Eckhardt, L.D. Lee, A theoretical basis for the analysis of multiversion software subject to coincident errors. IEEE Trans. Softw. Eng. **11**(12), 1511–1517 (1985)
26. D.E. Eckhardt, A.K. Caglavan, J.C. Knight, L.D. Lee, D.F. McAllister, M.A. Vouk, J.P. Kelly, An experimental evaluation of software redundancy as a strategy for improving reliability. IEEE Trans. Softw. Eng. **17**(7), 692–702 (1991)
27. M. Ege, M.A. Eyler, M.U. Karakas, Reliability analysis in N-version programming with dependent failures, in *Proceedings of the 27th Euromicro Conference*, Warsaw, 2001
28. K.N. Fleming, N. Mosleh, R.K. Deremer, A systematic procedure for incorporation of common cause events into risk and reliability models. Nucl. Eng. Des. **93**, 245–273 (1986)

29. J.E. Freund, A bivariate extension of the exponential distribution. J. Am. Stat. Assoc. **56**, 971–977 (1961)
30. O. Gaudoin, J.L. Soler, Statistical analysis of the geometric de-eutrophication software-reliability model. IEEE Trans. Reliab. **41**(4), 518–524 (1992)
31. O. Gaudoin, C. Lavergne, J.L. Soler, A generalized geometric de-eutrophication software-reliability model. IEEE Trans. Reliab. **43**(4), 536–541 (1994)
32. B.V. Gnedenko, Y.K. Belyayev, A.D. Solovyev, *Mathematical Methods of Reliability Theory* (Academic, New York, 1969)
33. A.L. Goel, Software reliability models: assumptions, limitations, and applicability. IEEE Trans. Softw. Eng. **11**(12), 1411–1423 (1985)
34. A.L. Goel, K. Okumoto, Time-dependent error-detection rate model for software reliability and other performance measures. IEEE Trans. Reliab. **28**(3), 206–211 (1979)
35. S.S. Gokhale, P.N. Marinos, K.S. Trivedi, Important milestomes in software reliability modeling, in *Proceedings of Software Engineering and Knowledge Engineering (SEKE) '96*, Lake Tahoe, 1996
36. K. Goševa-Popstojanova, K.S. Trivedi, Architecture-based approaches to reliability assessment of software systems. Perform. Eval. **45**, 179–204 (2001)
37. M.H. Halstead, *Elements of Software Science* (North-Holland, Amsterdam, 1977)
38. P. Hokstad, M. Rausand, Common cause failure modeling: status and trends, in *Handbook of Performability Engineering*, ed. by K.B. Misra (Springer, London, 2008)
39. C.Y. Huang, W.C. Huang, Software reliability analysis and measurement using finite and infinite server queueing models. IEEE Trans. Reliab. **57**(1), 192–203 (2008)
40. C.Y. Huang, M.R. Lyu, S.Y. Kuo, A unified scheme of some nonhomogenous Poisson process models for software reliability estimation. IEEE Trans. Softw. Eng. **29**(3), 261–269 (2003)
41. L. Huang, Q. Xu, Lifetime reliability for load-sharing redundant systems with arbitrary failure distributions. IEEE Trans. Reliab. **59**(2), 319–330 (2010)
42. Z. Jelinski, P.B. Moranda, Software reliability research, in *Statistical Computer Performance Evaluation*, ed. by W. Freiberger (Academic, New York, 1972)
43. H. Joe, *Multivariate Models and Dependence Concepts* (Chapman & Hall, London, 1997)
44. P.K. Kapur, A. Gupta, P.C. Jha, Reliability growth modeling and optimal release policy under fuzzy environment of an N-version programming system incorporating the effect of fault removal efficiency. Int. J. Autom. Comput. **4**(4), 369–379 (2007)
45. M. Kijima, *Markov Processes for Stochastic Modeling* (Chapman & Hall, London, 1997)
46. B.A. Kitchenham, B. Littlewood, *Measurement for Software Control and Assurance* (Elsevier, London, 1989)
47. J.C. Knight, N.G. Leveson, An experimental evaluation of the assumption of independence in multiversion programming. IEEE Trans. Softw. Eng. **12**(1), 96–109 (1986)
48. K.A.H. Kobbacy, D.N.P. Murthy, *Complex System Maintenance Handbook* (Springer, London, 2008)
49. W. Kremer, Birth-death and bug counting. IEEE Trans. Reliab. **32**(1), 37–47 (1983)
50. P.H. Kvam, E.A. Peña, Estimating load-sharing properties in a dynamic reliability system. J. Am. Stat. Assoc. **100**, 262–272 (2005)
51. N. Langberg, N.D. Singpurwalla, A unification of some software reliabilitymodels. SIAM J. Sci. Stat. Comput. **6**, 781–790 (1985)
52. J. Ledoux, Software reliability modeling, in *Handbook of Reliability Engineering*, ed. by H. Pham (Springer, London, 2003)
53. H.H. Lin, K.H. Chen, R.T. Wang, A multivariate exponential shared-load model. IEEE Trans. Reliab. **42**(1), 165–171 (1993)
54. B.H. Lindqvist, On the statistical modeling and analysis of repairable systems. Stat. Sci. **21**(4), 532–551 (2006)
55. B.H. Lindqvist, G. Elvebakk, K. Heggland, The trend-renewal process for statistical analysis of repairable systems. Technometrics **45**, 31–44 (2003)
56. B. Littlewood, Theories of software reliability: how good are they and how can they be improved? IEEE Trans. Softw. Eng. **6**(5), 489–500 (1980)

57. B. Littlewood, Stochastic reliability-growth: a model for fault-removal in computer-programs and hardware-designs. IEEE Trans. Reliab. **30**(4), 313–320 (1981)
58. B. Littlewood, Rationale for a modified Duane model. IEEE Trans. Reliab. **33**(2), 157–159 (1984)
59. B. Littlewood, D. Miller, Conceptual modeling of coincident failures in multiversion software. IEEE Trans. Softw. Eng. **15**(12), 1596–1614 (1989)
60. B. Littlewood, J.L. Verrall, A Bayesian reliability growth model for computer software. J. R. Stat. Soc. C **22**, 332–346 (1973)
61. B. Littlewood, J.L. Verrall, A Bayesian reliability model with a stochastically monotone failure rate. IEEE Trans. Reliab. **23**(2), 108–114 (1974)
62. H. Liu, Reliability of a load-sharing k-out-of-n:G system: non-iid components with arbitrary distributions. IEEE Trans. Reliab. **47**(3), 279–284 (1998)
63. J.C. Lu, Weibull extensions of the Freund and Marshall-Olkin bivariate exponential models. IEEE Trans. Reliab. **38**(5), 615–619 (1989)
64. M.R. Lyu, *Handbook of Software Reliability Engineering* (McGraw-Hill, New York/IEEE Computer Society, Los Angeles, 1996)
65. M.R. Lyu, Y. He, Improving the n-version programming process through the evolution of a design paradigm. IEEE Trans. Reliab. **42**(2), 179–189 (1993)
66. M.R. Lyu, Z. Huang, K.S. Sze, X. Cai, An empirical study on testing and fault tolerance of software reliability engineering, in *Proceedings of the 14th International Symposium on Software Reliability Engineering (ISSRE 2003)*, Denver, 2003
67. A.W. Marshall, I. Olkin, A multivariate exponential distribution. J. Am. Stat. Assoc. **62**, 30–44 (1967)
68. T.J. McCabe, A compexity measure. IEEE Trans. Softw. Eng. **2**(4), 308–320 (1976)
69. W.Q. Meeker, L.A. Escobar, *Statistical Methods for Reliability Data* (Wiley, New York, 1998)
70. D.R. Miller, Exponential order statistic models of software reliability growth. IEEE Trans. Softw. Eng. **12**(1), 12–24 (1986)
71. K.B. Misra, *Handbook of Performability Engineering* (Springer, London, 2008)
72. P.B. Moranda, Predictions of software reliability during debugging, in *Proceedings Annual Reliability and Maintainability Symposium*, Washington, DC, 1975
73. P.B. Moranda, Event-altered rate models for general reliability analysis. IEEE Trans. Reliab. **28**(5), 376–381 (1979)
74. J.D. Musa, A.F. Ackerman, Quantifying software validation: when to stop testing? IEEE Softw. **6**, 19–27 (1989)
75. J.D. Musa, K. Okumoto, A logarithmic Poisson execution time model for software reliability measurement, in *Proceedings 7th International Conference on Software Engineering*, Orlando, 1984
76. J.D. Musa, A. Iannino, K. Okumoto, *Software Reliability: Measurement, Prediction, Application* (McGraw-Hill, New York, 1987)
77. A. Myers, *Complex System Reliability: Multichannel Systems with Imperfect Fault Coverage* (Springer, London, 2010)
78. T. Nakagawa, *Maintenance Theory of Reliability* (Springer, London, 2005)
79. NEA, International common-cause failure data exchange. ICDE general coding guidelines. Technical note NEA/CSNI/R(2004)4. Nuclear Energy Agency (2004)
80. V.F. Nicola, A. Goyal, Modelling of correlated failures and community error recovery in multiversion software. IEEE Trans. Softw. Eng. **16**(3), 350–359 (1990)
81. E. Parzen, *Stochastic Processes* (Holden Day, San Francisco, 1962)
82. E.A. Peña, Dynamic modeling and statistical analysis of event times. Stat. Sci. **21**(4), 487–500 (2006)
83. H. Pham, *Software Reliability* (Springer, New York, 2000)
84. H. Pham, *Handbook of Reliability Engineering* (Springer, London, 2003)
85. H. Pham, *Recent Advances in Reliability and Quality in Design* (Springer, London, 2008)
86. H. Pham, L. Nordmann, X. Zhang, A general imperfect-software-debugging model with s-shaped fault-detection rate. IEEE Trans. Reliab. **48**(2), 169–175 (1999)

87. P.T. Popov, L. Strigini, J. May, S. Kuball, Estimating bounds on the reliability of diverse systems. IEEE Trans. Softw. Eng. **29**(4), 345–359 (2003)
88. P. Pukite, J. Pukite, *Modeling for Reliability Analysis* (IEEE, New York, 1998)
89. L. Rade, Expected time to failure of reliability systems. Math. Sci. **14**, 24–37 (1989)
90. A.E. Raftery, Inference and prediction for a general order statistic model with unknown population size. J. Am. Stat. Assoc. **82**, 1163–1168 (1987)
91. S. Ramani, S.S. Gokhale, K.S. Trivedi, in *Computer Performance Evaluation: Modelling Techniques and Tools, SREPT: Software Reliability Estimation and Prediction Tool,* ed. by R. Puigjaner et al. Lecture Notes in Computer Science (LNCS 1469) (Springer, New York, 1998)
92. M. Rausand, A. Høyland, *System Reliability Theory: Models, Statistical Methods, and Applications* (Wiley, Hoboken, 2004)
93. S.M. Ross, *Introduction to Probability Models*, 9th edn. (Academic, London, 2007)
94. E.M. Scheuer, Reliability of an m-out-of-n system when component failure induces higher failure rates in survivors. IEEE Trans. Reliab. **37**(1), 73–74 (1988)
95. G.J. Schick, R.W. Wolverton, Assessment of software reliability, in *Proceedings of the Operations Research* (Physica, Wirzberg-Wien, 1973). September 1972 in Hamburg
96. G.J. Schick, R.W. Wolverton, An analysis of competing software reliability models. IEEE Trans. Softw. Eng. **4**, 104–120 (1978)
97. J. Shao, L.R. Lamberson, Modeling a shared-load k-out-of-n:G system. IEEE Trans. Reliab. **40**(2), 205–209 (1991)
98. K. Sharma, R. Garg, C.K. Nagpal, R.K. Garg, Selection of optimal software reliability growth models using a distance based approach. IEEE Trans. Reliab. **59**(2), 266–276 (2010)
99. N.D. Singpurwalla, Foundational issues in reliability and risk analysis. SIAM Rev. **30**(2), 264–282 (1988)
100. N.D. Singpurwalla, *Reliability and Risk: A Bayesian Perspective* (Wiley, New York, 2006)
101. N.D. Singpurwalla, R. Soyer, Non-homogeneous autoregressive processes for tracking (software) reliability growth, and their Bayesian analysis. J. R. Stat. Soc. B **54**, 145–156 (1992)
102. N.D. Singpurwalla, S.P. Wilson, Software reliability modeling. Int. Stat. Rev. **62**, 289–317 (1994)
103. D.L. Snyder, M.I. Miller, *Random Point Processes in Time and Space*, 2nd edn. (Springer, New York, 1991)
104. H.S. Son, M.C. Kim, Software faults and reliability, in *Reliability and Risk Issues in Large Scale Safety-Critical Digital Control Systems*, ed. by P.H. Seong (Springer, London, 2009)
105. Z. Tang, J.B. Dugan, An integrated method for incorporating common cause failures in system analysis, in *Proceedings of the 50th Annual Reliability and Maintainability Symposium*, Los Angeles, 2004
106. X. Teng, H. Pham, A software-reliability growth model for N-version programming systems. IEEE Trans. Reliab. **51**(3), 311–321 (2002)
107. K. Tokuno, S. Yamada, An imperfect debugging model with two types of hazard rates for software reliability measurement and assessment. Math. Comput. Model. **31**, 343–352 (2000)
108. U.S. Nuclear Regulatory Commission, *Reactor Safety: An Assessment of Accident Risks in U.S. Commercial Nuclear Power Plants.* WASH-1400, NUREG-75/014 (U.S. Nuclear Regulatory Commission, Washington, DC, 1975)
109. E.A. van Doom, On the α-classification of birth-death and quasi-bith-death processes. Stoch Model **22**, 411–421 (2006)
110. J.K. Vaurio, An implicit method for incorporating common-cause failures in system analysis. IEEE Trans. Reliab. **47**(2), 173–180 (1998)
111. A.K. Verma, S. Ajit, M. Kumar, *Dependability of Networked Computer-Based Systems* (Springer, London, 2011)
112. W.L. Wagoner, The final report on a software reliability measurement study. TOR-0074(4112)-1. Aerospace Corporation, El Segundo (1973)

113. R. Wang, A mixture and self-exciting model for software reliability. Stat. Probab. Lett. **72**, 187–194 (2005)
114. R.T. Wang, A reliability model for multivariate exponential distributions. J. Multivar. Anal. **98**, 1033–1042 (2007)
115. R.T. Wang, A dependent model for fault tolerant software systems during debugging. IEEE Trans. Reliab. **61**(2), 504–515 (2012)
116. M. Xie, *Software Reliability Modelling* (World Scientific, Singapore, 1991)
117. L. Xing, Fault-tolerant network reliability and importance analysis using binary decision diagrams, in *Proceedings of the 50th Annual Reliability and Maintainability Symposium*, Los Angeles, 2004
118. S. Yamada, Software reliability growth models incorporating imperfect debugging with introduced faults. Electron. Commun. Jpn. Part 3 **81**(4), 33–40 (1998)
119. S. Yamada, S. Osaki, Software reliability growth modeling: models and applications. IEEE Trans. Softw. Eng. **11**(12), 1431–1437 (1985)
120. S. Yamada, M. Ohba, S. Osaki, S-shaped reliability growth modeling for software error detection. IEEE Trans. Reliab. **32**(5), 475–478 (1983)
121. S. Yamada, M. Ohba, S. Osaki, S-shaped software reliability growth models and their applications. IEEE Trans. Reliab. **33**(4), 289–292 (1984)
122. K.Z. Yang, An infinite server queueing model for software readiness assessment and related performance measures. Ph.D. Dissertation, Syracuse University, 1996
123. S. Zack, *Introduction to Reliability Analysis: Probability Models and Statistical Methods* (Springer, New York, 1992)
124. P. Zeephongsekul, G. Xia, S. Kumar, Software reliability growth models: primary failures generate secondary-faults under imperfect debugging. IEEE Trans. Reliab. **43**(3), 408–413 (1994)
125. T. Zhang, M. Horigome, Availability and reliability of system with dependent components and time-varying failure and repair rates. IEEE Trans. Reliab. **50**(2), 151–158 (2001)

Chapter 3
Energy-Efficient Design Techniques

Rong Ye and Qiang Xu

3.1 Introduction

While the relentless scaling of CMOS technology has brought digital IC designs with enhanced functionality and improved performance in every new generation, at the same time, the associated ever-increasing on-chip power and temperature densities make them suffer from more severe reliability threats [6, 93]. For example, as demonstrated in [94], the average mean-time-to-failure (MTTF) of a contemporary superscalar processor drops by about 4× from 180 to 65 nm technology node. In fact, the failure rates for today's electrical systems can be quite high, e.g., as high as 16.4 % for the Microsoft Xbox 360 within 10 months [90].

On the one hand, from the circuit level up to the system level, numerous power- and energy-efficient techniques have been proposed in the literature to achieve high energy efficiency. On the other hand, these solutions have high impact on system reliability, which needs to be addressed together with system performance and energy consumption in a holistic manner. For example, one of the most widely-used power minimization techniques, dynamic voltage and frequency scaling (DVFS) reduces power dissipation by scaling down supply voltage and operational frequency at runtime. By trading off performance for power, high energy efficiency can be achieved. As DVFS facilitates to reduce circuit temperature, it is also helpful for the lifetime reliability of the system. However, lowering down supply voltage inevitably leads to the increase of soft error rate due to the reduction of critical charge.

The remainder of this chapter is organized as follows. In Sect. 3.2, we present the preliminaries and background knowledge of this chapter. Various circuit-level and system-level energy-efficient design techniques are then introduced in Sects. 3.3 and 3.4, respectively. Next, Sect. 3.5 discusses the impact of power minimization

R. Ye • Q. Xu (✉)
The Chinese University of Hong Kong, Hong Kong, China
e-mail: rye,qxu@cse.cuhk.edu.hk

J. Mathew et al. (eds.), *Energy-Efficient Fault-Tolerant Systems*,
DOI 10.1007/978-1-4614-4193-9_3, © Springer Science+Business Media New York 2014

techniques on system reliability and Sect. 3.6 focuses on some emerging energy-efficient design solutions. Finally, Sect. 3.7 concludes this chapter.

3.2 Preliminaries

This section reviews the background knowledge in energy-efficient designs to make clear the definitions and models used in this domain.

3.2.1 Optimization Objectives

Before introducing the various sources of IC power dissipation, it is worth spending some efforts discussing the typical metrics used to evaluate the quality of a circuit design. The optimization objective during the design of complex chips has undergone a series of revolutions in the past several decades. As shown in Fig. 3.1, hardware area, circuit performance, and power dissipation are three important design factors that IC designers are familiar with. To be specific, for a circuit, its silicon cost is proportional to its size. Path delay distribution (in particular, critical path delay) determines its operational frequency, the most intuitive and commonly-used metric to evaluate circuit performance. As for power dissipation, it is not only related to the battery endurance of portable electronic devices but also affecting the reliability of the circuit. Therefore, IC design optimization is to explore the solution space to achieve an optimized tradeoff among the above three factors.

Thanks to the continuous scaling of CMOS technology, billions of transistors can be integrated onto a single chip nowadays, rendering hardware area no longer a serious concern. However, many other challenges emerge, among which reliability is probably the most critical one due to the ever-decreasing transistor feature size and the ever-increasing power and temperature density of the circuit. In fact, it is imperative for designers to explore the tradeoff among power, performance and reliability nowadays, as shown in Fig. 3.2.

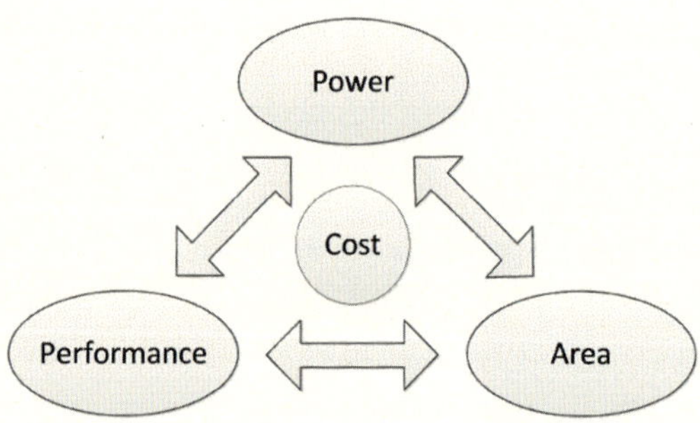

Fig. 3.1 Design tradeoff in conventional IC design

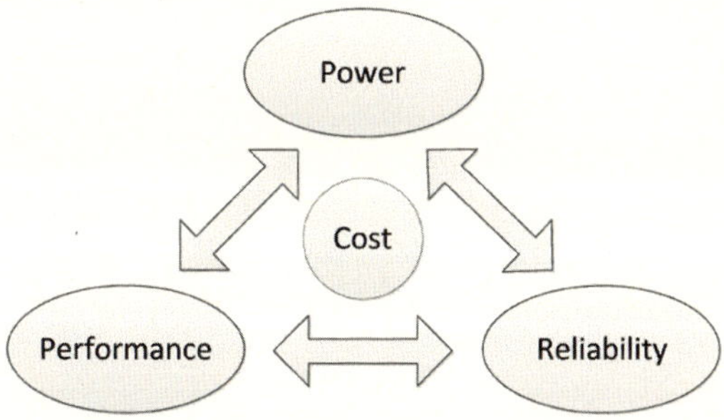

Fig. 3.2 Design tradeoff in today's IC design

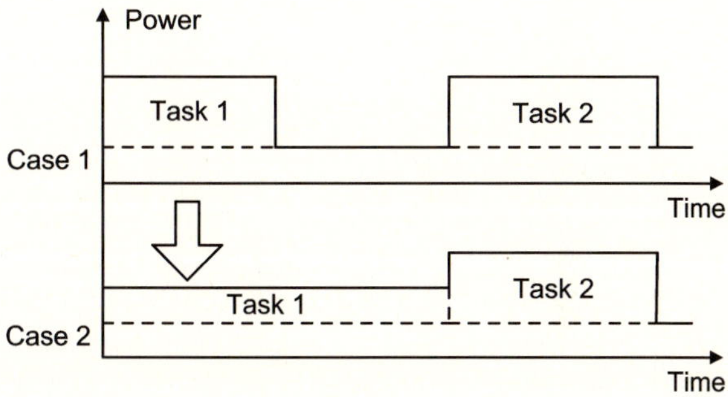

Fig. 3.3 Power versus energy

3.2.2 Power Versus Energy

It is essential to understand the difference between power and energy. Simply put, energy is the integral of power over time. As shown in Fig. 3.3, *power* is represented by the curve stretching with respect to *time* while *energy* is the area under the *power* curve. With low-power design techniques, we can lower the height of the *power* curve of *Task 1* from *Case 1* to *Case 2* (see Fig. 3.3), but the energy consumptions in these two cases may be the same. For example, suppose the power consumptions of *Task 1* in *Case 1* and *Case 2* are 8 and 4 W respectively and the times spent on executing it in these two cases are 1 and 2 s, respectively, the energy consumption in both *Case 1* and *Case 2* would be 8 J.

As for the impacts of power and energy, there are many factors and each of them has its specific role. For example, average power dissipation largely determines circuit operational temperature; peak power dissipation is one of the most significant factors to be considered when designing IC power distribution network. Energy consumption, on the other hand, determines the operating time of battery-powered portable device. Nevertheless, many low-power design techniques are indeed helpful to improve system energy efficiency, and in the rest of this chapter, we do not differentiate them much and the terms (e.g., power-efficient and energy-efficient) are often used interchangeably.

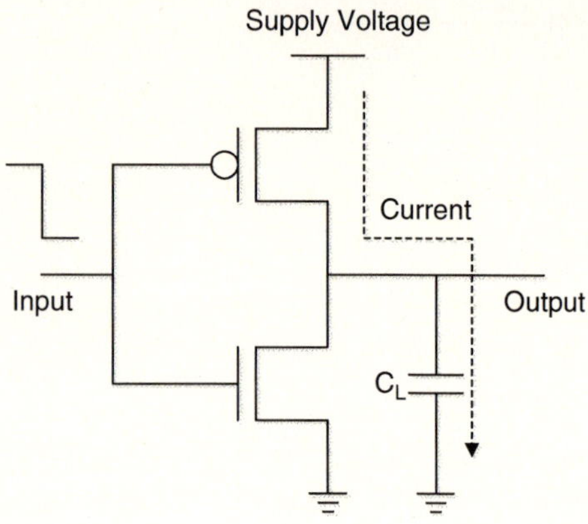

Fig. 3.4 Dynamic power due to signal transition

3.2.3 *Power Models*

Modeling the power consumption of circuits and systems is essential towards establishing measures to compare the quality of alternative designs in terms of the power needed for its operation. Generally speaking, the total power consumption of an electronic system consists of two components: dynamic power and static power [67, 82, 114]. Dynamic power is the power dissipation when devices are switching, while static power is caused by leakage currents (primarily caused by subthreshold leakage current) that are present even when no logic operations are performed.

3.2.3.1 Dynamic Power Dissipation

In a digital CMOS circuit, there are two sources of dynamic power dissipation. The first one results from charging and discharging of the load capacitance with signal transitions, wherein current flows through transistor channels and electrical energy gets converted into heat and dissipated away, as shown in Fig. 3.4. Such dynamic power is called switching power dissipation.

Instantaneous switching power dissipation $P_{sw}(t)$ can be modeled as

$$P_{sw}(t) = V_{dd} \cdot I_{dd}(t), \tag{3.1}$$

where V_{dd} is the supply voltage and I_{dd} is the current without considering leakage current. As for the average dynamic power dissipation, we can first have the energy consumed in each transition E_{pt} as follows:

$$E_{pt} = C_L \cdot V_{dd}^2, \tag{3.2}$$

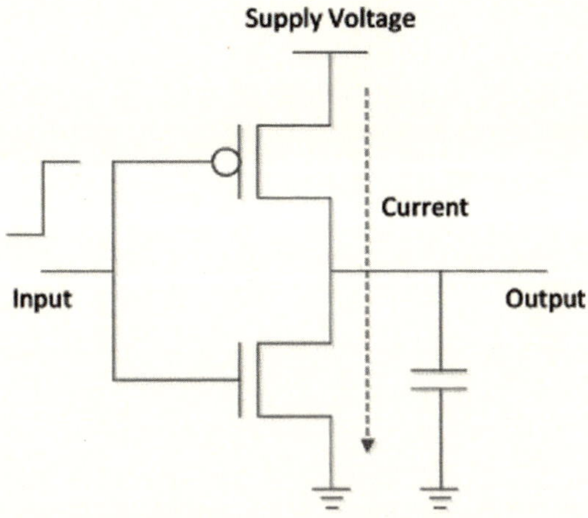

Fig. 3.5 Dynamic power due to short-circuit currents

where E_{pt} is the energy per transition and C_L is the load capacitance. Finally, we have

$$P_{sw} = \alpha \cdot C_L \cdot V_{dd}^2 \cdot f, \tag{3.3}$$

where α is the switching activity and f is the operational clock frequency.

The second source of dynamic power dissipation comes from the short-circuit currents as shown in Fig. 3.5, flowing directly from the power supply to the ground when the n-subnetwork and p-subnetwork of a CMOS gate are in on-state simultaneously. With input to the gate stable at either logic level, only one of the two subnetworks conducts and no short-circuit flows. However, when the output of a gate switches in response to changing inputs, both subnetworks are in on-state simultaneously for a short interval, resulting in short-circuit current flows. The duration of the interval depends on the input and the output transition times and hence the short-circuit dissipation also depends on it.

Both of the above sources of power dissipation in CMOS circuits are related to transitions at gate outputs and therefore collectively referred to as dynamic power dissipation. Because the ramp time of the input signal is usually quite short, the short-circuit current lasts for only a short time interval in each transition. Therefore, the switching power is usually the dominating one in dynamic power dissipation while short-circuit power is only of interest in some cases.

3.2.3.2 Static Power Dissipation

Static power is caused by leakage currents, which flow even when the input and output of a gate are not changing. One of the main reasons why CMOS technology becomes the dominating semiconductor technology is due to the fact that static power dissipation in CMOS circuits is much lower than that in other technologies.

However, as supply voltage reduces with technology scaling, transistors with low threshold voltages have to be used to maintain performance and reliability, yet the lower the threshold voltage, the greater the standby leakage current. In fact, for some high-performance circuits, the contribution of static power dissipation already has a dominant share of the total power dissipation.

In the near future, subthreshold leakage and gate leakage will be the dominant types of leakage current [64, 76]. Consequently, static power dissipation can be roughly modeled as follows:

$$P_{leakage} = I_{sr} \cdot T^2 \cdot e^{\frac{\mu \cdot V_{dd} + \gamma \cdot V_{bs} + \theta}{T}} \cdot + |V_{bs}| \cdot I_{ju}, \tag{3.4}$$

wherein I_{sr} is the reference leakage current at reference temperature, T is temperature, V_{bs} is the body bias voltage, I_{ju} is the junction leakage current, and μ, γ and θ are all curve fitting parameters depending on fabrication technology.

3.2.4 Power Minimization Methodology

As power minimization is one of the most important optimization objectives in today's IC designs, numerous energy-efficient design techniques at various steps of the design flow have been proposed in the literature and realized in industry designs, targeting at dynamic and/or static power dissipation. Many design parameters can be adjusted to achieve power reduction. Some of them are continuous, such as supply voltage and threshold voltage; while some others are discrete, such as different logic styles, topologies, and micro-architectures.

Ideally, designers would like to consider all parameters at the same time, and to define a single optimization problem to minimize power from the viewpoint of the entire system. However, the complexity of such an optimization problem with all the parameters included would be practically unsolvable. Consequently, IC design methodologies rely on some important concepts to help manage complexity: abstraction and hierarchy. Abstraction is used to hide the details, while hierarchy is for building larger entities through a composition of smaller ones. The abstraction and hierarchy in design process of ICs is shown in Fig. 3.6. There have been many techniques employed to perform energy-efficient computing at these different levels. For example, at device level, we can employ multi-threshold devices to reduce power dissipation of certain logic gates; at circuit level, clock gating can be used to turn off the clock tree for inactive parts of the circuit to reduce dynamic power; at logic level, low-power logic synthesis can be used to reduce power dissipation; at architecture level, some techniques like architecture synthesis and transformation can be used; and at system level, designers can perform task scheduling and allocation to optimize power consumption for processors. In this chapter, we mainly discuss power optimization techniques at circuit level and system level.

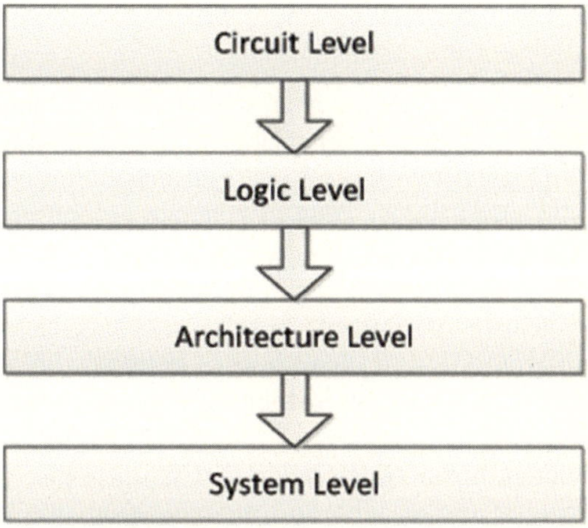

Fig. 3.6 Abstraction and hierarchy in design process of ICs

3.3 Circuit Level Power Minimization Techniques

In this section, we discuss some representative low-power design techniques at circuit level.

3.3.1 Transistor Sizing

Transistor sizing, the operation to manipulate the width of the channel of a transistor, is an effective technique to improve the timing performance and/or power efficiency of CMOS circuits [10]. That is, by increasing the width of the channel and hence the transistor size, the driving capability of the transistor increases with reduced signal rise/fall times at the gate output.

Earlier works in transistor sizing focus on minimizing the area of the circuit subject to a certain delay constraint [14, 78]. With aggressive technology scaling, silicon area is of less concern and sizing transistors for power minimization has been popular [32, 54].

Theoretical analysis for transistor sizing usually assumes a continuous sizing model, which is only possible for full custom design. For standard cell based IC designs, transistor sizes are pre-determined in the cell library and with some discrete values only. In the early days, the number of library cells were quite small, ranging between 50 and 100 cells. Further optimization needs for power-performance tradeoff have changed this situation substantially. With the need for more sizing options for each logical cell, today's libraries usually have up to 1,000 cells.

3.3.2 *Technology Mapping*

Technology mapping is an optimization technique that selects library cells for the implementation of a given logic function, serving as the final step of logic synthesis. The logic network, resulting from technology-independent optimizations, is mapped with given library cells to generate an optimized circuit while satisfying design constraints.

Technology mapping is the step in IC design flow where transistor/gate sizing is actually performed. Besides choosing between the same kind of cells with different sizes, it also chooses between different gate mappings: either simple cells with small fan-in logic or complex cells with large fan-in logic. Generally speaking, simple gates are good from a performance perspective due to the quadratic relationship between delay and fan-in logic, while complex gates with smaller intrinsic capacitance are attractive from the energy perspective. Consequently, designers tend to replace simple gates with complex ones in non-critical timing paths for power reduction.

Technology mapping is usually formulated as a directed acyclic graph (DAG) covering problem [99, 100]. To be specific, the Boolean function to be mapped is represented in canonical form as a subject DAG. Similarly, gates in the library are represented as gate DAGs. Technology mapping is then to find an optimized covering of the nodes of the subject DAG using the available gate DAGs.

While conventional technology mapping focuses on minimizing area and/or circuit delay, the graph covering formulation of technology mapping has been extended to integrate power optimization as an important objective [26, 62, 99–101]. For instance, in [99,100], the subject DAG is created for the Boolean function to be mapped using a basis function consisting of two-input NAND/NOR gates and inverters, and the overall algorithm flow of technology mapping for low power follows along the algorithm for area and delay.

3.3.3 *Supply Voltage Scaling*

Supply voltage scaling is probably the most effective technique used for power reduction, due to the quadratic dependency of dynamic power on supply voltage [114]. Moreover, it also facilitates to reduce leakage power since leakage currents decrease as supply voltage scaling down. Both static voltage scaling (SVS) and dynamic voltage scaling (DVS) have been widely used in today's IC designs, which achieve power reduction with small performance penalty.

With SVS, we have a multiple supply voltage (MSV) design, wherein circuits are partitioned into multiple "voltage islands" and each island operates at a specified supply voltage that satisfies its own performance requirement. In such designs, to meet the timing requirement of each voltage island, the corresponding supply voltage has to be high enough to drive the most timing-critical cell, even though the rest of cells may have much more relaxed timing requirements. Consequently, the voltage assignment to a certain island is dominated by the required voltage of the most timing-critical path, and designers tend to group cells with similar timing requirement together as voltage islands. Numerous works have been presented to achieve an optimized MSV design at various design stages, e.g., floorplanning [67–69, 82], post-floorplanning [60, 61], placement [34, 63], and post-placement [17, 55, 106–108]. Note that, wherever an output from a low V_{dd} island needs to drive an input to a high V_{dd} island, a level shifter is required at the interface, resulting in extra power overhead. Because of this, more islands do not always imply better energy efficiency, as shown in [69].

With DVS, supply voltage is adaptively changed according to runtime performance demands [38, 45, 88]. When relatively low performance is sufficient, supply voltage and clock frequency are both lowered at runtime, delivering reduced performance with substantial power reduction. Similar to most control systems, there are three key components to implement a DVS system: a targeted system that supports operation at multiple voltages, a management centre that generates commands to control the behavior of the targeted system, and a control loop that delivers the commands and collects system status at runtime.

3.3.4 Clock Gating

Up to 50 % or even more of dynamic power is spent in clock distribution network, driving clock buffers with extremely high toggle rates. Clock gating is an effective technique to address this problem, which disables a portion of inactive clock tree and its relevant logic elements. The dynamic power consumption is hence dramatically reduced (leakage power dissipation remains) [72].

Clock gating is usually implemented at register-transfer level (RTL) [57], which identifies the clusters of FFs that share a common enable control signal, and uses this enable signal to control such a clock gating circuit that is connected to the clock ports of all these FFs. Consequently, as long as the enable signal is de-asserted, these FFs under control are clock-gated to have no switching activities and consume no dynamic power. Generally speaking, EDA tools are able to identify fine-grained clock gating opportunities while module-level clock gating is conducted manually.

3.3.5 Input Vector Control

The input vector control (IVC) problem is to find input vector combinations that minimize the leakage current of a circuit, taking advantage of *stacking effect*, the phenomenon that subthreshold leakage current flowing through a stack of series connected transistors is greatly reduced if some of them are turned off. Simply speaking, the subthreshold leakage through a logic gate depends on the applied

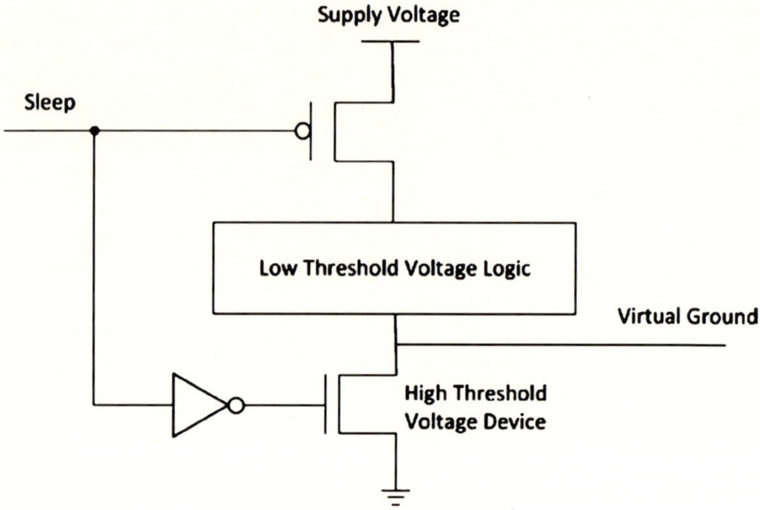

Fig. 3.7 An example to show MTCMOS circuit structure

input vector, making the total leakage current of a circuit depend on primary input combinations. Consequently, by applying proper input vectors, the circuit can work in a low leakage state.

The most straightforward method to find low-leakage input vectors is to enumerate all combinations of primary inputs. Clearly, such an exhaustive method is only applicable to those circuits with a small number of primary inputs. For large circuits, a more efficient and effective method is to use some probabilistic metaheuristic algorithms that can exploit historical information in order to speculate on new search points to obtain a near optimal solution. With the ever-increasing impact of leakage power in nanometer devices, IVC techniques have attracted lots of attention (e.g., [1,30,47,71]). For example, a stack transistor insertion technique is presented in [47], wherein the authors first identify circuit input vectors, putting most of the circuit into a low-leakage state, and then insert transistors to enable leakage control of those leakage paths that are not able to turn off series transistors.

3.3.6 Multi-threshold Designs

Multiple threshold CMOS (MTCMOS) circuit, which has both high and low threshold transistors in a single chip, can be used to tradeoff circuit performance and leakage power.

A common implementation of MTCMOS for reducing power makes use of sleep transistors. That is, low V_t transistors are used as logic cells to provide high performance while high V_t transistors with low leakage serve as sleep transistors, disconnecting logic cells from power supply network. As shown in Fig. 3.7,

MTCMOS involves high V_t transistors to gate the power supply of a logic block constructed using low V_t transistors [53]. When high V_t sleep transistors are turned on, low V_t logic works as usual, and switching is performed through fast low V_t devices. When the circuit is in inactive mode, high V_t transistors are turned off, disconnecting the gates from the ground and resulting in low subthreshold leakage current.

Compared to MSV designs, which require level shifters and special layout strategies, introducing multiple thresholds has relatively smaller impact on the design flow. Existing designs can be modified into MTCMOS blocks by simply adding high V_t transistors as power supply switches. However, by introducing an extra series device to the power supply network, MTCMOS circuits incur a performance penalty when compared to regular CMOS circuits, if these devices are not sized properly. Moreover, retention registers need to be added to hold the data of main register of the power-gated block.

3.4 System Level Power Minimization Techniques

Managing power dissipation at system level is able to considerably decrease energy consumption because we have a better view on application needs at this level. Various energy-efficient design techniques, such as dynamic power management (DPM), dynamic voltage scaling (DVS) and task migration, have been explored in the literature and applied in state-of-the-art low-power system designs.

DPM for electronic systems, which trades off performance for power savings in a controlled fashion, is one of the most successful techniques used for energy-efficient computing [7]. By taking system workloads into account, DPM reduces power dissipation via selectively shutting down (or lowering the performance of) inactive system components. For example, a microprocessor can be put in sleep mode for power reduction when it is idle for some time and it is waken up when new tasks arrive. Consequently, the system can be modeled as a power state machine [110]. The transitions between states are controlled by commands issued by a power manager (PM) that observes the workload of the system and decides when and how to force power state transitions. The power manager makes state transition decisions according to the power management policy implemented in the system a priori. The choice of DPM policy that minimizes power under performance constraints (or maximizes performance under power constraint) is a constrained optimization problem.

To be specific, a power-manageable system can be modeled as shown in Fig. 3.8 [7,97], which includes three components: service requester (*SR*), service queue (*SQ*) and service provider (*SP*). SR issues requests as the event source, while SP processes requests. SQ buffers requests that cannot be processed at once, if SP is too busy. SR has several operational states to represent different service request rates, which can roughly be considered as the possible request number issued in a time unit. SP can have different modes, such as run mode to process requests, and sleep mode to save

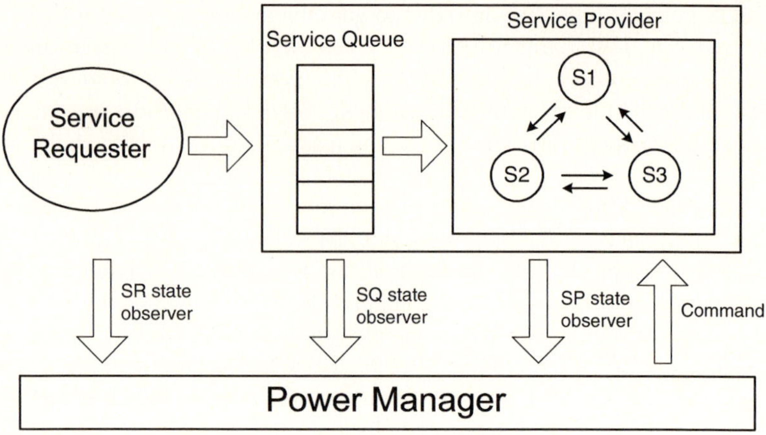

Fig. 3.8 Abstract structure of DPM system model

power in case of no requests. As for SQ, its state can be simply described as the stored request number. The power manager observes system states (consisting of SR, SQ and SP states), and controls the behavior of SP, to achieve power savings at certain performance penalty. This is a general framework and in practice designers should detail this framework to make it more applicable and efficient.

System level DPM effectively reduces power consumption by shutting down during the idle periods between application workloads. DVS, on the other hand, reduces energy consumption by changing processor speed and voltage at run-time depending on the needs of the executed applications [88]. Early DVS algorithms simply set processor speed based on the processor utilization of fixed intervals; while more sophisticated voltage scaling techniques consider the individual requirements of running tasks. The approaches presented in [45] assume that all tasks run at their worst case execution time (WCET). The workload variation slack times are exploited on a task-by-task basis in [84], and are fully utilized in [59]. Pering et al. [73] introduces a voltage scheduler that determines the operating voltage by analyzing application requirements. The scheduling is done at task level, by setting processor frequency to the minimum value needed to complete all tasks. For applications with high frame-to-frame variance, such as MPEG video, schedule smoothing is done by scheduling tasks to complete twice the amount of work in twice the allocated time. In most DVS approaches presented in the literature, scheduling is done at the task level, assuming multiple threads. The prediction of task execution times is done either using worst case execution times, or heuristics. Such approaches neglect that DVS can be done within a task or for single-application devices. For instance, in MPEG decoding, the variance in execution time on frame basis can be very large.

Note that, conducting system level power optimization can be quite different in different types of targeted systems. In this section, we introduce system level power

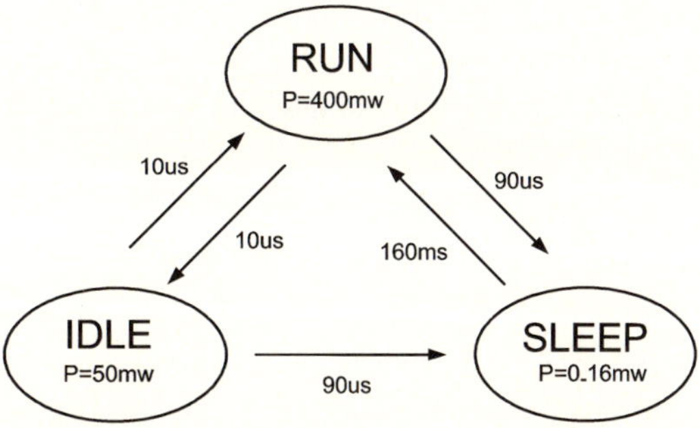

Fig. 3.9 Power states of StrongARM SA1100

management technique based on two important types of computing devices: general purpose processors and embedded systems. Generally speaking, in general purpose processors, the arrivals of tasks are unknown and hence how to predict task arrivals is quite important for effective DPM, while in embedded systems some preliminary information about tasks is assumed to be acquired in advance.

3.4.1 General Purpose Processors

Due to the unknown task characteristics in general purpose processors, state transitions in DPM usually involve non-trivial performance penalty and power cost. An eager power management policy that turns off system components as soon as they are idle may even increase the system power dissipation and degrades its performance at the same time. Consequently, how to optimize the DPM policy, the procedure that takes decision on the state of the system components, is a rather complex constrained optimization problem, especially considering the fact that a component may have multiple operational modes with different power benefits and transition costs [18]. For example, a processor in deep sleep state has lower power consumption but requires more transition time and transition power when waken up, compared to that in light sleep state.

For a complex electronic system that supports DPM, we can model it as a finite-state machine with multiple power modes [18], as shown in Fig. 3.9, taking StrongARM SA-1100 processor as an example. This processor has three operational modes: run mode, idle mode, and sleep mode. In each mode, the processor has different power dissipation, and the mode transition would induce both performance penalty and power cost. As systems typically experience non-uniform workloads (manifested as idle periods among tasks), we can still obtain power savings from selective shutdown even after compensating the non-trivial transition cost.

There are numerous related works in dynamic power management in the literature. From the aspect of conducting effective power state transitions, generally speaking, existing DPM policies can be classified into two categories [27]: heuristic policies and stochastic policies. Time-out policy [56] is one of the most widely-used heuristic policies, which simply turns off a component when the duration time, for which the component has been in idle period, exceeds a pre-defined time interval. Time-out policies are simple and robust, but they may be too fast or too slow to react. Stochastic policies, on the other hand, model system state changes and request arrivals as stochastic processes. Markov decision process [7] and Semi-Markov decision process [87] are often adopted to derive an optimal DPM policy according to these models.

In the above works, DPM policies are determined at design stage and they may not work well with varying workload characteristics and environment conditions. Learning-based DPM solutions are thus attractive since they are able to adapt to varying system conditions and workloads. Srivastava et al. [91] explored a shutdown mechanism to predict the length of idle time based on real-life traces and recent computation history. In [43], Hwang et al. predicted the current idle period length using exponential average approach based on previous idle periods. In [96], Steinbach proposed reinforcement learning-based DPM policy to perform mid-level power management in wireless network cards. Theocharous et al. [98] considered user annoyance as a performance constraint and presented a user-based adaptive power management technique. In [27], Dhiman and Rosing proposed to dynamically select the best DPM policy from a set of candidate policies. In [97], Tan et al. presented an approach for system-level power management in a partially observable environment, based on model-free constrained reinforcement learning.

There are also some recent works that consider DPM in multi-core processors, which can be categorized into per-core approach [44, 49, 50] and chip-wide approach [31, 38]. In [44], Canturk et al. proposed an approach to set the power mode of each core to meet a power budget. Jung et al. [49, 50] presented a supervised learning-based DPM framework for multi-core processors. Their approach, however, determines power management actions for each core based on their individual workload prediction and hence is not a "true" multi-core power management scheme. In [38], Sebastian et al. utilized a control theory based controller to apply DVFS technique, but the task-to-core allocation is fixed in their approach. In [31], Mohammad et al. proposed a hierarchical DPM framework under given throughput constraint, which employs core consolidation, coarse-grained DVFS and task allocation at the CMP level and fine-grained DVFS based on closed-loop feedback control at the individual core level. This work required to obtain task characteristic a priori for task allocation.

3.4.2 Embedded Systems

In response to today's competitive electronics market, when designing complex embedded systems, it is increasingly popular to employ pre-designed multiprocessor system-on-a-chip (MPSoC) platforms and map applications onto them to reduce design risk and achieve short time-to-market [77]. Various platforms with specific functionalities reflecting the need of the expected application domain have been developed in the industry recently, e.g., ARM PrimeXsys platform [2].

When building platform-based embedded systems, a basic issue is to conduct task allocation and scheduling for applications, in which the allocation of tasks is to effectively utilize the available processors while scheduling is to meet various requirements. One major trend in embedded system design is towards energy-efficient computing based on the concept of performance on demand, by dynamically adjusting the operational voltage and frequency of processors based on instantaneous processing requirement. There is a rich literature on energy-efficient design methodologies (e.g., [46]), which mainly resort to DVS and slack reclaiming to cut down the energy consumption of embedded processors. In particular, Schmitz et al. [81] proposed an energy-efficient co-synthesis framework for multi-mode embedded systems under the consideration of mode execution probabilities, in which a single execution mode occupies the entire MPSoC at a time.

Both DVS and DPM can be used to achieve energy-efficient scheduling for embedded systems. When both solutions are available, it is advocated to consider DVS first [46]. The input specification of low power system scheduling problem is usually given in terms of a set of task graphs. A typical problem formulation based on MPSoCs can be developed as follows:

Problem: Given

- The floorplan of the platform-based MPSoC embedded system that consists of ℓ processor cores;
- n execution modes. Each mode i is represented by a directed acyclic task graph $\mathbf{G}_i = (\mathbf{V}_i, \mathbf{E}_i)$, wherein each node in $\mathbf{V}_i$ indicates a task in $\mathbf{G}_i$, and $\mathbf{E}_i$ is the set of directed arcs that represent precedence constraints;
- The joint probability density function[1] that the system is in various modes $f_{Y_1,Y_2,\cdots,Y_n}(y_1, y_2, \cdots, y_n)$, where y_i represents the probability that the system is in execution mode i;
- The execution time $w_{i,j,k}$ of task j of mode i on processor k under maximum supply voltage V_{dd};
- The power consumption $P_{i,j,k}$ of task j of mode i on processor k under maximum supply voltage V_{dd};
- Deadline $d_{i,j}$ of task j of mode i, meaning that task j in $\mathbf{G}_i$ should be finished before $d_{i,j}$;
- Other design constraints, e.g., target service life L and the corresponding reliability requirement $\eta\%$;
- System parameters, e.g., failure mechanism parameters (e.g., activation energy E_a of electromigration) and the corresponding failure distributions;

to determine a periodical task allocation and schedule on the given MPSoC platform for each execution mode such that the expected energy consumption is minimized,

[1]The mode execution probabilities can be estimated as in [81].

under the performance constraints that real-time tasks are finished before deadlines and some other design constraints.

3.5 Impact of Power Minimization Techniques on Reliability

With technology scaling, reliability has emerged as a major design constraint for high-performance IC designs. As energy-efficient design techniques change the power/temperature densities of a circuit, it is imperative to take their impact on reliability into consideration. In this section, we discuss the above issue from three aspects: hard errors that affect circuit lifetime reliability (Sect. 3.5.1); soft errors that cause flipping of values in circuit storage elements (Sect. 3.5.2); and timing errors that appear on circuit critical paths (Sect. 3.5.3).

3.5.1 Hard Error

No doubt to say, designers need to ensure the lifetime reliability of IC products so that their failure rates would not exceed customers' expectations. This has become a challenging task with relentless technology scaling [93]. On the one hand, some well-known failure mechanisms such as time dependent dielectric breakdown (TDDB) in the gate oxides and electromigration (EM) on interconnects have increasing adverse effects due to shrinking feature size. On the other hand, degradation of device parameters over the circuit's lifetime has emerged as a major threat to system reliability. In particular, circuit wearout resulting from negative bias temperature instability (NBTI) and random telegraph noise (RTN) that cause electrical parameter shift (e.g., transistor threshold voltage increase) is of particular concern with technology scaling and it is shown that they could result in significant performance degradation of the circuit over its service life [90].

Energy-efficient design techniques generally are helpful to improve circuit lifetime reliability as they facilitate to reduce circuit power/temperature density. However, as functional blocks in a system age differently, without explicitly taking their aging rates into consideration, IC designs may still not meet their lifetime reliability requirement. As discussed earlier, DVS reduces power consumption by scaling down voltage and frequency of operation at runtime. As reducing circuit power facilitates to lower device temperature while most circuit failure mechanisms exacerbate themselves under high temperature, circuit lifetime reliability is usually improved implicitly. On the other hand, aggressive power management policies may also decrease circuit lifetime reliability because of the degradation effect that temperature cycles have on modern IC materials [92]. Rosing et al. [75] plot the tradeoff between mean time to failure (MTTF) and power saving in Fig. 3.10, wherein they present three common failure mechanisms: EM, TDDB, and thermal cycling (TC). These values are obtained from a particular test core that has only one

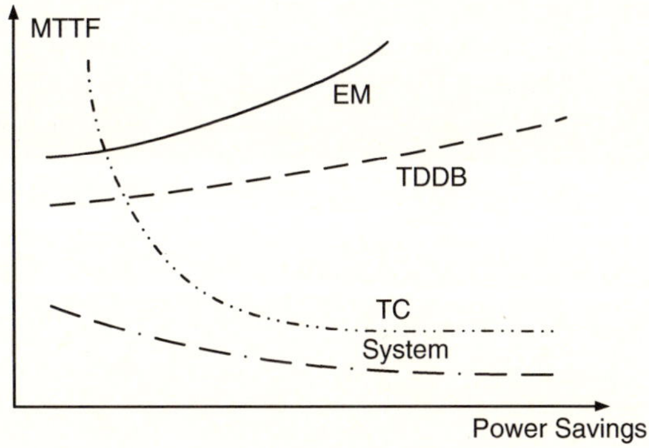

Fig. 3.10 MTTF vs. power savings

sleep state but no DVS capability with 95 nm technology. Although more aggressive power management policies improve MTTF due to EM and TDDB, the circuit suffer from significant thermal cycling due to frequent shutdowns. As a result, the overall MTTF in fact decreases as more power savings are obtained.

Many widely-accepted reliability models for the above failure mechanisms at device and circuit level have been proposed and empirically validated by academia and industry [8, 9, 48, 95]. They are usually quite complex and involve a number of parameters. Let us take MTTF due to EM as an example [48]:

$$MTTF_{EM} = \frac{A_{EM}}{J^n} \cdot e^{\frac{E_{aEM}}{\kappa T}}, \tag{3.5}$$

where A_{EM} is a constant determined by the physical characteristics of the metal interconnect, J is the current density, E_{aEM} is the activation energy of electromigration, n is an empirically-determined constant, κ is Boltzmann's constant, and T is the temperature.

Circuit lifetime reliability can be improved at various design levels. At device/gate level, we can employ gate resizing and threshold voltage tuning to make device less vulnerable [21,74,109]. At circuit level, we can give more design margin to those frequently-stressed circuit paths. For example, in [15], a variable-latency adder design is presented for NBTI tolerance, wherein the proposed VL-adder can automatically shift data capturing clock edge to tolerate NBTI-induced delay degradation on critical timing paths. In [105], a novel input vector control technique is proposed to minimize NBTI effects. In [16], the minimum cost reliability driven routing is considered and an iterative rounding-based integer linear programming algorithm is proposed to mitigate EM effects. At micro-architectural level, block-level redundancy can be introduced to protect those functional units that are prone to wearout failures. In [25], a tool capable of evaluating NBTI vulnerabilities early in the design cycle was developed to facilitate architectural level aging analysis, which can be used to guide redundancy allocation. In [23], by judiciously binding and scheduling applications onto the buses in the microarchitecture, the activities on some critical wires can be reduced to mitigate EM degradation. In [86], the

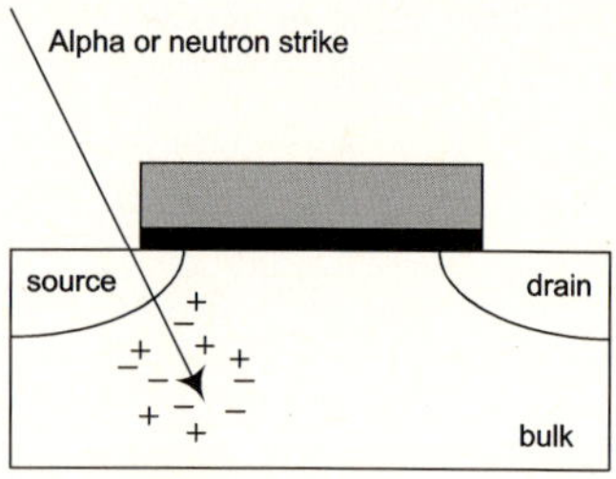

Fig. 3.11 Interaction of an alpha particle or a neutron with silicon crystal

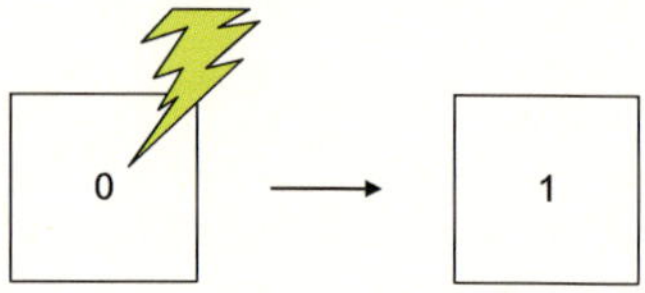

Fig. 3.12 Strike on a storage device can flip the bit stored

authors proposed to use the inherent redundancy available in microarchitectures to handle hard failures and enable graceful performance degradation in fail-in-place systems. Finally, at architectural and OS level, reliability-driven task allocation and scheduling techniques and dynamic reliability management (DRM) policies can be employed to balance the stress on different processor cores on an MPSoC design. In [39–42], several task allocation and scheduling techniques are presented to tradeoff lifetime reliability and energy consumption for MPSoC designs.

From the above, we can see that although energy-efficient design techniques can lower system power dissipation and thus implicitly increase circuit lifetime reliability for some failure mechanisms, its impact to the overall system lifetime reliability is not certain. Without a systematic solution, today's IC designs either take a rather conservative approach that employs many of the above solutions to mitigate reliability threats, which results in non-trivial hardware/performance overhead, or simply rely on power/thermal control techniques to improve reliability, which may not be sufficient to meet lifetime requirement [41].

3.5.2 *Soft Error*

Radiation-induced transient faults arise from energetic particles, such as alpha particles from packaging material and neutrons from the atmosphere, generating electron-hole pairs (directly or indirectly) as they pass through a semiconductor devices, as demonstrated in Fig. 3.11. Transistor source and diffusion nodes can collect these charges. A sufficient amount of accumulated charge may invert the state of a logic device, such as a latch, static random access memory (SRAM) cell, or gate, thereby introducing a logical fault into the circuit's operation (Fig. 3.12). Because this type of fault does not reflect a permanent malfunction of the device, it is termed *soft* or *transient*.

As a fundamental parameter to reflect soft error vulnerability of an electronic device, critical charge (Q_{crit}) describes the minimum charge that must be deposited by particle strike to cause a device to malfunction. It is usually computed using integrated circuit simulators, such as SPICE, by injecting current pulses into the sensitive nodes of device. These pulses represent the current generated from electron-hole pairs created by an alpha particle or neutron strike. Because the charge value has the following quadratic relationship with respect to the supply voltage V_{dd}:

$$Q_{charge} = C \times V_{dd}^2, \tag{3.6}$$

where C is capacitance, it is clear to observe that the soft error vulnerability of a device, represented by critical charge Q_{crit}, will also quadratically increase with respect to V_{dd}. For example, Hazucha et al. [36] showed that reducing the supply voltage from 3.3 to 2.2 V for an SRAM cell in a 0.6 μm technology decreases the Q_{crit} from 91.4 to 51.5 fC. As many energy-efficient techniques achieve power saving by reducing supply voltage, it is inevitable that they are associated with higher soft error rate [83].

In order to mitigate the impact of soft errors, we have to rely on redundancy, i.e., either by duplicating hardware component or by performing redundant computation. With redundancy, computational results are compared, and the disagreement between redundant components indicates the occurrence of soft error. Clearly, without a careful vulnerability analysis, this redundancy-based protection would lead to significant area/performance overhead and cause more energy consumption. Consequently, when pursuing power minimization with reduced supply voltage, the increase of soft error rate and its associated overhead should be carefully taken into account.

3.5.3 *Timing Error*

In a synchronous design, the delays of combinational logic paths must not exceed the operational clock period; otherwise, timing errors occur. To be specific, in a synchronous circuit, let S_{ij} represent the timing slack of a certain path between two flip-flops (FF_i and FF_j), we have

$$S_{ij} = T_{cp} - T_{setup} - D_{ij}, \tag{3.7}$$

where T_{cp} is the clock period, T_{setup} is the setup time of FFs, and D_{ij} is the maximum path delay. If the timing slack S_{ij} is not positive, we say timing error occurs on this path.

With the continuous downscaling of transistor feature size, there is an increasing uncertainty for the timing behavior of today's ICs, often manifesting themselves as infrequent timing errors on speed-paths, i.e., critical or near-critical paths. There are multiple factors that contribute to this effect: (i) inevitable static process variation caused by manufacturing imperfection leads to the mismatch of timing performance between the designed value and the actual one; (ii) dynamic variations in supply voltage, temperature, and multiple-input switching cause varying circuit delay at runtime; (iii) circuit aging mechanisms such as NBTI lead to gradual increase of circuit delay over its lifetime. Considering these variation effects, the path delay becomes a random variable, therefore we have

$$\tilde{S}_{ij} = T_{cp} - T_{setup} - \tilde{D}_{ij}, \tag{3.8}$$

where $\tilde{S}_{ij}$ is the random variable of timing slack and $\tilde{D}_{ij}$ is the random variable of the maximum path delay. The probability for $\tilde{S}_{ij}$ being negative is considered as the timing error probability. Traditionally, a large timing guard band has to be reserved to tolerate the above variation effects.

With supply voltage scaling down to reduce power consumption, there are some other side-effects, e.g., the delay increase of logic gates, thereby increasing the possibility of timing violation. To be specific, we have the delay model as follows:

$$D = \frac{K_i \cdot V_{dd}}{(V_{dd} - V_{th})^{\alpha}}, \tag{3.9}$$

where D is gate delay, K_i and α are fitting parameters as defined in [82]. That is to say, power consumption is reduced with voltage scaling down at the expense of reduced timing performance or increased timing error probability. From this perspective, the tradeoff between power and performance must be carefully considered when conducting power minimization. For example, in the region-based multi-supply voltage (MSV) design, one of the most critical problems is to identify those non-critical circuit blocks and group them together as an individual voltage island, so that low supply voltage can be applied to these blocks for power reduction while timing performance of the entire system would not be reduced.

3.6 Emerging Energy-Efficient Design Techniques

The quest for better energy efficiency is never ending, even after decades of research and practice. In this section, we discuss three emerging energy-efficient design techniques: timing speculation (TS) at circuit level, approximate computing at logic level and adaptive power management at system level. Both timing speculation and approximate computing achieve power savings by allowing errors to occur, while adaptive power management is a robust power management technique that can work under variable system environment (e.g., workloads and usage patterns). In Sect. 3.6.1, we introduce timing speculation and show how it enables the tradeoff among performance, power and reliability by allowing infrequent timing errors to occur. In Sect. 3.6.2, we discuss approximate computing techniques, which are able

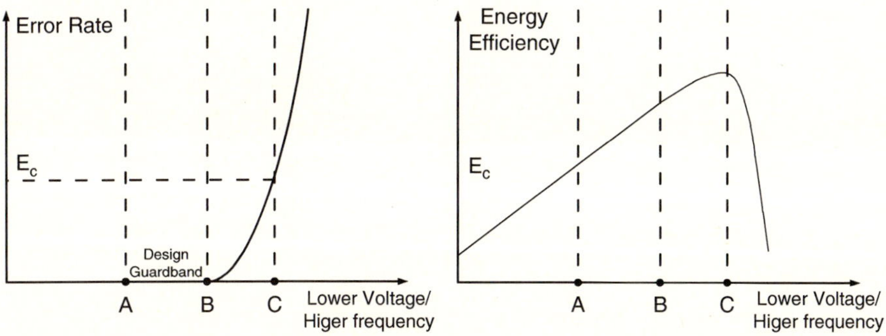

Fig. 3.13 Motivation for timing speculation

to trade off computation quality (e.g., accuracy) and computational effort (e.g., energy consumption). In Sect. 3.6.3, we introduce adaptive power management techniques. A case study on learning-based power management is shown in Sect. 3.6.4.

3.6.1 Timing Speculation

Conventional IC designs try all means to achieve error-free computation, even under worst-case combinations of process, voltage, and temperature (PVT) variations and wearout effects [12, 29]. As the above circuit non-idealities inevitably worsen with technology scaling [11], more design guardband has to be incorporated to ensure IC timing correctness. Consequently, such worst-case design methodology results in pessimistic designs with considerable power and performance overheads [5], lessening the benefits provided by technology scaling. As can be seen in Fig. 3.13, even though a particular circuit may operate at point B without timing errors, during the design phase, we have to conservatively let the circuit work at point A with lower frequency and/or higher supply voltage to ensure its timing correctness throughout its service life. To address the above problem, better-than-worst-case (BTWC) design methodology that allows reliability to be traded off against power and performance was proposed to dramatically improve the energy efficiency of computation [4,37]. The basic idea behind BTWC design methodology is that, since circuit non-idealities mainly manifest themselves as infrequent timing errors on critical paths of the circuit (if sufficient design guardband is not incorporated) [79], we can over-clock the chip and/or reduce the supply voltage of the chip to a point that timing errors occur and achieve resilient computation (instead of error-free computation) by performing timing error detection and correction. This approach is generally referred to as timing speculation. As can be seen in Fig. 3.13, a timing speculative circuit can operate at point C with much higher frequency or much less supply voltage, thus greatly improving the circuit's energy efficiency. Due to this

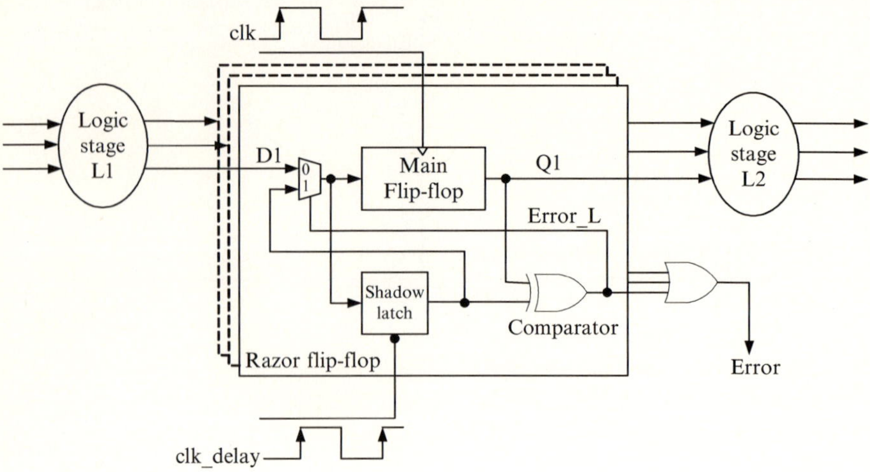

Fig. 3.14 The design of Razor-FF

significant benefit, TS techniques have attracted lots of research interests from both academia and industry [3].

Without loss of generality, let us discuss one of the most representative TS techniques, Razor [22, 28], to illustrate how resilient computation can be achieved with timing speculation. To detect timing errors on critical paths, the receiving ends of critical paths, referred to as suspicious flip-flops, are replaced with Razor flip-flops (Razor-FFs), which includes a main flip-flop, an additional shadow latch and some control logic (see Fig. 3.14). The main flip-flop latches the output signal at the clock edge with possible timing error, while the shadow latch, controlled by a delayed clock signal, latches the signal a fraction of a cycle later, which guarantees to receive the correct value. Consequently, when the shadow latch and the main FF values do not agree, indicated by the comparator, the timing error is detected. For microprocessors, timing error recovery can be achieved with microarchitectural support [89]. That is, when a timing error is detected, the processor pipeline is flushed and the correct result from the shadow latch is returned back into the pipeline. Then, by replaying instructions (at possibly lower frequency), the processor is able to recover from the timing error [13].

Timing error recovery inevitably incurs some performance loss and extra energy consumption. As can be observed in Fig. 3.13, further increase of frequency and/or decrease of voltage beyond point C will lead to too many rollbacks and hurt system performance/energy efficiency. Therefore, it is essential to reduce timing error rate (TER) to optimize timing speculative circuits [24]. However, there is usually a "wall of critical paths" in the final implementation of the circuit from the traditional worst-case design flow. This is due to the nature of today's IC design and optimization flow, e.g., gates on those initially non-critical paths are often downsized to tradeoff for power and area, making many such paths to be critical too. This suggests that, given a fixed circuit design, the effectiveness of timing speculation techniques is limited by a fixed threshold beyond which the circuit will become unusable.

To address the above issue, various optimization techniques were presented for timing speculative circuits in the literature. The key issue in this optimization problem is to reshape the path delay distribution of the circuit so that those frequently-exercised timing paths are optimized with more timing slack while other paths are allowed to have timing errors. EVAL [80] proposes a so-called high-dimensional dynamic adaptation technique that trades error rate for processor frequency by tilting, shifting, or reshaping the path distributions of various functional units. Blueshift [33] identifies and optimizes the most frequently exercised critical paths by on-demand selective biasing and path constraint tuning. DynaTune [104] optimizes the most frequently-sensitized critical paths of the circuit by assigning low threshold voltage to those critical gates that are strongly related to the occurrence of timing errors. Kahng et al. [52] proposed a slack redistribution strategy to increase the level of voltage overscaling under a given TER constraint to minimize the power consumption. Ye et al. conducted pre-silicon clock skew scheduling [112] and post-silicon clock skew tuning [111] to manipulate timing slacks at circuit level to reduce timing error rate in timing-speculative circuits. In [113], the authors studied the voltage island generation problem to achieve better energy efficiency in timing-speculative circuits.

The above techniques are helpful for TER reduction, but one common limitation is that they conduct optimization on top of a given circuit netlist and hence are not capable of manipulating the logic structure of the circuit. In [19], the authors attempted to conduct logic synthesis for BTWC designs. They constructed a simple timing error probability model and used it to guide the "balance" logic optimization step, which is a logic decomposition method initially used for delay minimization [20]. The effectiveness of this solution, however, is not very impressive from their experimental results, likely due to the lack of accuracy of the unvalidated timing error model and the simple strategy to include timing errors into optimization cost function only. In [66], the proposed logic synthesis technique manipulates circuit structures from the ground up and dramatically reduces timing error rates. In [65], cost-efficient re-synthesis solutions are proposed to reduce the number of suspicious FFs and pad the short paths.

3.6.2 Approximate Computing

A large and growing number of applications are inherently error-tolerant, which do not require "strict" correctness but rather approximate correctness. Applications of such kind include multimedia, DSP, wireless communication, data mining and synthesis. They may process noisy data sets and the associated algorithms are stochastic or involve a human interface with limited perceptual capability. For these applications, approximate computing, being able to trade off computation quality

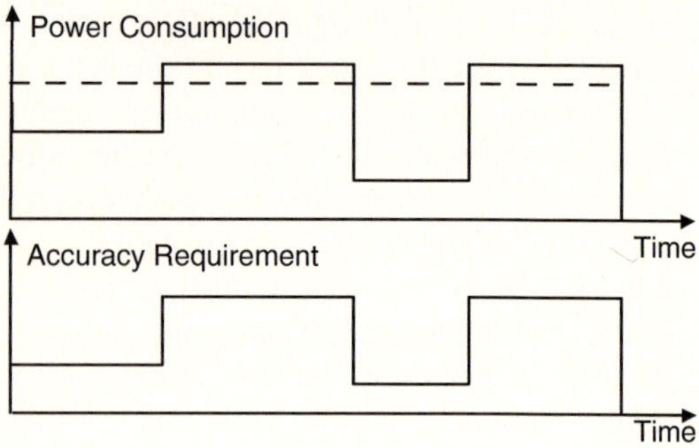

Fig. 3.15 A motivational example for approximate computing

(e.g., accuracy) and computational effort (e.g., energy), has attracted lots of attention recently (e.g., [51, 58, 85, 102, 103]).

Generally speaking, approximate hardware designs implement a slightly different yet more energy-efficient and/or faster Boolean function. Various approximate designs for specific arithmetic components were presented in the literature, taking advantage of the structural properties of these components. In [35], the authors proposed to design approximate full adder cells with reduced complexity at the transistor level and utilized them to design approximate multi-bit adders. In [58], the authors presented a 2×2 approximate multiplier block by elaborately revising the K-map of 2×2 multiplier. By using this inaccurate multiplier with simpler logic design, significant power savings can be achieved. The above works try to substitute original computation components with approximate ones at design stage, while in [51], the authors illustrated how power benefits can be obtained at runtime with an accuracy-configurable design. As shown in Fig. 3.15, the dashed line represents the case with accurate computation while the solid line represents the case with approximate computing. In accurate computation, we always have a relatively high power consumption while in approximate computing we can adjust the system setting and hence consume less power according to varying accuracy requirement of different system workloads. Motivated by the above, [51] proposes a pipelined adder and an error correction mechanism. By selectively turn on/off the correction component in different pipeline stages of the approximate adder, the computation accuracy can be controlled to satisfy system requirement.

Instead of designing certain arithmetic components, [102] presents a systematic methodology for logic synthesis of functionally approximate circuits. As shown in Fig. 3.16, the concept "quality constraint circuit" is used to formulate the problem of approximate synthesis, which consists of three components: the original circuit, the approximate circuit and the quality function. The original circuit represents the structural description of circuit specification that needs to be approximated while the quality function defines the acceptable accuracy constraint. By using the primary outputs of both original circuit and approximate circuit, the quality function can output one bit Q-value to indicate whether the accuracy constraint is satisfied. Consequently, by taking primary inputs into the original circuit, we can utilize quality function to obtain the output range of approximate circuit that would not

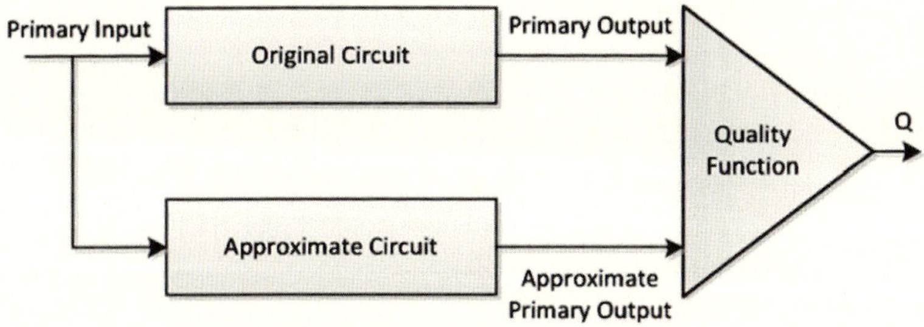

Fig. 3.16 Logic synthesis framework of quality constraint circuit

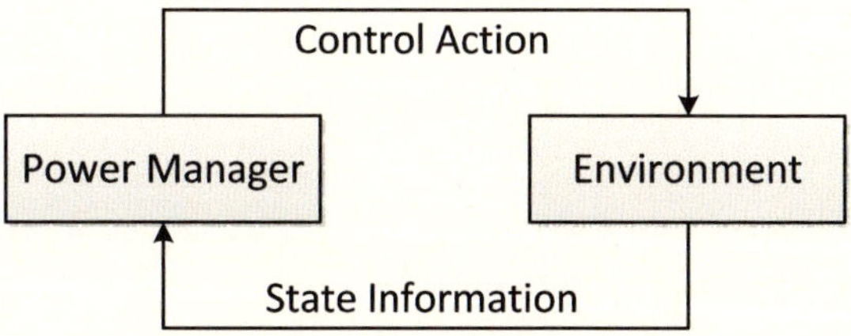

Fig. 3.17 Conceptual framework of adaptive power management

affect the value of Q. These input combinations can be used to obtain a simplified approximate design to achieve power savings or performance improvement.

3.6.3 Adaptive Power Management

Adaptive power management (APM) techniques reduce power dissipation by turning off certain idle components adaptively according to the characteristics of runtime workloads. As discussed in Sect. 3.4, when designing power management policies, the main challenge is that it is difficult, if not impossible, to know which components should be shut down and when is the opportune moment.

APM is an example of autonomic system, where autonomic control is applied for regulation of processes without direct human intervention. Conceptually, *autonomics* is derived from human body's autonomic nervous system that controls individual organ function. The objective of autonomic computing is to create systems that are self-managing, self-healing, and self-protecting. It promises to reduce expenditures associated with operations, and to significantly improve the end user's experience [98].

As shown in Fig. 3.17, an APM system [98] consists of two components at least: power manager and environment. Usually, power manager monitors the environment states and uses this information as inputs to determine a power control action that adapts to current environment condition. After applying the generated action to environment, the environment under control can tune itself to meet varying condition. From this perspective, an APM system needs to be able to reason about

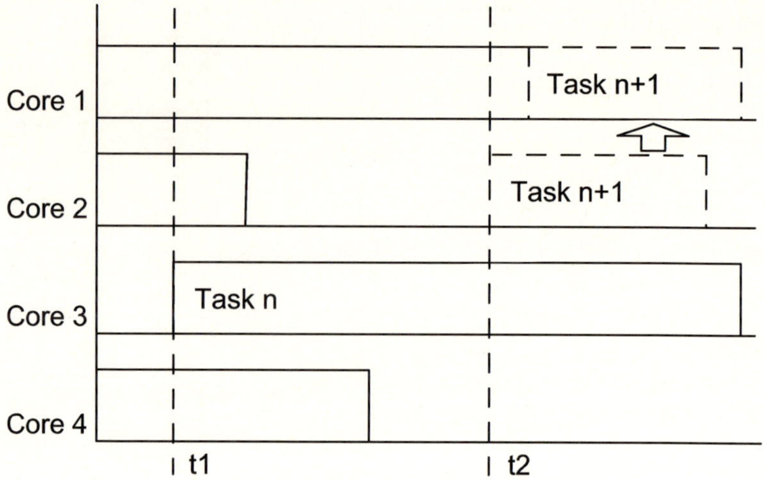

Fig. 3.18 A motivational example for task allocation on multi-core processors

the environment uncertainty, due to the fact that user context cannot be directly observed from sensors in a completely accurate manner, such as temperature and current sensors.

To deal with the above uncertainties, artificial intelligence and machine learning techniques can be employed in APM systems, which facilitate to understand the environmental behavior and then take proper management actions under a self-improved control policy.

3.6.4 Case Study

In this subsection, we take a learning-based power management technique on multi-core processors as an example to show how to utilize machine learning algorithm to realize APM.

3.6.4.1 Motivation

In multi-core processors, we have the flexibility to assign a task to any processor core and hence the idle periods on processor cores become partially controllable, which can be exploited for power savings. Note that, for the sake of simplicity, we assume that each task is executed on only one core and there is no dependency between tasks. In addition, we mainly consider dynamic power consumption for task execution.

Figure 3.18 presents the motivational example. In this 4-core processor, when $task_{n+1}$ arrives, allocating it to different cores for processing may lead to very

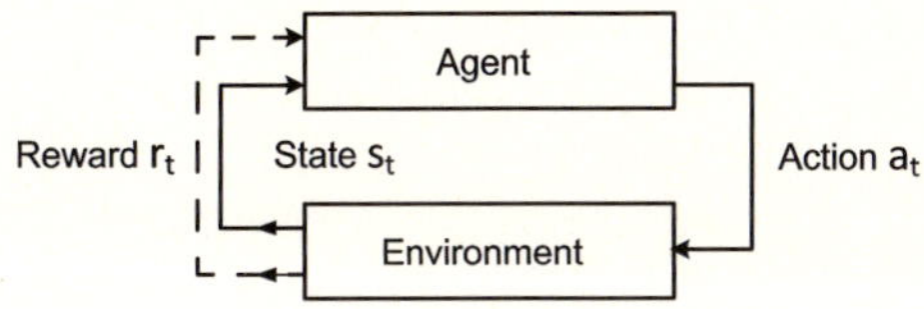

Fig. 3.19 Q-learning cycle [70]

different results. Suppose the task is assigned to *core* 2. Since this core has been idle for some time, it might be in sleep mode at this time point, and we have to wake it up to process this task, causing extra power dissipation and performance penalty. If, however, the task is assigned to *core* 1, we are able to save the above cost without incurring much performance penalty since it is about to finish the task assigned to it earlier. Ideally, if we can assign a new task to a processor core that has just finished its earlier-assigned task at that time point, we do not need to suffer from any cost.

Motivated by the above, we develop a novel learning-based DPM framework for multi-core processors that judiciously allocate tasks on processor cores to achieve a better tradeoff between power dissipation and system performance, as discussed in the following.

3.6.4.2 Background on Q-Learning

Q-learning, as one of the prevalent reinforcement learning algorithms, has been applied in many scientific and engineering fields. Since it is also used in our proposed DPM solution, we briefly introduce it in the following.

The basic idea of Q-learning [70] is to decide on what action to take based on current system state information in order to maximize the expected *reward* in the future by mapping states to actions. In standard Q-learning framework (as shown in Fig. 3.19), an agent is connected to its environment via perception and actions. In each step t, the agent observes the system state s_t, chooses an action a_t to perform, and then receives $r(s_t, a_t)$ from the environment and observes new state s_{t+1}. Formally, the model consists of

- A discrete set of environmental states, $S = \{s_t\}$;
- A discrete set of agent actions, $A = \{a_t\}$;
- A reward function $R=\{r(s_t, a_t)\}$: $S \times A \rightarrow R$.

In each state, there is a Q-value associated with each action. The definition of Q-value is the sum of the reinforcements received when performing the associated action and then following the given policy thereafter. Given the definition, it is easy to derive the equivalent of the Bellman equation for Q-learning:

$$Q(s_t, a_t) = r(s_t, a_t) + \gamma \cdot \max_{a_{t+1}} Q(s_{t+1}, a_{t+1}), \tag{3.10}$$

which is the objective to be maximized in Q-learning. According to this definition, when receiving *reward* in each learning cycle, we update Q-value according to the following equation:

$$Q^{new}(s_t, a_t) = Q(s_t, a_t) + \mu \cdot [r(s_t, a_t) + \gamma \cdot \max_{a_{t+1}} Q(s_{t+1}, a_{t+1}) - Q(s_t, a_t)]. \quad (3.11)$$

Here $r(s_t, a_t)$ is the *reward* received in state s_t with action a_t taken; μ is learning rate; and γ is discount rate. It should be noted that

- The learning rate μ determines what extent the newly acquired information will override the old information to, while the discount rate γ determines the importance of future rewards;
- The number of possible system states and actions must be finite, and as the number of states and actions increases, the Q-table gets bigger and thus the learning accuracy deteriorates quickly;
- If the agent always just takes the action with the highest Q-value for a given state, it might end up in a local maximum, because one action might be repeatedly taken without exploring new actions.

3.6.4.3 System Framework

As discussed earlier, we can model the targeted power-manageable system as shown in Fig. 3.8 with three components: SR, SQ and SP. The power manager observes system states (consisting of SR, SQ and SP states), and controls the behavior of SP, to achieve power savings at certain performance penalty. Based on the above, we setup our Q-learning model for DPM problem in multi-core processors.

State Space

In our Q-learning model, system states are composed of the states of SQ and SP only, because the state of SR is unknown a priori. To simplify the problem, we firstly consider how to describe the state space of a single-core system, and then extend it to multi-core processors.

For single-core processors, we use a vector with two dimensions to describe its state (s_t, q_t). Therein, s_t stands for the processor power state, e.g., run mode or sleep mode. q_t represents the queue status, which indicates how many task requests are stored in queue to wait for processing. Suppose we consider $q_t = 0, 1$, and 2 respectively for the cases that the number of requests in the queue is 0, 1 and larger than 1. There are as many as $(n_c \cdot n_q)$ states, where n_c is the number of power states, and n_q is the number of queue states. Let (s_{t+1}, q_{t+1}) represent the next state and a_t represent the taken action, we have $(s_t, q_t) \xrightarrow{a_t} (s_{t+1}, q_{t+1})$.

To represent system states in multi-core processors with n cores, we can extend the state vector from two dimensions to $2n$ dimensions. Hence, we have the state representation $(s_{t1}, s_{t2}, \ldots, s_{tn}; q_{t1}, q_{t2}, \ldots, q_{tn})$, wherein s_{ti} and q_{ti} are the core power state and the waiting queue state for core i, respectively. With the above representation, however, the size of the state space increases to $(n_c \cdot n_q)^n$. Such a huge state space is a critical problem for learning-based approaches, because in this case many more training samples are needed, and learning accuracy deteriorates quickly. To solve this problem, we utilize neural network to approximate Q-values.

Action Space

In our model, we sample the system state at each time point when a task request arrives. The power manager then observes the current system state, and determines an action for SP to operate. As shown in Fig. 3.18, when $task_{n+1}$ arrives, its arrival time can determine the time point t_2. At that time point, power manager samples system state, and chooses an action to apply. The action is composed of two components: not only the core that $task_{n+1}$ is assigned to, but also the power state of the assigned core after finishing this task. In other words, the power manager presets the power modes for all the cores. If idle time slots appear in the cores, they will transfer to the appointed power modes.

The action can be represented as ($core_t$,$mode_t$). The variable $mode_t$ stands for the preset mode for assigned core, and $core_t$ is the core index to indicate which core to assign this task to. In this case, the action space size is $(n_c \cdot n)$, where n_c is the number of power modes for each core and n is the number of cores.

Reward

The objective of DPM techniques is usually to achieve the maximum power savings at slight performance penalty cost. To achieve a tradeoff between the two items, the *reward* function used in our Q-learning model is expressed as below:

$$R(\mathbf{s_t}, \mathbf{a_t}) = -(P(\mathbf{s_t}, \mathbf{a_t}) + \beta \cdot RT(\mathbf{s_t}, \mathbf{a_t})), \tag{3.12}$$

where R is reward, P is mean power dissipation, RT is response time and β is the coefficient to trade off power and performance. If β-value is changed, the weights of mean power and response time in *reward* function are adjusted to satisfy system demand. A larger β-value means that response time is more important to our concern.

At the time point with system state $\mathbf{s_t}$, the power manager chooses action $\mathbf{a_t}$. Then the system state transfers from $\mathbf{s_t}$ to $\mathbf{s}_{t+1}$, and corresponding reward value can be received.

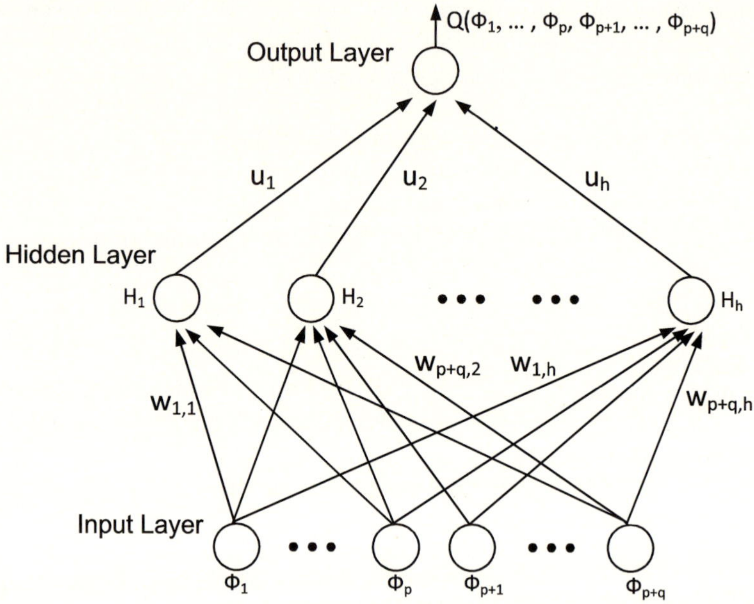

Fig. 3.20 Multi-layer sigmoid approximation neural network

3.6.4.4 Learning Algorithm

Q-Function Approximation

One of the most challenging issues in our work is the huge system space size ($(n_c \cdot n_q)^n \cdot (n_c \cdot n)$), which is exponentially increased with respect to processor core number n. Q-learning at its simplest version uses tables to store Q-values. This not only costs insufferable memory, but also requires a huge amount of training samples to learn the Q-table accurately. For example, if we describe core with up to 3 power states and 2 queue states, a 8-core processor would have $(3 \cdot 2)^8$ system states and $(3 \cdot 8)$ actions. Suppose that 5 training samples are needed for each table cell, $((3 \cdot 2)^8 \cdot (3 \cdot 8) \cdot 5) = 201,553,920$ training samples are required, which is almost impossible.

To address the above issue, we use neural network (NN) [70] to approximate the Q-function. Hence, the key task in the Q-learning for our problem becomes how to estimate the mapping $Q(\mathbf{s}, \mathbf{a}) : \mathbf{s} \times \mathbf{a} \rightarrow Q$. We adopt the feedforward neural network to model the mapping $Q(\mathbf{s}, \mathbf{a})$ and represent the value function of state-action pair (s_t, a_t). There are a variety of neural networks that are applicable to function approximation, and we consider back propagation neural network (BPNN) [70], one of the most prevailing neural algorithms in dealing with function approximation.

As shown in Fig. 3.20, there are three layers in the used neural network, namely, input, hidden, and output layers, respectively. In the input layer, the input vector is the composite of state vector ($\mathbf{s_1},\ldots,\mathbf{s_n}$) and action vector $\mathbf{a}$. We use binary encoding scheme to denote the input vectors for every possible system state and action. In the hidden layer, the hidden nodes H_i employs the following sigmoid function,

$$H_j = 1/(1 + e^{-\sum_{i=1}^{p+q} \Phi_i \cdot w_{i,j}}), \tag{3.13}$$

where $w_{i,j}$ and u_i are the parameters of neural network, Φ_i denotes one bit of binary input, p denotes the bit number of binary input for system state, while q denotes the bit number of binary input for action. In the output layer, the approximated Q-value function is given by

$$Q(\mathbf{s}, \mathbf{a}) = \sum_{i=1}^{h} H_i \cdot u_j. \tag{3.14}$$

As a whole, this neural network describes a non-linear mapping $Q(\mathbf{s}, \mathbf{a})$. At each time t, the parameters of the network ($w_{1,1},\ldots, w_{p+q,h}$; $u_1,\ldots,u_h$) are updated in a gradient manner with the help of the back-propagation algorithm [70]. The errors propagate backwardly from the output nodes to the inner nodes to adjust the network's weights. When it is applied to Q-learning, the input of back propagation neural network is the state-action pair and its output is the Q-value corresponding to the state-action pair.

Action Selection

The action selection mechanism is an important component of Q-learning. There are two problems to tackle in our action selection phase.

First, if the agent always takes the action with the highest Q-value for a given state, it might end up in a local maximum, because one action might be repeatedly taken without exploring new actions, which prevents us from finding other solution. In other words, action selection may greatly affect learning effectiveness, due to the tradeoff between exploitation and exploration. To balance these two aspects, we employ ε-greedy method for action selection, so that the agent can reinforce the evaluation of the known actions to be good and also explore unknown actions, which helps in avoiding local maximum. We gives the action that owns the highest Q-value a high selected probability $(1 - \varepsilon)$, and all the actions equally share the remaining probability ε. The probability for choosing a certain action a_i is presented as below:

$$P_i = \begin{cases} (1-\varepsilon) + (\varepsilon/num) & \text{if the Q value of action } a_i \text{ is the highest;} \\ \varepsilon/num & \text{otherwise.} \end{cases} \tag{3.15}$$

```
1.  Initialization()
2.      Set neural network parameters to random values
            between 0 and 1.0
3.      Initialize system state s
4.  Repeat for each step of the current episode
5.      Select action a using ε-greedy
6.      Take action a
7.      Observe next state and receive reward R
8.      Update back propagation neural network parameters
            using gradient descent algorithm
9.      Set system state s ← s′
10. Until there are no more episodes
```

Fig. 3.21 Q-learning algorithm based on back propagation neural network

We consider $\varepsilon = 10\,\%$ here, and *num* is action number. That means, we have the probability of 10 % to select another action instead of the action with highest Q-value, to void local maximum.

Second, since one of the motivations is to reduce unnecessary power state transitions to avoid transition costs, our algorithm has the trend to allocate tasks successively to certain cores. This may induce temperature stress on certain cores and cause reliability concerns. To tackle this problem, our action selection mechanism is further modified. Each time, if the temperature of the core chosen is higher than a pre-defined temperature threshold, we would give up this action and try to select another action from the remaining cores by ε-greedy again.

Overall Flow

To sum up, the Q-learning algorithm is illuminated in Fig. 3.21, which starts with initialization (Line 1). The procedure is repeated until there are no more episodes. For every episode, the Q-values are computed via back propagation neural network. We then select an action in the ε-greedy manner (Line 5), and take the action to transfer system state from $\mathbf{s}$ to $\mathbf{s}'$ and receive *reward* value (Line 7). When we get the *reward* as feedback, we update the parameters of back propagation neural network using gradient descent algorithm (Line 8), and update state $\mathbf{s}$ using next state $\mathbf{s}'$ (Line 9).

Note that, an off-line training phase with a convergence criterion (e.g., the normalized error of approximated Q-value is less than 5 %) can be used to improve the solution quality during the beginning of task execution, if necessary. We perform online training to both Q-learning and neural network by considering each task as one training sample.

3.7 Conclusion

The quest for high energy efficiency is never ending. A large amount of works in this area have been presented in the literature and implemented in industrial designs. In this chapter, we first introduce background knowledge in low power designs. We then discuss some classic energy-efficient design techniques and their impact on circuit reliability. Finally, some emerging energy-efficient design techniques are introduced.

References

1. A. Abdollahi, F. Fallah, M. Pedram, Leakage current reduction in CMOS VLSI circuits by input vector control. IEEE Trans. Very Large Scale Integr. Syst. **12**(2), 140 (2004)
2. ARM, ARM11 PrimeXsys Platform, http://www.jp.arm.com/event/images/\forum2002/02-print_arm11_primexsys_platform_ian.pdf
3. T. Austin, Diva: a reliable substrate for deep submicron microarchitecture design, in *Proceedings of the International Symposium on Microarchitecture (MICRO)*, Haifa, 1999
4. T. Austin, V. Bertacco, Deployment of better than worst-case design: solutions and needs, in *Proceedings of the International Conference on Computer Design (ICCD)*, San Jose, 2005
5. T. Austin, V. Bertacco, D. Blaauw, T. Mudge, Opportunities and challenges for better than worst-case design, in *Proceedings of the IEEE Asia South Pacific Design Automation Conference (ASP-DAC)*, Shanghai, 2005
6. T. Austin, V. Bertacco, S. Mahlke, Y. Cao, Reliable systems on unreliable fabrics. IEEE Des. Test Comput. **25**(4), 322–332 (2008)
7. L. Benini, A. Bogliolo, G. Paleologo, G. De Micheli, Policy optimization for dynamic power management. IEEE Trans. Comput. Aided Des. **18**(6), 813–833 (2001)
8. S. Bhardwaj, W. Wang, R. Vattikonda, Y. Cao, S. Vrudhula, Predictive modeling of the NBTI effect for reliable design, in *Proceedings of the Custom Integrated Circuits Conference*, San Jose, 2006
9. J.R. Black, Electromigration – a brief survey and some recent results. IEEE Trans. Electron Devices **16**(4), 338–347 (1969)
10. M. Borah, R.M. Owens, M.J. Irwin, Transistor sizing for low power CMOS circuits. IEEE Trans. Comput. Aided Des. Integr. Circuits Syst. **15**(6), 665–671 (1996)
11. S. Borkar, Designing reliable systems from unreliable components: the challenges of transistor variability and degradation, in *Proceedings of the International Symposium on Microarchitecture (MICRO)*, Barcelona, 2005
12. S. Borkar, T. Karnik, S. Narendra, J. Tschanz, A. Keshavarzi, V. De, Parameter variations and impact on circuits and microarchitecture, in *Proceedings of the ACM/IEEE Design Automation Conference (DAC)*, Anaheim, 2003
13. K. Bowman, J. Tschanz, N.S. Kim, J.C. Lee, C.B. Wilkerson, S.-L. Lu, T. Karnik, V.K. De, Energy-efficient and metastability-immune timing-error detection and instruction-replay-based recovery circuits for dynamic-variation tolerance, in *Proceedings of the International Solid State Circuits Conference (ISSCC)*, San Francisco, 2008
14. H.Y. Chen, S.M. Kang, iCOACH: a circuit optimization aid for CMOS high-performance circuits, in *Proceedings of the International Conference on Computer-Aided Design (ICCAD)*, Santa Clara, 1988
15. Y. Chen, H. Li, C. Koh, G. Sun, J. Li, Y. Xie, K. Roy, Variable-latency adder (VLAdder) designs for low power and NBTI tolerance. IEEE Trans. Very Large Scale Integr. Syst. **18**(11), 1621–1624 (2010)

16. X. Chen, C. Liao, T. Wei, S. Hu, An interconnect reliability-driven routing technique for electromigration failure avoidance. IEEE Trans. Dependable Secur. Comput. **9**, 770–716 (2010)
17. R. Ching, E. Young, K. Leung, C. Chu, Post-placement voltage island generation, in *Proceedings of the International Conference on Computer-Aided Design (ICCAD)*, San Jose, 2006
18. E. Chung, L. Benini, G. Micheli, Dynamic power management using adaptive learning tree, in *Proceedings of the International Conference on Computer-Aided Design (ICCAD)*, San Jose, 1999
19. J. Cong, K. Minkovich, Logic synthesis for better than worst-case designs, in *Proceedings of the International Symposium on VLSI Design, Automation and Test*, Hsinchu, 2009
20. J. Cortadella, Timing-driven logic bi-decomposition. IEEE Trans. Comput. Aided Des. **22**(6), 675–685 (2003)
21. M.B. da Silva, V.V.A. Camargo, L. Brusamarello, G.I. Wirth, R. da Silva, NBTI-aware technique for transistor sizing of high-performance CMOS gates, in *10th Latin American Test Workshop*, Rio de Janeiro, 2009
22. S. Das, C. Tokunaga, S. Pant, W.-H. Ma, S. Kalaiselvan, K. Lai, D.M. Bull, D.T. Blaauw, RazorII: in situ error detection and ccorrection for PVT and SER tolerance. IEEE J. Solid State Circuits **44**(1), 32–48 (2009)
23. A. Dasgupta, R. Karri, Electromigration reliability enhancement via bus activity distribution, in *Proceedings of the Design Automation Conference (DAC)*, Las Vegas, 1996
24. M. de Kruijf, S. Nomura, K. Sankaralingam, Unified model for timing speculation: evaluating the impact of technology scaling, CMOS design style, and fault recovery mechanism, in *Proceedings of the IEEE/IFIP International Conference on Dependable Systems and Networks (DSN)*, Chicago, 2010
25. M. Debole, W. Wang, Y. Wang, Y. Xie, V. Nayaranan, Y. Cao, A framework for estimating NBTI degradation of microarchitectural components, in *Proceedings of the Asia-South Pacific Design Automation Conference (ASP-DAC)*, Yokohama, 2009
26. S. Devadas, S. Malik, A survey of optimization techniques targeting low power VLSI circuits, in *Proceedings of the Design Automation Conference (DAC)*, San Francisco, 1995
27. G. Dhiman, T. Rosing, Dynamic power management using machine learning, in *Proceedings of the International Conference on Computer-Aided Design (ICCAD)*, San Jose, 2006
28. D. Ernst, N.S. Kim, S. Das, S. Pant, R. Rao, T. Pham, C. Ziesler, D. Blaauw, T. Austin, K. Flautner, T. Mudge, Razor: a low-power pipeline based on circuit-level timing speculation, in *Proceedings of the International Symposium on Microarchitecture (MICRO)*, San Diego, 2003
29. D. Frank, R. Puri, D. Toma, Design and CAD challenges in 45 nm CMOS and beyond, in *Proceedings of the International Conference on Computer-Aided Design (ICCAD)*, San Jose, 2006
30. F. Gao, P. Hayes, Exact and heuristic approaches to input vector control for leakage power reduction. IEEE Trans. Comput. Aided Des. Integr. Circuits Syst. **25**(11), 2564 (2006)
31. M. Ghasemazar, E. Pakbaznia, M. Pedram, Minimizing the power consumption of a chip multiprocessor under an average throughput constraint, in *Proceedings of the International Symposium on Quality Electronic Design (ISQED)*, San Jose, 2010
32. L.A. Glasser, L.P. Hoyte, Delay and power optimization in VLSI circuits, in *Proceedings of the Design Automation Conference (DAC)*, Albuquerque, 1984
33. B. Greskamp, L. Wan, U.R. Karpuzcu, J.J. Cook, J. Torrellas, D. Chen, C. Zilles, Blueshift: designing processors for timing speculation from the ground up, in *Proceedings of the International Symposium on High Performance Computer Architecture (HPCA)*, Shanghai, 2009
34. L. Guo, Y. Cai, Q. Zhou, X. Hong, Logic and layout aware voltage island generation for low power design, in *Proceedings of the Asia and South Pacific Design Automation Conference (ASP-DAC)*, Yokohama, 2007

35. V. Gupta, D. Mohapatra, S.P. Park, A. Raghunathan, K. Roy, IMPACT: IMPrecise adders for low-power approximate computing, in *Proceedings of the International Symposium on Low Power Electronics and Design (ISLPED)*, Fukuoka, 2011
36. P. Hazucha, C. Svensson, S.A. Wender, Cosmic-ray soft error rate characterization of a standard 0.6 μm CMOS process. IEEE J. Solid State Circuits **35**(10), 1422–1429 (2000)
37. R. Hegde, N.R. Shanbhag, Energy-efficient signal processing via algorithmic noise-tolerance, in *Proceedings of the International Symposium on Low Power Electronics and Design (ISLPED)*, San Diego, 1999
38. S. Herbert, D. Marculescu, Analysis of dynamic voltage/frequency scaling in chip-multiprocessors, in *Proceedings of the International Symposium on Low Power Electronics and Design (ISLPED)*, Portland, 2007
39. L. Huang, Q. Xu, Energy-efficient task allocation and scheduling for multi-mode MPSoCs under lifetime reliability constraint, in *Proceedings of the Design, Automation, and Test in Europe (DATE)*, Dresden, 2010
40. L. Huang, F. Yuan, Q. Xu, Lifetime reliability-aware task allocation and scheduling for MPSoC platforms, in *Proceedings of the Design, Automation, and Test in Europe (DATE)*, Nice, 2009
41. L. Huang, F. Yuan, Q. Xu, On task allocation and scheduling for lifetime extension of platform-based MPSoC designs. IEEE Trans. Parallel Distrib. Syst. **22**(12), 2088–2099 (2011)
42. L. Huang, R. Ye, Q. Xu, Customer-aware task allocation and scheduling for multi-mode MPSoCs, in *Proceedings of the Design Automation Conference (DAC)*, San Diego, 2011
43. C. Hwang, A. Wu, A predictive system shutdown method for energy saving of event-driven computation, in *Proceedings of the International Conference on Computer-Aided Design (ICCAD)*, San Jose, 1997
44. C. Isci, A. Buyuktosunoglu, C. Cher, P. Bose, M. Martonosi, An analysis of efficient multi-core global power management policies: maximizing performance for a given power budget, in *Proceedings of the of International Symposium on Microarchitecture (MICRO)*, Orlando, 2006
45. T. Ishihara, H. Yasuura, Voltage scheduling problem for dynamically variable voltage processors, in *Proceedings of the International Symposium on Low Power Electronics and Design (ISLPED)*, Monterey, 1998
46. N.K. Jha, Low power system scheduling and synthesis, in *Proceedings of the International Conference on Computer-Aided Design (ICCAD)*, San Jose, 2001
47. M.C. Johnson, D. Somasekhar, K. Roy, Leakage control with efficient use of transistor stacks in single threshold CMOS, in *Proceedings of the Design Automation Conference (DAC)*, New Orleans, 1999
48. J.S.S.T. Association, *Failure Mechanisms and Models for Semiconductor Devices (JEP122-B)* (JEDEC, Solid State Technology Association), Arlington, VA 22201–2107 United States (2009)
49. H. Jung, M. Pedram, Improving the efficiency of power management techniques by using Bayesian classification, in *Proceedings of the International Symposium on Quality of Electronic Design (ISQED)*, San Jose, 2008
50. H. Jung, M. Pedram, Supervised learning based power management for multicore processors. IEEE Trans. Comput. Aided Des. **29**(9), 1395–1408 (2010)
51. A.B. Kahng, S. Kang, Accuracy-reconfigurable adder for approximate arithmetic design, in *Proceedings of the Design Automation Conference (DAC)*, San Francisco, 2012
52. A.B. Kahng, S. Kang, R. Kumar, J. Sartori, Slack redistribution for graceful degradation under voltage overscaling, in *Proceedings of the Asia South Pacific Design Automation Conference (ASP-DAC)*, Taipei, 2010
53. J. Kao, A.P. Chandrakasan, Dual-threshold voltage techniques for low-power digital circuits. IEEE J. Solid State Circuits **35**(7), 1009–1018 (2000)
54. J. Kao, A. Chandrakasan, D. Antoniadis, Transistor sizing for low power CMOS circuits, in *Proceedings of the Design Automation Conference (DAC)*, Anaheim, 1997
55. Z. Karimi, M. Sarrafzadeh, Fine-grained post placement voltage assignment considering level shifter overhead, in *Proceedings of the IEEE/IFIP VLSI System on Chip Conference (VLSI-SoC)*, Madrid, 2010

56. A. Karlin, M. Manesse, L. McGeoch, S. Owicki, Competitive randomized algorithms for nonuniform problems. Algorithmica **11**, 542–571 (1994)
57. J. Kathuria, M. Ayoubkhan, A. Noor, A review of clock gating. Int. J. Electron. Commun. Eng. **1**(2), 106–114 (2011)
58. P. Kulkarni, P. Gupta, M. Ercegovac, Trading accuracy for power with an underdesigned multiplier architecture, in *Proceedings of the Annual Conference on VLSI Design*, Chennai, 2011
59. S. Lee, T. Sakurai, Run-time voltage hopping for low-power real-time systems, in *Proceedings of the International Symposium on Low Power Electronics and Design (ISLPED)*, Rapallo, 2000
60. W. Lee, H. Liu, Y. Chang, An ILP algorithm for post-floorplanning voltage-island generation considering power-network planning, in *Proceedings of the International Conference on Computer-Aided Design (ICCAD)*, San Jose, 2007
61. W. Lee, D. Marculescu, Y. Chang, Post-floorplanning power/ground ring synthesis for multiple-supply-voltage designs, in *Proceedings of the International Symposium on Physical design (ISPD)*, San Diego, 2009
62. B. Lin, H.D. Man, Low-power driven technology mapping under timing constraints, in *Proceedings of the International Conference on Computer Design (ICCD)*, Cambridge, 1993
63. B. Liu, Y. Cai, Q. Zhou, X. Hong, Power driven placement with layout aware supply voltage assignment for voltage island generation in dual-vdd designs, in *Proceedings of the Asia and South Pacific Design Automation Conference (ASP-DAC)*, Yokohama, 2006
64. Y. Liu, H. Yang, R. Dick, H. Wang, L. Shang, Thermal vs energy optimization for dvfs-enabled processors in embedded systems, in *Proceedings of the International Symposium on Quality Electronic Design (ISQED)*, San Jose, 2007
65. Y. Liu, F. Yuan, Q. Xu, Re-synthesis for cost-efficient circuit-level timing speculatio, in *Proceedings of the Design Automation Conference (DAC)*, San Diego, 2011
66. Y. Liu, R. Ye, F. Yuan, R. Kumar, Q. Xu, On logic synthesis for timing speculation, in *Proceedings of the International Conference on Computer-Aided Design (ICCAD)*, San Jose, 2012
67. Q. Ma, E. Young, Multivoltage floorplan design. IEEE Trans. Comput. Aided Des. Integr. Circuits Syst. **29**(4), 607–617 (2010)
68. Q. Ma, Z. Qian, E. Young, H. Zhou, MSV-driven floorplanning. IEEE Trans. Comput. Aided Des. Integr. Circuits Syst. **30**(8), 1152–1162 (2011)
69. W.K. Mak, J.W. Chen, Voltage island generation under performance requirement for SoC designs, in *Proceedings of the Asia and South Pacific Design Automation Conference*, Yokohama, 2007
70. S. Marsland, *Machine Learning: An Algorithmic Perspective* (CRC, Boca Raton, c2009)
71. S. Mukhopadhyay, C. Neau, R.T. Cakici, A. Agarwal, C.H. Kim, K. Roy, Gate leakage reduction for scaled devices using transistor stacking. IEEE Trans. Very Large Scale Integr. (VLSI) Syst. **11**(4), 716–730 (2003)
72. P.R. Panda, B.V.N. Silpa, A. Shrivastava, K. Gummidipudi, *Power-efficient System Design* (Springer New York Dordrecht Heidelberg London, 2010)
73. T. Pering, T. Burd, R. Brodersen, Voltage scheduling in the IpARM microprocessor system, in *Proceedings of the International Symposium on Low Power Electronics and Design (ISLPED)*, Rapallo, 2000
74. Z. Qi, R.S. Mircea, NBTI resilient circuits using adaptive body biasing, in *Proceedings of the ACM Great Lakes Symposium on VLSI (GLSVLSI)*, Orlando, 2008
75. T.S. Rosing, K. Mihic, G.D. Micheli, Power and reliability management of SoCs. IEEE Trans. Very Large Scale Integr. (VLSI) Syst. **15**(4), 391 (2007)
76. K. Roy, S. Mukhopadhyay, H. Mahmoodi-Meimand, Leakage current mechanisms and leakage reduction techniques in deep-submicrometer CMOS circuits. Proc. IEEE **91**(2), 305–327 (2003)
77. A. Sangiovanni-Vincentelli, et al., Benefits and challenges for platform-based design, in *Proceedings of the Design Automation Conference (DAC)*, San Diego, 2004

78. S.S. Sapatnekar, V.B. Rao, P.M. Vaidya, S.M. Kang, An exact solution to the transistor sizing problem for CMOS circuits using convex optimization. IEEE Trans. Comput. Aided Des. Integr. Circuits Syst. **12**, 1621–1634 (1993)
79. S.R. Sarangi, B. Greskamp, R. Teodorescu, J. Nakano, A. Tiwari, J. Torrellas, VARIUS: a model of process variation and resulting timing errors for microarchitects. IEEE Trans. Semicond. Manuf. **21**(1), 3–13 (2008)
80. S. Sarangi, B. Greskamp, A. Tiwari, J. Torrellas, EVAL: utilizing processors with variation-induced timing errors, in *Proceedings of the International Symposium on Microarchitecture (MICRO)*, Lake Como, 2008
81. M.T. Schmitz, B.M. Al-Hashimi, P. Eles, Cosynthesis of energy-efficient multimode embedded systems with consideration of mode-execution probabilities. IEEE Trans. Aided Des. Integr. Circuits Syst. **24**(2), 153–169 (2005)
82. D. Sengupta, R.A. Saleh, Application-driven voltage-island partitioning for low-power system-on-chip design. IEEE Trans. Comput. Aided Des. Integr. Circuits Syst. **28**(3), 316–326 (2009)
83. N. Shanbhag, Reliable and efficient system-on-chip design. IEEE Trans. Comput. **37**(3), 42–50 (2004)
84. Y. Shin, K. Choi, Power conscious fixed priority scheduling for hard real-time systems, in *Proceedings of the Design Automation Conference (DAC)*, New Orleans, 1999
85. D. Shin, S.K. Gupta, Approximate logic synthesis for error tolerant applications, in *Proceedings of the Design, Automation, and Test in Europe (DATE)*, Dresden, 2010
86. P. Shivakumar, S.W. Keckler, C.R. Moore, D. Burger, Exploiting microarchitectural redundancy for defect tolerance, in *Proceedings of the International Conference on Computer Design (ICCD)*, San Jose, 2003
87. T. Simunic, L. Benini, G. Micheli, Event-driven power management of portable systems, in *Proceedings of the International Symposium on System Synthesis*, San Jose, 1999
88. T. Simunic, L. Benini, A. Acquaviva, P.W. Glynn, G. DeMicheli, Dynamic voltage scaling for portable systems, in *Proceedings of the Design Automation Conference (DAC)*, Las Vegas, 2001
89. R. Sproull, I. Sutherland, C. Molnar, The counterflow pipeline processor architecture. IEEE Des. Test Comput. **11**(3), 48 (1994)
90. SquareTrade, Report on Xbox 360 failure rates (2008), http://blog.squaretrade.com/2008/02/xbox-fail-rates.html
91. M. Srivastava, A. Chandrakasan, R. Brodersen, Predictive system shutdown and other architectural techniques for energy-efficient programmable compuation. IEEE Trans. Very Large Scale Integr. Syst. **4**(1), 42–55 (1996)
92. J. Srinivasan, S.V. Adve, P. Bose, J. Rivers, C.K. Hu, *RAMP: A Model for Reliability Aware Microprocessor Design* (IBM, Poughkeepsie, 2003)
93. J. Srinivasan, S.V. Adve, P. Bose, J.A. Rivers, The case for lifetime reliability-aware microprocessors, in *Proceedings of the International Symposium on Computer Architecture (ISCA)*, Munich, 2004
94. J. Srinivasan, S.V. Adve, P. Bose, J.A. Rivers, Lifetime reliability: toward an architectural solution, in *Proceedings of the International Symposium on Microarchitecture (MICRO)*, Barcelona, 2005
95. J.H. Stathis, Reliability limits for the gate insulator in CMOS technology. IBM J. Res. Dev. **46**(2/3), 265–283 (2002)
96. C. Steinbach, *A Reinforcement Learning Approach to Power Management*. AI technical report, M.Eng thesis, Artificial Intelligence Laboratory, Massachusetts Institute of Technology, 2002
97. Y. Tan, W. Liu, Q. Qiu, Adaptive power management using reinforcement learning, in *Proceedings of the International Conference on Computer-Aided Design (ICCAD)*, San Jose, 2009
98. G. Theocharous, et al., Machine learning for adaptive power management. Intel Technol. J. **10**, 4 (2006)

99. V. Tiwari, P. Ashar, S. Malik, Technology mapping for low power, in *Proceedings of the Design Automation Conference (DAC)*, Dallas, 1993
100. V. Tiwari, P. Ashar, S. Malik, Technology mapping for low power in logic synthesis. VLSI J. Integr. **20**(3), 243–268 (1996)
101. C. Tsui, M. Pedram, A.M. Despain, Technology decomposition and mapping targeting low power dissipation, in *Proceedings of the Design Automation Conference (DAC)*, Dallas, TX, USA, 1993
102. S. Venkataramani, A. Sabne, V. Kozhikkottu, K. Roy, A. Raghunathan, SALSA: systematic logic synthesis of approximate circuits, in *Proceedings of the Design Automation Conference (DAC)*, San Francisco, 2012
103. R. Venkatesan, A. Agarwal, K. Roy, A. Raghunathan, MACACO: modeling and analysis of circuits for approximate computing, in *Proceedings of the International Conference on Computer-Aided Design (ICCAD)*, San Jose, 2011
104. L. Wan, D. Chen, Dynatune: circuit-level optimization for timing speculation considering dynamic path behavior, in *Proceedings of the International Conference on Computer-Aided Design (ICCAD)*, San Jose, 2009
105. Y. Wang, H. Luo, K. He, R. Luo, Y. Xie, H. Yang, Temperature-aware NBTI modeling and the impact of input vector control on performance degradation, in *Proceedings of the International Conference on Design Automation and Test in Europe (DATE)*, Nice, 2007
106. H. Wu, M. Wong, Incremental improvement of voltage assignment. IEEE Trans. Comput. Aided Des. Integr. Circuits Syst. **28**(2), 217–230 (2009)
107. H. Wu, M. Wong, I. Liu, Timing-constrained and voltage-island-aware voltage assignment. In *Proceedings of the Design Automation Conference (DAC)*, San Francisco, 2006
108. H. Wu, M. Wong, I. Liu, Y. Wang, Placement-proximity-based voltage island grouping under performance requirement. IEEE Trans. Comput. Aided Des. Integr. Circuits Syst. **26**(7), 1256–1269 (2007)
109. X. Yang, K. Saluja, Combating NBTI degradation via gate sizing, in *Proceedings of the International Symposium on Quality Electronic Design (ISQED)*, San Jose, 2007
110. R. Ye, Q. Xu, Learning-based power management for multi-core processors via idle period manipulation, in *Proceedings of the Asia and South Pacific Design Automation Conference (ASP-DAC)*, Sydney, 2012
111. R. Ye, F. Yuan, Q. Xu, Online clock skew tuning for timing speculation, in *Proceedings of the International Conference on Computer-Aided Design (ICCAD)*, San Jose, 2011
112. R. Ye, F. Yuan, H. Zhou, Q. Xu, Clock skew scheduling for timing speculation, in *Proceedings of the Design, Automation, and Test in Europe (DATE)*, Dresden, 2012
113. R. Ye, F. Yuan, Z. Sun, W.-B. Jone, Q. Xu, Post-placement voltage island generation for timing-speculative circuits, in *Proceedings of the Design Automation Conference (DAC)*, Austin, 2013
114. K. Yeo, K. Roy, *Low-Voltage, Low-Power VLSI Subsystems*. McGraw-Hill Professional Engineering, McGraw-Hill, Inc., New York, NY, USA, 2004

Chapter 4
Error Correction Coding for Electronic Circuits

Juan A. Maestro, Pedro Reviriego, and Mark F. Flanagan

4.1 Introduction

Digital electronic circuits are subject to many types of error. Considering the effect of such errors on the circuit functionality, they can be classed as permanent, transient or intermittent. Permanent (or "hard") errors disrupt the functionality of the circuit for its entire lifetime [1]. An example of a hard error is a stuck at one/zero fault in a logical gate, in which the output is fixed to a logical value regardless of the values of the inputs. Permanent errors can be caused, for example, by manufacturing defects, aging, or radiation effects. Transient (or "soft") errors only affect the functionality of the circuit for a short time. An example of a transient error is a radiation-induced soft error, in which a particle impacts the circuit and changes the logical value of one circuit node [2]. Transient errors can also be caused by noise or crosstalk. The circuit functions correctly after the error event, but if an incorrect value is stored in a register or memory then the system state can be erroneous. Intermittent errors are those which affect a circuit node in such a way as to cause errors frequently but not constantly [3]. These errors are commonly caused by marginal or unstable behavior, which may or may not cause an error, depending on the conditions.

A wide range of techniques can be used to deal with errors in electronic circuits. Permanent faults which occur during the manufacturing process can be identified in the test phase, and the defective parts can be discarded [1]. In some circuits, it may be more cost-effective to include redundant elements that are used to repair the errors detected in the test phase. This is commonly the case for memories, where redundant rows/columns are used to replace those with permanent errors [4].

J.A. Maestro (✉) • P. Reviriego
Universidad Antonio de Nebrija, Madrid, Spain
e-mail: jmaestro@nebrija.es; previrie@nebrija.es

M.F. Flanagan
University College Dublin, Dublin, Ireland
e-mail: mark.flanagan@ieee.org

J. Mathew et al. (eds.), *Energy-Efficient Fault-Tolerant Systems*,
DOI 10.1007/978-1-4614-4193-9_4, © Springer Science+Business Media New York 2014

For permanent errors caused by aging, discarding the defective part is not an option as the part is already working in the system. Therefore, preventive actions are taken in the design phase to ensure that the parts will be operational for the specified lifetime. One example of aging effects is Negative Bias Temperature Instability (NBTI), which tends to degrade the circuit speed [5]. Therefore, in the design phase some additional delay margin is added so that when the speed degrades over time, the circuit still meets the delay requirements. Intermittent errors are typically a symptom of a marginal failure which can degenerate into a permanent error. Therefore, the same techniques as those described for permanent errors can be used, the main challenge being the identification of the errors during the testing phase. To mitigate transient errors, some techniques can be used during the manufacturing phase to improve noise immunity or to reduce the probability of particle impacts causing soft errors. However, ensuring that no errors will occur may not be possible or may have an unacceptable cost. When that is the case, some transient errors will occur. In addition, in many cases the error is caused by an external source that can affect any circuit node and therefore no screening can be done at the test phase. To mitigate the effects of these transient errors, redundancy can be added to the design so that the errors are detected and corrected [6]. A typical solution is to replicate the design; with duplication, an error can be detected, and with triplication, corrected. Replication requires a large overhead in terms of circuit area and energy consumption, which limits its application. In some specific applications it is possible to design ad-hoc redundancy schemes that provide efficient protection at a lower cost. One example is that of signal processing circuits whose algorithmic properties can be used to implement fault-tolerance [7]. For memories, registers and interconnections, another option is to use Error Correction Codes (ECCs) [8]. While replication can be viewed as a primitive ECC, ECCs in general represent a more sophisticated way of adding redundancy, enabling more powerful error correction capabilities with a lower overhead. There are different types of ECCs; one important category is that of *block codes* [9]. In a block code, a set of k bits is used to generate a larger block of n bits. The n-k additional bits, called *parity* bits, enable the detection and correction of errors. The operation of generating the block of n bits is known as *encoding* and the operation of *decoding* extracts the k original bits from the n coded bits. Another type of code, known as a *convolutional code*, operates on a sequence of bits for which there is no fixed length. The overheads associated with an ECC are mainly the number of added parity bits (n-k), the encoder, and the decoder. The purpose of this chapter is to provide an overview of the use of ECCs to protect electronic circuits from errors.

Most existing work on error correction coding has been carried out in the area of communications. Error Correction Codes have in fact played a fundamental role in the development of modern communication systems [9]. The use of error correction enables transmission over noisy channels and makes it possible to approach the capacity of the channel. The use of ECCs in electronic circuits started decades ago, focusing on the protection of memories [8]. In this case, the design requirements for the ECC are different from those in communication systems. This has led to different solutions being used in memories to those used in communications. The same reasoning applies to other electronic circuits. For example, in communications the received signal is typically quantized with more than two levels. This provides additional information (called *soft information*) on the *reliability* of each of the bits, which is then used in the decoding process. For electronic circuits, this additional information is available only in a few cases, as for example in some Flash memories [10]. In most electronic circuits, the values read from a memory or register are simply equal to zero or one. Therefore, the decoding algorithm is different. Another difference is that in electronic circuits, the decoding latency typically has to be smaller. For example, in a processor register file or cache, the access to a value has to be done typically in one clock cycle, or else in a few clock cycles. In communications, even if the link speed is high, larger decoding latencies are allowed as the main concern is to ensure that the decoder throughput meets the link speed. These latency requirements influence the algorithms and architectures for the encoder and decoder. In electronic circuits, the data divides naturally into blocks such as a memory word or a register which stores a value. In communications, data is typically received serially and the concept of block, when used, is introduced somewhat artificially. Therefore, in this context serial decoding algorithms are commonly used. However, in circuits the decoding of a block is typically done in parallel as the complete block is accessed in a single operation. From this discussion, the specifics of the use of ECCs to protect electronic circuits become apparent. The objective of this chapter is to provide an overview of ECCs used for circuit protection, focusing on their implementation and describing which codes are most suited for each circuit type. To this end, the rest of the chapter is divided into two parts. In the first part, the requirements of different types of electronic circuit are described. This analysis focuses on the protection requirements in terms of the types of error that the circuit suffers and also on the performance requirements in terms of circuit energy consumption, area and delay. The second part of the chapter then describes the different ECCs that are commonly used to protect circuits, as well as some which have been recently proposed in research papers. The description provides a basic introduction to the codes used, and then focuses on describing the error correction capabilities and the implementation overheads. This is then linked at the end of the chapter with the first part by highlighting, for each type of circuit, the ECCs which are most suitable.

4.2 Protection Requirements of Electronic Circuits

There are many different types of electronic circuits and when considering protection against errors, each one has its own set of requirements. In the following, these requirements are discussed for the circuits in which Error Correction Codes (ECCs) are commonly used. The parameters considered are the type of error that needs to be corrected, and the area and latency constraints for the ECCs.

4.2.1 Registers

In digital circuits, it is common to find a set of flip-flops that are grouped to store a value. This is the case, for example, in Finite State Machines (FSMs) where a register is used to store the state [11]. A soft error that affects a register can cause a system malfunction, requiring a reset to recover the correct functionality. Error Correction Codes are used to protect registers, so that even when a soft error affects the register, the correct value can be recovered. The most common assumption is that errors will affect only a single bit in the register. This can be reasonable, as flip-flops have a larger size than memory cells and therefore it is less likely that a particle impact affects more than one register bit. However, as technology scales, multiple bit errors may become an important issue. The requirements in terms of latency are typically such that encoding and decoding must take place in a fraction of a clock cycle. Larger latencies would make it impossible to decode, use a value, and encode the new value within a clock cycle. The duration of a clock cycle is design-specific and can vary significantly. For a very high speed circuit, the use of ECCs may not even be possible, while for a low speed design, complex ECCs may not impact the overall latency. In any case, the complexity of the encoder and decoder also has to be low, since otherwise it may be more cost-effective to use replication. The same reasoning applies to the number of parity bits. Another important parameter for an ECC is the block size. In the case of registers, the block can have any number of bits depending on the specific design. In many cases, the number of bits in the block should be a power of two, and typically does not exceed 64 bits. All these requirements restrict, in most cases, the use of ECCs in registers to simple codes. The requirements are summarized below:

- Correct single bit errors.
- Latency equal to a fraction of a clock cycle.
- Complexity of encoder/decoder equal to or smaller than that of the register.
- Number of added bits (parity bits) is smaller than the number of bits in the register.
- Variable register length, typically equal to or smaller than 64 bits.

4.2.2 Register Files

Another structure that is commonly found in digital circuits is a set of registers, or a *register file*. This is the case for example in processors and controllers [12]. Many other circuits also have a set of registers for configuration and monitoring. The ECC requirements of a register file are similar to those of a single register, but there are two main differences. The first difference is that in register files, the bit width is almost always a power of two. The second difference deals with the complexity of the ECC, and may be explained as follows. For a register file, the area cost of the encoder and decoder can be larger, as it is shared among a set of registers. On the other hand, the number of parity bits has a larger impact as the bits are added to each of the registers. This means that ECCs with more complex encoding/decoding can be used if they reduce the number of parity bits. Finally, regarding the type of errors which affect the register file, in most studies single bit errors are considered, but multi-bit errors are also starting to become an issue [13]. The requirements are summarized below:

- Correct single bit errors (multiple bit errors starting to become an issue).
- Latency equal to a fraction of a clock cycle.
- Complexity of encoder/decoder should be a small fraction of that of the register file.
- Number of parity bits is significantly smaller than the number of bits in the register.
- Block size is a power of two; typical values are 16, 32 and 64 bits.

4.2.3 Caches

Memory caches are an important element in modern processors and computing systems [14]. Since they are intended to improve the speed of access to data stored in the memory, they work at high speed. This imposes tight constraints on the latency of any ECC. The size of caches becomes larger with every new technology generation and is commonly in the order of Megabytes. This means that the encoder and decoder complexity can be significant and yet still have a small impact on the circuit area of the cache. The number of parity bits, on the other hand, should be minimized as it impacts directly the size of the cache (this is because these parity bits are added to each cache entry). The block sizes in caches are typically larger than those in register files, with values that are also powers of two. Block sizes of 512, 1024 and 2048 bits are commonly used. It is important to note that in addition to radiation-induced errors, there is another source of transient errors in caches that is important, which may be explained as follows. Energy consumption is a large issue in caches, and one way to reduce it is to lower the voltage supply; this can cause some random transient errors that are then corrected by the ECC [14]. This type of error, as well as radiation-induced errors, can affect single or multiple bits in caches [15]. The requirements for caches are summarized below:

- Correct single bit errors and multiple bit errors.
- Latency equal to a fraction of a clock cycle.
- Significant complexity of encoder/decoder can be acceptable.
- Number of parity bits significantly smaller than the number of bits in an entry.
- Block size is a power of two; typical values are 512, 1024 and 2048 bits.

4.2.4 SRAM Memories

The transient errors in Static Random Access Memories (SRAMs) can affect a single bit or multiple bits in a memory word. As technology scales, the proportion of multiple bit errors increases [16]. This means that the ECC must be able to correct multiple bits unless interleaving is used. Interleaving consists in placing the bits that belong to the same logical word physically apart. This ensures that a particle impact, even if it affects multiple memory cells, will only corrupt one bit per memory word. Interleaving can however have an impact on memory area and power consumption [17]. Latency has to be a fraction of the memory access time to ensure that the impact of the ECC on the memory speed is small. The complexity that can be afforded in the encoder and decoder depends largely on the size of the memory. For a large memory, a complex decoder can represent only a small fraction of the memory area, while for a small memory the overhead may be unacceptable. The impact of the number of redundant bits is independent of the memory size, as the bits are added in each word. The word size of the memory is in most cases a power of two with common values being 16, 32 and 64 bits. This value coincides, in most cases, with the data bus and register file width. The requirements for SRAMs are summarized below:

- Correct single bit errors and multiple bit errors.
- Latency equal to a fraction of a clock cycle (although larger than in the case of register files and caches).
- Acceptable complexity of encoder/decoder depends on the memory size.
- Number of parity bits is significantly smaller than the number of bits in a word.
- Block size is a power of two; typical values are 16, 32 and 64 bits.

4.2.5 DRAM Memories

Dynamic Random Access Memories (DRAMs) suffer from errors of a specific type: these are known as *data retention failures*. In DRAM memories, the contents of the cells have to be refreshed periodically to avoid data loss. To minimize the probability of data loss, frequent refreshes can be used. This however increases power consumption, which is an important issue in modern computing systems. This leads to a tradeoff between reliability and power consumption. DRAMs also suffer radiation-induced soft errors which can affect multiple bits as in SRAMs. A recent study on a large number of DRAMs operating in computing systems shows that permanent errors are also frequent during the device lifetime [18]. Therefore, Error Correction Codes in this case have to deal with both transient and permanent errors. DRAM devices are commonly grouped in memory modules with larger bit widths. In this case, the correction of errors due to the failure of one of the devices is also an important issue. This can be achieved by using interleaving as proposed in [19] or more sophisticated non binary codes like for example Reed-Solomon (RS) codes.

The latency of the ECCs used has to be a fraction of a clock cycle, but in this case the memories are slower than caches, so that larger latencies can be tolerated. The size of the DRAM modules is typically large, and therefore significant complexity can be allowed in the encoder/decoder with little impact on the cost of the module. The data bit width of the modules is typically a power of two, with values of 64 bits, 128 bits or more. The total number of bits in the module is typically larger to allow the implementation of ECCs. The requirements for DRAMs are summarized below:

- Correct single bit errors and multiple bit errors. Multiple bit errors include failure of an entire DRAM device in a memory module.
- Latency equal to a fraction of a clock cycle.
- Significant complexity of encoder/decoder is acceptable.
- Number of parity bits is significantly smaller than the number of bits in a word. In many cases, the number of parity bits is restricted by the configuration of the memory module.
- Block size is a power of two; typical values are 64 and 128 bits.

4.2.6 Content Addressable Memories (CAMs)

Content Addressable Memories (CAMs) are a special type of memory used in a variety of applications in computing and communications [20]. A CAM stores keys, together with their associated values, in pairs (key, value), and when a search key is presented to the memory, the address of the key that matches the search key is obtained. Using that address, the associated value can be read from a standard memory where the values are stored. A CAM includes additional logic to compare the stored keys with the key presented to the memory. This is done in parallel for all the memory words, using comparators for each cell and merging the results of all bits in a word to determine if there is a match. The need to merge the results of the individual cells of a word limits the use of interleaving to deal with multiple errors. Another important consideration is that errors in CAMs have different effects than in standard memories. Errors in a CAM can cause two effects: *false positives* and *false negatives* [21]. A false positive occurs when the key stored in the word affected by errors matches a search key presented to the memory. In this case, an incorrect value is returned by the CAM. A false negative occurs when the search key presented to the memory should match the key stored in the word, but does not since the latter contains errors. In this case, the corrupted entry will not match the search key and no match will be found. Error correction codes can be used to avoid false positives by storing the key and the additional parity bits in each CAM entry. Also, to avoid false negatives, the circuit which merges the results of the individual bits can be modified to allow for mismatches in some bits [22].

Content Addressable Memories (CAM) can suffer single as well as multiple bit errors. Latency is not a critical issue for CAMs because the decoder is not needed (the comparison is done directly with the coded key). Therefore, only the encoder is used when accessing the CAM, to encode the presented key and compare it with the encoded keys stored in the CAM. The same reasoning applies to the encoder/decoder complexity, as the most complex block is typically the decoder, which is not used in CAMs. The number of parity bits has to be small, since it impacts directly the power consumption of CAMs as well as the area. Power consumption is an important issue for CAMs as they perform many comparisons in parallel which results in large power consumption. The bit width of CAM memories depends on the application, but the most common values are powers of two. The requirements for CAMs are summarized below:

- Detect and/or correct single and multiple bit errors.
- Latency equal to a fraction of a clock cycle (only applies for the encoder).
- Complexity is only that of the encoder.
- Number of parity bits is significantly smaller than the number of bits in a word.
- Block size is a power of two.

4.2.7 Flash Memories

Flash memories provide non-volatile storage and are used in a wide range of applications. They are characterized by the use of large block sizes of several thousands of bits. This enables the use of more complex ECCs [23, 24]. Also, the information retrieved when the memory is read is not simply a binary value, but an analog voltage. The value of the voltage provides an indication of the reliability of the bit read, and can be used to improve the performance of the ECCs [25, 26]. This is similar to the case of communications [9]. Therefore, the use of ECCs in Flash memories is quite different from the case of the other circuits considered. In this case, advanced codes such as Low-Density Parity-Check (LDPC) are used [10, 27] and the decoding resembles that used in a communication system. At the same time, there are some salient differences regarding data representation and coding for this application which make it an active area of current research [26]. The use of ECCs in Flash memories is not considered in the rest of this chapter. For further reading on this topic, the reader is referred to [25].

4.2.8 Interconnections

As the complexities of electronic circuits increase, an Integrated Circuit (IC) typically incorporates more and more blocks. Interconnections are needed to connect those blocks and also to connect the IC with other ICs in the system [28].

These interconnections can suffer errors due to noise and crosstalk, and therefore ECCs are used to protect them [29]. As with memories, interconnections can suffer from single and multiple bit errors. The number of bits is also typically a power of two, with values of 32 and 64 being common. Most of the time, this is also the width of a memory word. The latency of the ECC is important as it impacts the speed of the interconnection. The same is true of the complexity of the encoder and decoder, which impacts area and power consumption. Therefore, both latency and complexity should be small. The number of parity bits should be a fraction of the number of data bits, but this number is not as critical as in memories.

Apart from noise and crosstalk (mentioned previously), combinational logic provides another source of errors affecting interconnections. An error in a logic gate would produce an incorrect output which would be transmitted forward by an interconnection. All combinational modules (multiplexers, gates, arithmetic modules, etc.) are subject to error, and therefore they are a potential hazard for the reliability of the system.

Moreover, errors may be produced in the combinational logic which forms the encoding and decoding processes associated with the ECCs themselves, which would produce a major breakdown of the protection system.

There are many mechanisms which prevent errors in combinational logic, but these are outside the scope of this chapter.

The requirements for interconnections are summarized below:

- Correct single bit errors and multiple bit errors.
- Latency should be small.
- Complexity of encoder/decoder should be small.
- Number of parity bits is smaller than the number of data bits.
- Block size is typically a power of two, with values of 32 and 64 being common.

4.3 Error Correction Codes for Electronic Circuits

There are two main types of Error Correction Code: *convolutional codes* and *block codes*. Both types are widely used in digital communications. Convolutional codes have been traditionally preferred due to the availability of efficient decoding algorithms which can use soft-input decoding [30]. Soft-input decoding uses not only the value of the received bit, but also the value of the received signal as an indication of the *reliability* of that bit. As mentioned previously, this improves the performance of the decoder. In recent years, algorithms for block and convolutional codes, which are based on the principle of processing soft inputs and producing soft outputs, have led to extremely advanced codes such as Turbo Codes and Low Density Parity Check (LDPC) codes; these codes are now commonly used in communication systems. For circuits, although the use of convolutional codes has been proposed to protect memories [31] and state machines [32], in most cases block codes are used [6, 8]. This is because in memories and circuits, typically no soft information is available, i.e., a register or memory cell takes a value of zero or one with no indication of its reliability. The concept of block is also natural to memories and flip-flops which are typically organized in words and registers. Therefore, in what follows, convolutional codes are not considered.

There are many different types of block code. Most codes used to protect circuits are linear block codes and many of them are also binary. A binary (n,k) linear block code takes k data bits and produces (by encoding) a larger block of n bits such that each of the n bits is equal to the exclusive-OR (XOR, or modulo-2 sum) of some subset of the k original data bits. This means that the coded block can be obtained by multiplying the original data block by a Generator Matrix $\boldsymbol{G}$. If $\boldsymbol{u}$ is the block (or vector) containing the k data bits, then the coded block $\boldsymbol{v}$ is obtained as follows:

$$\boldsymbol{v} = \boldsymbol{u}\boldsymbol{G} \tag{4.1}$$

The *Hamming distance* between two binary blocks is defined as the number of bits that are different in the two blocks. Also, the *minimum distance* of the code, written $d_{\min}$, is defined as the minimum over all Hamming distances between pairs of different valid coded blocks. If a code has minimum distance $d_{\min}$, then by adding the additional *n-k parity* bits, the code ensures that any two coded blocks have different values in at least $d_{\min}$ positions. For a code with minimum distance $d_{\min}$, errors affecting up to $d_{\min}$-1 bits will be detected. To correct errors, the optimal approach is to assume that the original data was that corresponding to the valid coded block *closest* (in Hamming distance) to the block read. A code with minimum distance $d_{\min}$ can correct errors that affect up to $(d_{\min}$-1)/2 bits. Therefore, the main parameters of the code are n, k and $d_{\min}$. In circuits, k is typically given by the bit width of the memory word or register that needs to be protected. The values of n-k and $d_{\min}$ are correlated, so that generally speaking, to achieve larger minimum distances, more parity bits need to be added to the original k data bits. The relative overhead to achieve a given minimum distance decreases with larger block sizes, lowering the protection cost. As discussed previously, other important parameters in the selection of an ECC for a circuit are the latency and complexity of the encoder and decoder, and the ability to correct multiple errors, especially *correlated* multiple errors. In the rest of this section, different ECCs used for circuit protection are described and their features are related to the needs of each circuit type described in the previous section. The results of the analysis are summarized at the end of the section in a table (Table 4.7) which lists suitable ECCs for each circuit category.

4.3.1 Single Parity Check (SPC) Codes

The simplest type of linear code is a Single Parity Check (SPC) code which computes the exclusive-or (XOR, or modulo-2 sum) of the k data bits, and appends this single parity bit to the data block to form the overall coded block. In this case,

n-$k = 1$ and $d_{\min} = 2$. A Single Parity Check code can detect single errors and the encoding and error checking procedures are trivial. The ***G*** matrix is formed by an identity matrix of size k to which a column containing ones in all positions is added. In spite of its simplicity, a Single Parity Check is used in many circuits in which error detection is sufficient. Those include register files in processors [33], state machines [34] and arithmetic circuits [35, 36]. In the latter case, parity prediction can be used to detect errors in the combinational logic that computes the arithmetic operation [35]. Parity bits can also be used to correct errors in memories when they are combined with other techniques such as Built-in Current Sensors (BICS). In this case, the parity bit identifies the word in error and the BICS identifies the bit affected [37]. A Single Parity Check can be also used in Content Addressable Memories to avoid false positives due to a single bit error. The main features of SPC codes are summarized below:

- Detect single errors.
- The number of additional (parity) bits is always equal to one.
- There are no restrictions on the size of the data block (k).
- Encoding and decoding latency is approximately equal to the delay of $\log_2(n)$ two-input XOR gates.
- The area of the encoder/decoder is approximately that of n two-input XOR gates.

4.3.2 Single Error Correction Double Error Detection (SEC-DED) Codes

When error detection is not sufficient, codes that can correct errors can be used. Single Error Correction (SEC) codes which can correct a single bit error in a block are commonly used to protect memories and circuits [6, 8]. There are different types of SEC codes, a famous example being the class of *Hamming codes* [38]. SEC codes have a minimum distance of three. This means, however, that a double error can be miscorrected into another valid coded word. For this reason, Single Error Correction Double Error Detection (SEC-DED) codes are preferred [8]. SEC-DED codes have a minimum distance of four, guaranteeing that a word with two errors cannot be at a Hamming distance of one from a valid coded word. A SEC-DED code can be constructed from a Hamming code by adding a parity bit; this parity bit is used to identify single errors and avoid miscorrection (this is often referred to as an *extended Hamming code* [9]). Other SEC-DED codes have been proposed over the years to reduce the implementation cost [39] or to minimize the probability of miscorrection when a triple error occurs [40]. Another extension of SEC-DED codes are SEC-DED-DAEC codes that can also correct double errors when they are

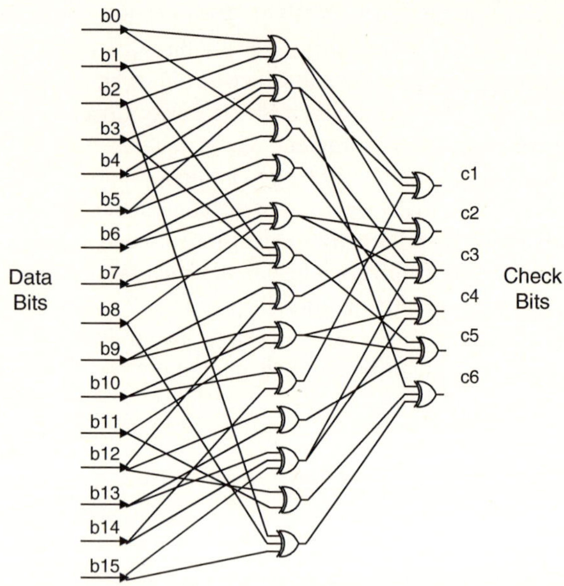

Fig. 4.1 Encoder for a SEC-DED code with $n = 22$ and $k = 16$

adjacent [41]. This is interesting to protect against correlated errors such as Multiple Cell Upsets (MCUs). This approach has recently been extended in order to develop SEC codes that also correct double/triple adjacent errors and double almost-adjacent errors [42].

For a block of k data bits, the number of redundant bits (n-k) required by typical SEC-DED codes is $\log_2(k) + 2$. Therefore, the relative overhead is smaller for larger data blocks. SEC-DED codes can be used to protect blocks of any size by using code shortening, but they are typically used for block sizes which are equal to a power of two. The encoder for a SEC-DED code is simply a set of parity equations, each covering some subset of the data bits. As an example, the **G** matrix of a SEC-DED code with $n = 22$ and $k = 16$ proposed by Hsiao in [39] is shown in Eq. 4.2. Note that here the original data bits appear in the coded word unchanged; such an encoder is called *systematic*. The final six columns contain the parity-check equations for each of the n-$k = 6$ parity bits. One possible implementation of the encoder is presented in Fig. 4.1; it can be observed that some XOR gates can be shared in the computation of several parity bits. The number of two-input XOR gates required to implement the encoder is related to the number of ones in the **G** matrix and to the amount of sharing between parity bits. As an example, the number of required XOR gates for the SEC-DED codes proposed by Hsiao is 48, 96 and 208 for $k = 16$, 32 and 64 respectively [41].

$$
\mathbf{G} = \left[\begin{array}{ll}
1\,0\,0\,0\,0\,0\,0\,0\,0\,0\,0\,0\,0\,0\,0\,0 & 1\,1\,1\,0\,0\,0 \\
0\,1\,0\,0\,0\,0\,0\,0\,0\,0\,0\,0\,0\,0\,0\,0 & 1\,1\,0\,0\,1\,0 \\
0\,0\,1\,0\,0\,0\,0\,0\,0\,0\,0\,0\,0\,0\,0\,0 & 1\,1\,0\,0\,0\,1 \\
0\,0\,0\,1\,0\,0\,0\,0\,0\,0\,0\,0\,0\,0\,0\,0 & 1\,0\,0\,0\,1\,1 \\
0\,0\,0\,0\,1\,0\,0\,0\,0\,0\,0\,0\,0\,0\,0\,0 & 1\,0\,1\,0\,0\,1 \\
0\,0\,0\,0\,0\,1\,0\,0\,0\,0\,0\,0\,0\,0\,0\,0 & 1\,0\,0\,1\,0\,1 \\
0\,0\,0\,0\,0\,0\,1\,0\,0\,0\,0\,0\,0\,0\,0\,0 & 0\,1\,1\,1\,0\,0 \\
0\,0\,0\,0\,0\,0\,0\,1\,0\,0\,0\,0\,0\,0\,0\,0 & 0\,1\,1\,0\,1\,0 \\
0\,0\,0\,0\,0\,0\,0\,0\,1\,0\,0\,0\,0\,0\,0\,0 & 0\,1\,1\,0\,0\,1 \\
0\,0\,0\,0\,0\,0\,0\,0\,0\,1\,0\,0\,0\,0\,0\,0 & 0\,1\,0\,1\,1\,0 \\
0\,0\,0\,0\,0\,0\,0\,0\,0\,0\,1\,0\,0\,0\,0\,0 & 1\,0\,0\,1\,1\,0 \\
0\,0\,0\,0\,0\,0\,0\,0\,0\,0\,0\,1\,0\,0\,0\,0 & 0\,0\,0\,1\,1\,1 \\
0\,0\,0\,0\,0\,0\,0\,0\,0\,0\,0\,0\,1\,0\,0\,0 & 0\,1\,0\,0\,1\,1 \\
0\,0\,0\,0\,0\,0\,0\,0\,0\,0\,0\,0\,0\,1\,0\,0 & 0\,0\,1\,1\,1\,0 \\
0\,0\,0\,0\,0\,0\,0\,0\,0\,0\,0\,0\,0\,0\,1\,0 & 1\,0\,1\,1\,0\,0 \\
0\,0\,0\,0\,0\,0\,0\,0\,0\,0\,0\,0\,0\,0\,0\,1 & 0\,0\,1\,1\,0\,1
\end{array}\right] \tag{4.2}
$$

The decoding of a systematic SEC-DED code starts by re-computing the parity bits from the data bits read, and checking these against the parity bits read. The vector obtained is known as the *syndrome* [39]. This process is similar to that of the encoder and is illustrated in Fig. 4.2. Whenever there are nonzero-valued bits in the syndrome, at least one error must have occurred. In this case, the value of the syndrome is used to identify if the error has affected a single bit or two bits. This is done by comparing the value of the syndrome with each one of the values that correspond to single bit errors. If one of those patterns match, the error is corrected; otherwise an uncorrectable error is signaled. The structure of the decoder is illustrated in Fig. 4.3. The syndrome decoding consists of a set of k AND gates each having n-k inputs. Each gate checks if the syndrome matches the syndrome caused by one of the k possible single bit errors. The syndrome decoding and correction make the decoder more complex than the encoder – this is typically the case for ECCs in general.

The latencies of both the encoder and decoder grow with the data block size (k), but the increments are small. They are typically measured using the logic depth in number of gates. As a summary of the complexity of SEC-DED codes, the gate count and logic depth are presented in Table 4.1. This is a simplification, as the area and delay of different gate types can vary significantly. However it provides a useful first estimate. It can be observed that the gate count grows approximately linearly with k while the delay has a much smaller increment. Comparing with a Single Parity Check, both the gate count and the delay are increased significantly.

The main features of SEC-DED codes are summarized below:

- Can correct single bit errors and detect double errors.
- The number of parity bits, n-k, is $\log_2(k) + 2$.
- They are designed for blocks where the number of data bits (k) is a power of two.

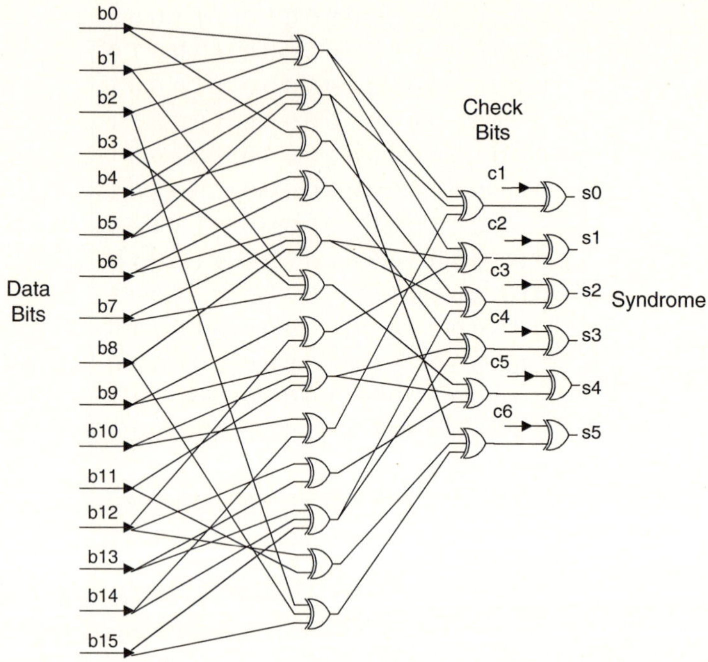

Fig. 4.2 Syndrome computation for a SEC-DED code with $n = 22$ and $k = 16$

- Encoding and decoding latency is moderate.
- The area of the encoder/decoder is also moderate.

4.3.3 Multi-bit Error Correction Codes

When single error correction is not enough, multi-bit error correction codes are needed. For electronic circuits, the appearance of soft errors which affect multiple bits is becoming increasingly common, and is expected to grow even further in the coming years [16]. Manufacturing defects or errors due to low-power/reliability tradeoffs can also cause multiple bit errors [14]. For environments in which the error arrival rate is high, multiple bit errors caused by the accumulation of single bit errors over time can also be a concern [43]. In some cases, such as a radiation-induced multiple error, the bits affected are correlated [44], while in others, such as the accumulation of single bit errors, they are not. In the first case, codes that can correct large bursts of errors are useful [9]. This will be discussed in more detail in the rest of this section. The main issues associated with multi-bit error correction codes in electronic circuits are related to their complexity. For the same data block size k, they require a larger number of additional parity bits. The decoding is also more complex and slower, which can result in large latencies.

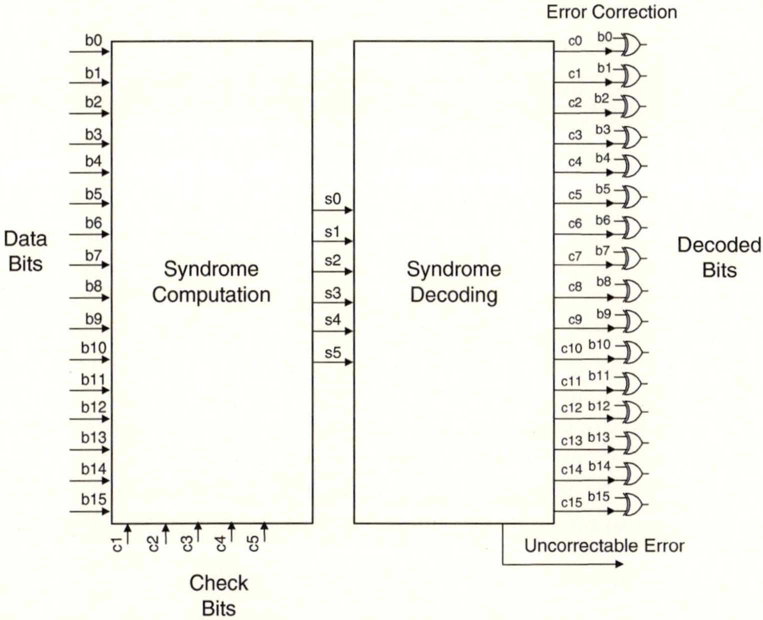

Fig. 4.3 Structure of the decoder for a SEC-DED code

Table 4.1 Complexity estimates for SEC-DED codes

n	k	Gate count	Logic depth
22	16	180	8
39	32	369	8
72	64	728	9
137	128	1705	11

As mentioned previously, advanced ECCs are commonly used in communication systems [9]. Those include Reed-Solomon (RS), Bose-Chaudhuri-Hocquenghem (BCH) and Low-Density Parity-Check codes. However, for electronic circuits the design parameters are different, as in most cases the latency and complexity has to be smaller; also, no soft information is available in the decoding. RS, BCH and LDPC codes have been proposed to protect some circuits like caches and memories [45–48]. The implementation of BCH decoders has also been optimized to the needs of memory protection [49]. Since RS, BCH and LDPC codes are well known and described in detail in many textbooks, we focus in this chapter on specific codes that have been proposed to protect electronic circuits. These include Orthogonal Latin Square (OLS) codes, Euclidean Geometry (EG) codes and Difference Set (DS) codes. In all of these cases, a simple decoding algorithm can be implemented with low latency which is an important feature for circuit applications. In the following, each of the codes is described in detail, focusing on the features of the code that are relevant for circuit protection. Finally, at the end of the section several enhancements in the use of multi-bit ECCs are discussed. These enable low power implementations, additional error detection capabilities and the correction of burst of errors.

4.3.3.1 Orthogonal Latin Square (OLS) Codes

Orthogonal Latin Square codes for memory protection were introduced in [50]. They are based on the concept of a *Latin Square*, which has seen many applications [51]. A Latin square of size m is an m-by-m matrix, each of whose rows and columns are permutations of the digits 0, 1, ..., $m-1$. Two Latin squares are *orthogonal* if, when one is superimposed on the other, every ordered pair of digits appears exactly once. An OLS code with $k = m^2$ is derived from a set of h mutually Orthogonal Latin Squares of size m as follows. Each Latin square is used to compute m parity check bits, such that each parity check involves the bits that have the same digit value in the Latin Square. This means that each parity check is computed from exactly m bits. The total number of parity checks is then equal to mh, and each bit participates in exactly h parity checks. Also, any two bits share at most one parity check equation. Therefore, using the h parity checks, h independent error checks can be done for each bit. Each other bit can only corrupt one of those checks. Therefore decoding can be done by inverting the bit when the majority of the h checks takes a value of one. This will correct errors affecting up to $t = h/2$ bits. This decoding algorithm is known as One-Step Majority Logic Decoding (OS-MLD) and can be used only for certain classes of codes [9]. Its simplicity results in a low latency which, as mentioned previously, is important in many circuits. An interesting property of OLS codes is that for a given data block size k, a code that can correct $t+1$ errors has a parity check matrix that contains that of a code that can correct t errors, and so on. This enables a modular construction and usage of these codes, which can be useful to provide schemes in which the error correction capability is adaptive [29].

The main parameters of an OLS code are the data block size $k = m^2$ and the number of parity bits $n\text{-}k = 2mt$. This number of parity bits is large compared with other codes, such as for example BCH codes for which $n\text{-}k \leq t \log_2(n)$. In many designs the block size is a power of two, and for these cases OLS codes are limited to values which are a power of four, e.g. $k = 16$, 64, 256, etc. For a code with data block size $k = m^2$ that can correct t errors, the decoding latency is $\log_2(m+1) \approx 0.5 \log_2(k)$ two-input XOR gates plus a $2t$-input majority gate, as well as the final XOR gate to perform correction. The number of gates required to implement the decoder is $2m^2t = 2kt$ two-input XOR gates plus k $2t$-input majority gates. The decoder structure is very regular and therefore amenable to implementation optimizations using full-custom or semi-custom designs. The implementation of the majority vote can also be optimized using a voltage sense amplifier instead of a logic design to reduce latency. In [14], a latency similar to that of a two-input XOR gate was mentioned for this optimized majority gate implementation. Assuming this result, the total decoding latency would be that of $\log_2(m+1)+2$ two-input XOR gates.

Table 4.2 Parameters of OS-MLD decodable Euclidean Geometry codes

n	k	J	t
15	7	4	2
63	37	8	4
255	175	16	8
1023	781	32	16

Finally, it is worth mentioning that OLS codes can also be extended to correct bursts of errors that affect multiple adjacent bits. This has been studied in [52], showing that for $k = 256$, large bursts can be corrected by adding a few additional parity check bits to the code.

The main features of OLS codes are summarized below:

- Can correct a variable number of errors t; furthermore, these codes can be extended to also correct bursts of adjacent errors.
- The code construction is modular, so that a code that can correct $t + 1$ errors includes a code that can correct t errors.
- The block sizes are typically a power of four, i.e., $k = 16, 64, 256$.
- The number of additional bits is $2\sqrt{kt}$.
- Encoding and decoding latency is moderate. With an optimized majority vote implementation, the decoding latency can be approximately that of $\log_2(\sqrt{k} + 1) + 2$ two-input XOR gates. Encoding latency is $\log_2(\sqrt{k})$ two-input XOR gates.
- The area of the encoder/decoder is moderate. The decoder requires $2kt$ two-input XOR gates plus k $2t$-input majority gates.

4.3.3.2 Euclidean Geometry (EG) Codes

Finite geometries have been used to derive many error-correcting codes [9]. One example is the class of Euclidean Geometry (EG) codes; these codes are based on the structure of Euclidean Geometries over a Galois Field. For circuit applications, it is interesting that among EG codes there is a subclass of codes that is one-step majority logic decodable (OS-MLD) [9]. As discussed in the context of OLS codes, this enables a simple decoder implementation with low latency. This subclass of EG codes has been proposed to protect memories [53–56].

The parameters of EG codes are limited to a small number of options, which are shown in Table 4.2 for block sizes up to 1023 bits. The parameter J is the number of MLD equations for each bit and t denotes the number of errors that the code can correct. It can be observed that as the data size k grows, so too does the number of correctable errors. It is also interesting to note that here k is not a power of two. Therefore, shortening should be used to adjust the data block size when it has to be a power of two. The construction of the codes is not modular as with OLS codes, but the number of parity bits is lower. The number of parity bits, n-k, is given by 3^s-1, where the parameter s is related to the coded block size via $n = 2^{2s}-1$ [9].

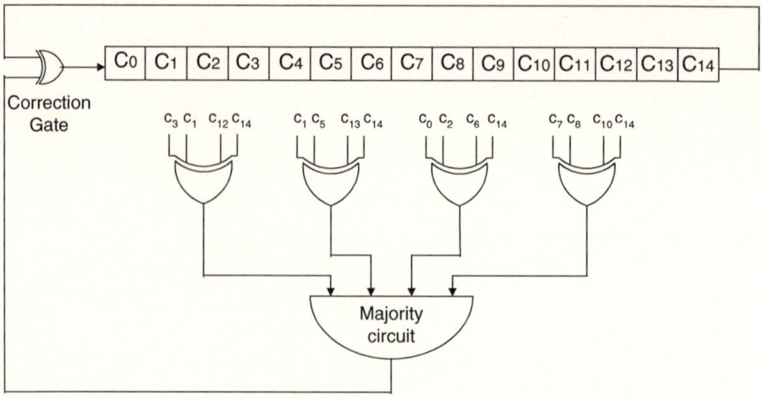

Fig. 4.4 Serial type-II one-step majority logic decoder for the ($n = 15$, $k = 7$) EG code

The decoding can be implemented using OS-MLD. For EG codes, there are two alternatives to implement OS-MLD, called type-I and type-II decoders [9]. In both cases, traditional implementations are serial, i.e., bits are decoded one at a time. This is suitable for communications systems, where the bits are received serially, but not for circuit or memory protection, where in most situations correction needs to be done for the complete block at one time. For circuit protection, the type-II decoder has been considered, as it is amenable to a parallel implementation [56]. The serial type-II decoder for the EG code with $n = 15$ and $k = 7$ is shown in Fig. 4.4. The logic needed to implement the decoder is very simple. However, the number of cycles required for serial decoding is equal to n, which results in a large latency. A parallel implementation can be obtained by replicating the combinational logic in Fig. 4.4 for each of the n bits [56]. The properties of EG codes enable an optimized implementation of the parallel decoder, as some equations are shared for different bits. This reduces the number of required MLD equations to n instead of nJ, as discussed in [57].

The decoding complexity is best expressed as a function of $J = 2^s$, where s is related to the block size via $n = 2^{2s} - 1 = J^2 - 1$. For large values of n, the approximation $J \approx \sqrt{n}$ can be used in the complexity estimations. The complexity of the combinational logic needed in a serial decoder is only $J\ (J - 1) + 1 \approx n - \sqrt{n}$ two-input XOR gates plus a J-input majority gate.[1] For a parallel decoder, the complexity increases to approximately nJ two-input XOR gates plus n J-input majority gates. The decoding latency is that of $\log_2(J) + 1$ two-input XOR gates plus a majority gate. As discussed previously in the context of OLS codes, it has been reported that majority gates can be implemented with low latency using a voltage sense amplifier [14]. In that case, latency would be approximately $\log_2(J) + 2$ two-input XOR gates. Given the reduced number of choices for the code parameters, the

[1] For EG codes, $J = \sqrt{n+1} = 2^s$.

Table 4.3 Number of two-input gates required to implement a parallel decoder for EG codes

n	Majority logic	Equations	Correction gate	Total
15	105	45	15	165
63	1,701	441	63	2,205
255	23,205	3,825	255	27,285

Table 4.4 Logic depth, in number of two-input gates, of a parallel decoder for EG codes

n	Majority logic	Equations	Correction gate	Total
15	3	2	1	6
63	6	3	1	10
255	10	4	1	15

complexity estimates are summarized in Tables 4.3 and 4.4. In this evaluation, the implementation of the majority gate was done using logic gates as proposed in [56]. Therefore, a detailed breakdown among majority gates, parity check equations and correction gates is provided to enable the evaluation of alternative majority gate implementations [14]. This is important, as in our evaluation majority gates are the major component in both area and latency, especially for large data blocks. Comparing the results with those for SEC-DED codes in Table 4.1, it can be observed that the circuit area is significantly larger but the increase in delay is small.

The main features of OS-MLD EG codes are summarized below:

- Can correct a variable number of errors t, depending on the block size.
- The block sizes are limited and the values are not a power of two.
- The number of parity bits for a coded block of size $n = 2^{2s} - 1$ is $3^s - 1$.
- Encoding and decoding latency is moderate. With an optimized majority vote implementation, the decoding latency can be approximately that of $s + 2$ two-input XOR gates.
- The area of the encoder/decoder is moderate and can be reduced by optimizing the majority gate implementation. The decoder requires $n2^s$ two-input XOR gates plus n 2^s-input majority gates.

4.3.3.3 Difference Set (DS) Codes

Another class of codes that are OS-MLD decodable is that of Difference Set (DS) codes. These codes are based on the concept of a *perfect difference set* [9]. Their use to protect memories has been recently studied [58, 59]. As with EG codes, the number of options is limited as regards block sizes and error correction capabilities. The available parameters for block sizes up to 1057 bits are shown in Table 4.5. The number of choices is the same as that for EG codes and the values of the parameters are also similar. DS codes have one more MLD equation than their EG counterparts for the same value of t (the value of the parameter J is greater by one). This can

Table 4.5 Parameters of difference set codes

n	k	J	t
21	11	5	2
73	45	9	4
273	191	17	8
1057	813	33	16

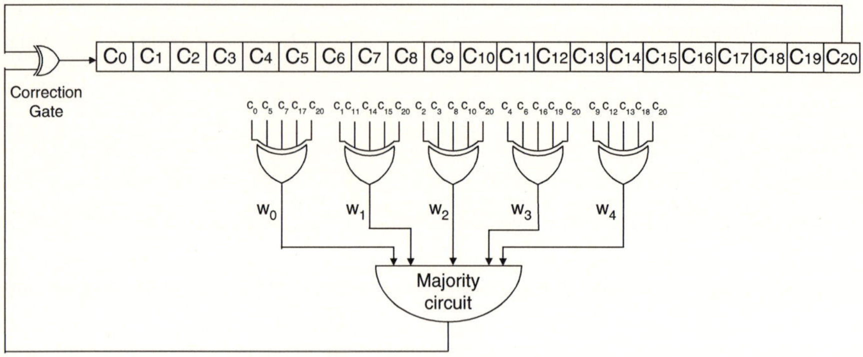

Fig. 4.5 Serial type-II one-step majority logic decoder for the ($n = 21$, $k = 11$) DS code

be used to provide additional error detection capabilities, as will be discussed in the next subsections. The number of parity bits, n-k, is given by $3^s + 1$ where s is related to the coded block size via $n = 2^{2s} + 2^s + 1$. This is similar to the case of EG codes.

The decoding of DS codes is also similar to that of OS-MLD EG codes. The type-II decoder for the code with $n = 21$ and $k = 11$ is illustrated in Fig. 4.5. An important difference is that in a DS code, every bit participates in exactly one of the MLD equations, except the bit being decoded (which participates in all of the equations). By contrast, in an EG code, some bits do not participate in any of the MLD equations at a given iteration.

The complexity and latency of the decoder for a DS code is similar to that of an EG code with corresponding parameters. The complexity is better expressed as a function of $J = 2^s + 1$ (recall that s is related to the block size via $n = 2^{2s} + 2^s + 1$). For large values of n, the approximation $J \approx \sqrt{n}$ can be used in the complexity estimations. The serial decoder requires $J*(J-1) + 1$ two-input XOR gates and a majority gate. The parallel decoder can be implemented with $nJ \approx n\sqrt{n}$ two-input XOR gates and n majority gates. The latency of decoding is that of $\log_2(J) + 1$ two-input XOR gates plus a majority gate. All these values are in line with those of EG codes. Therefore, DS codes provide additional code parameter choices with similar decoding complexity and latency.

However, Difference Set codes have some particular features which can be used to correct burst of errors. This has been recently studied in [60]. The results show that by judiciously placing the bits in the memory/circuit and correspondingly modifying the OS-MLD decoding algorithm, large bursts of adjacent errors can be corrected.

The main features of DS codes are summarized below:

- Can correct a variable number of errors t, depending on the block size. Burst error correction capabilities can also be implemented with no additional bits.
- The block sizes are limited and the values are not a power of two.
- The number of parity bits for a coded block of size $n = 2^{2s} + 2^s + 1$ is $3^s + 1$.
- Encoding and decoding latency is moderate. With an optimized majority vote implementation, the decoding latency can be approximately that of $\log_2(2^s + 1) + 2$ two-input XOR gates.
- The area of the encoder/decoder is moderate and can be reduced by optimizing the majority gate implementation. The decoder requires $n(2^s + 1)$ two-input XOR gates plus n majority gates each having $2^s + 1$ inputs.

4.3.3.4 Enhancing the Use of Multi-bit ECCs for Circuit Protection

This subsection presents several enhancements in the use of multi-bit ECCs for circuit protection. These are related to power consumption and latency optimizations, additional error detection capabilities, and the correction of burst of errors.

Reducing Decoding Power Consumption and Latency

The discussion of multi-bit ECC has so far focused on the use of OS-MLD codes which enable decoding with moderate complexity and low latency. This is because latency is a critical factor in many circuit and memory designs. Power consumption, which is also important, has however not been addressed so far. It could be argued that power consumption is directly related to encoder/decoder complexity and to the number of parity bits, and therefore complexity estimates could also provide a first indication of power consumption. However, there are some enhancements which can be used to reduce the power consumption of ECCs. One approach is to perform error detection first, and to then proceed with the rest of the decoding only when there are detected errors. The rationale is that in most cases the decoded block will be error-free, and therefore the rest of the decoding process can be spared. This will reduce power consumption and may also lower the latency when the block is error-free.

The method by which error detection is performed varies with the ECC code. Checking for any non-zero bit in the syndrome is a simple alternative which is especially suitable for ECCs in which the syndrome is computed as part of the decoding process. This is the case for BCH and OLS codes. For BCH codes, the use of this approach has been proposed to optimize the implementation of ECCs in caches [45]. For codes such as OLS codes which have a modular construction, this approach can be further optimized by noting that an ECC that can correct t errors contains in its parity check matrix, the parity check matrix of a code that corrects

$\lceil t/2 \rceil$ errors.[2] Therefore the syndrome for the code that corrects $\lceil t/2 \rceil$ errors can be used to detect errors affecting t or fewer bits. This optimization can also be used for BCH codes (which have a modular structure for the parity check matrix); this has been evaluated recently in [61].

For EG and DS codes, when the type-II OS-MLD decoder is used, the syndrome is not computed as part of the decoding. In these cases, it may be more interesting to perform error detection as part of the OS-MLD process. For a serial decoder, the first decoding cycles can be used to detect errors. This has been studied for DS codes in [57] and for EG codes in [62], showing that three decoding cycles can detect most errors. In fact, all errors affecting five or fewer bits were detected for the case of DS codes. For EG codes, all errors affecting four or fewer bits were detected. This method can also be used in a parallel decoder to reduce power consumption significantly.

Improving the Error Detection Capability

For memories and circuits, it is important to avoid undetected errors, as these may lead to silent data corruption. This is the main reason why SEC-DED codes are preferred over SEC codes. For multi-bit error correction codes, it is interesting that a code which corrects t errors can also detect $t+1$ or more errors. For $t+1$ errors this requires a minimum distance of $2t+2$, such that block with $t+1$ errors is at least at distance $t+1$ from any valid coded word and no miscorrection takes place. For the codes considered, DS codes have precisely a minimum distance of $2t+2$ and therefore can provide additional error detection. This is not the case for EG codes, while for OLS codes, since code construction is modular it can be achieved by adding more parity bits. Some techniques can be used to also detect some of the errors which affect more than $t+1$ bits. For example, the number of corrected bits can be checked to see whether it is larger than t.

The additional error detection for DS codes has been studied in [63], showing that detection of $t+1$ errors can be achieved by modifying the OS-MLD process. The error detection enhanced decoder is shown in Fig. 4.6 for the code with $n=21$ and $k=11$. The modifications are (i) that now correction is only done when there is a majority of four (in general, $t+2$) among the MLD equations and (ii) that an OR of all the MLD equations is used to detect errors. The modification of the majority vote threshold to $t+2$ avoids miscorrections when there are $t+1$ errors. Once the OS-MLD is completed, if there were $t+1$ errors there could still be errors in the coded block. Therefore those errors are detected by repeating the first OS-MLD iterations as described in the previous subsection. This scheme was also extended to detect more than $t+1$ errors by counting the number of corrected errors and checking for specific values in the majority votes.

[2]Here $\lceil x \rceil$ denotes the smallest integer greater than or equal to x.

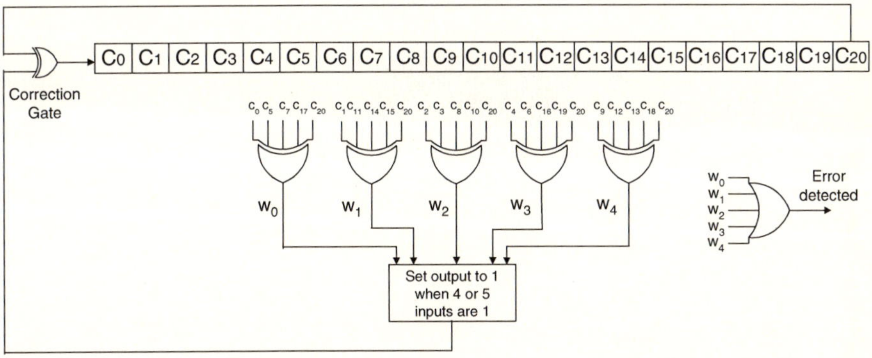

Fig. 4.6 Error detection enhanced type-II one-step majority logic decoder for the $(n = 21, k = 11)$ DS code

For OLS codes, error detection of $t + 1$ errors can be implemented by adding additional parity bits. For a data block of size k and a code that can correct t bits, $2t\sqrt{k}$ parity bits are required in an OLS code. By adding a further $\sqrt{k}$ bits the code will have a minimum distance of $2t + 2$, and $2t + 1$ MLD equations can be used to correct each bit. Then a similar approach as that proposed for DS codes can be used. In this case, the error detection after the modified OS-MLD decoding can be implemented using the syndrome computation. Therefore, enhanced error detection OLS codes can also be efficiently implemented.

Correction of Bursts of Errors

There are many different sources of errors in electronic circuits, which cause both random and correlated errors. An example of random errors is the accumulation of single radiation-induced soft errors, and an example of correlated errors is multiple radiation-induced soft errors. The codes described so far can correct t random errors, but when correlated errors are present, it is also important to consider whether the codes can correct bursts of adjacent or physically close errors. The correction of burst of errors has been considered for OLS and DS codes, while no study has yet been presented for EG codes.

For OLS codes, the correction of bursts of adjacent errors was considered in [52]. The proposed scheme adds a few additional bits to the coded block to achieve the correction of burst of errors. For example, the correction of triple adjacent errors for a double error correction OLS code is achieved with few additional bits. Correction of larger bursts is also studied, showing that only slightly more additional bits are needed. However, no analysis of the impact of the modifications on the decoder latency and complexity has been presented.

For DS codes, the correction of bursts of errors has been considered in [60]. In this case, no additional parity bits are added to the coded block. The scheme relies on placing bits that belong to the same MLD equation together, in order that bursts

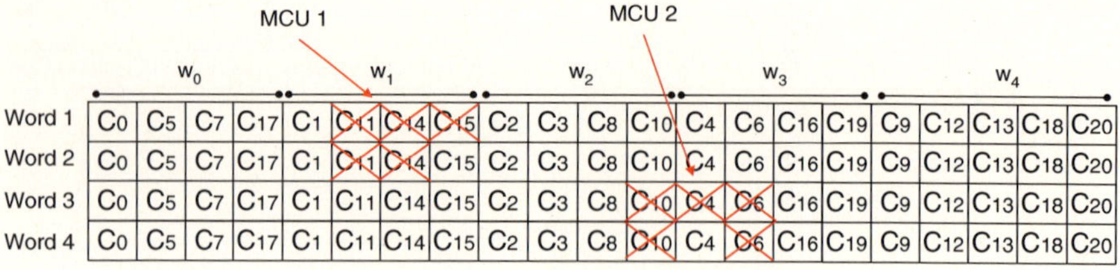

Fig. 4.7 Proposed bit placement for the bits in the $(n = 21, k = 11)$ DS code

of errors affect bits in only one or two MLD equations. This is illustrated in Fig. 4.7, where two examples of burst errors are shown. In this case, the errors are Multiple Cell Upsets (MCUs) caused by radiation. It can be observed that the first MCU affects only bits that belong to equation w_1, while errors in the second MCU affect both w_2 and w_3. The key observation is that errors which lie in one MLD equation when decoding a given bit will also be in one MLD equation when decoding each of the bits of the original equation. This is a consequence of the properties of perfect difference sets which are used to derive DS codes, and enables the correction of bursts of errors. To see how correction can be achieved, let us consider that there are t_b errors in a single MLD equation, for example w_1. Then when decoding the bits that formed part of w_1 in the beginning, all errors will be in one MLD equation. This means that the majority vote will take a value of at least $J-1$, and the errors will be corrected. However, previous to that other bits will be decoded and for these, errors can appear in different MLD equations causing the vote to take a value of up to t_b. This means that if a simple majority of $J/2$ is used for correction, miscorrection can take place when $t_b \geq J/2$. Therefore, to effectively correct bursts of errors, the decoding is done first with a large threshold for the majority vote ($J-1$), and then repeated with values decreasing down to $J/2$. The proposed algorithm is illustrated in Fig. 4.8. This modified decoding procedure has an impact on latency which, as mentioned previously, can be mitigated by first checking whether there are errors, in which case we proceed with the rest of the decoding stages.

The proposed scheme is able to correct up to $J-2$ errors in a single MLD equation and a variable number of additional errors in the other equations. If the number of errors in one equation is denoted by C and the number of the other errors by R, then the error correction capabilities of the proposed scheme are those shown in Table 4.6. For the code with $n = 1057$, only values of R up to 8 are presented. It can be observed that significant burst error correction capabilities are achieved, especially for codes with large n.

4.3.4 Emerging Codes and Error Correction Schemes

The codes described or mentioned so far are *linear* codes; this means that the computation of each of the parity check bits is achieved via modulo-2 summation

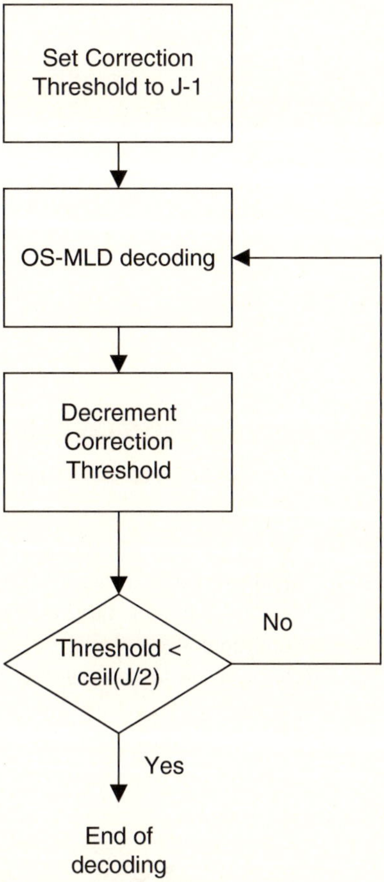

Fig. 4.8 Modified OS-MLD procedure to correct bursts of errors

Table 4.6 Error correction capabilities for the DS codes with the burst correction algorithm

N	J	$R=0$	$R=1$	$R=2$	$R=3$	$R=4$	$R=5$	$R=6$	$R=7$	$R=8$
21	5	$C=3$	$C=1$	$C=0$	–	–	–	–	–	–
73	9	$C=7$	$C=5$	$C=3$	$C=1$	$C=0$	–	–	–	–
273	17	$C=15$	$C=13$	$C=11$	$C=9$	$C=7$	$C=5$	$C=3$	$C=1$	$C=0$
1057	33	$C=31$	$C=29$	$C=27$	$C=25$	$C=23$	$C=21$	$C=19$	$C=17$	$C=15$

(or XOR operation) of some of the data bits. This ensures that when the receiver computes the syndrome to perform error detection and/or correction (as explained in Sect. 4.3.2), this syndrome does not depend on the *values* of the data bits stored, but only on the set of bit positions in which the errors have occurred. This property allows for low-complexity syndrome-based error detection and correction schemes. Recently, the use of *nonlinear* codes has been proposed to protect memories. For example, in [64] the use of nonlinear SEC-DED codes is analyzed, showing that they can reduce the number of undetected or miscorrected errors. The area and delay results presented suggest that these codes are more complex to implement than traditional linear SEC-DED codes. In any case, the use of nonlinear codes for circuit protection is an interesting area for future research.

Another topic that has gained interest in recent years is the development of adaptive ECC schemes that can provide a variable ECC solution depending on the observed error rate. One example is the adaptive ECC presented in [65] in which the block size is adapted dynamically. Another interesting idea was presented in [66] where the ECC is decoupled from the physical memory implementation, thus enabling greater flexibility. Taking advantage of that flexibility while maintaining performance in terms of area, power and speed is also an interesting research problem.

4.4 Selection of Error Correction Codes for Electronic Circuits

This chapter began by first presenting an overview of the main circuits for which error correction codes are used. The goal was to summarize, for each of these, the protection requirements and the implementation constraints. Subsequently, several error correction codes capable of meeting these requirements have been discussed. Now these parts are linked together by discussing which codes are most suitable to protect each circuit type.

4.4.1 Registers

In registers, encoding and decoding complexity is a limiting factor as the encoder and decoder are used to protect a single register and are not shared among many words as in the case of memories. Latency is also an important constraint. These factors in most cases limit the use of ECCs in registers to Single Parity Check (SPC) and SEC/SEC-DED codes. The use of multi-bit error correction codes has a large impact on performance and other alternatives like triplication (commonly known as Triple Modular Redundancy or TMR) may be more effective.

4.4.2 Register Files

Similarly to registers, register files impose tight latency constraints on the ECC. The limits to the complexity of the encoder and decoder are however relaxed, as they are shared among all the registers. This again makes SPC and SEC/SEC-DED codes good options for protection. When the cost of the additional bits is acceptable, OLS codes are also an attractive option which fit well with common register sizes (64 bits).

4.4.3 Caches

A distinct feature of caches is the use of large data blocks (512, 1024, etc.). Large blocks reduce the relative overhead of the ECC, making the use of multi-bit ECCs more attractive. At the same time, the encoder and decoder complexity is shared among a large number of blocks, reducing the impact on overall cost. Latency is an important issue as only low delay decoders can be used.

In addition to SPC and SEC/SEC-DED codes, OLS and BCH codes are good options to protect caches. The use of DS or EG codes for block sizes of 512 can also be of interest when the number of errors to correct is large (>10).

4.4.4 SRAM Memories

For memories, the cost of the encoder and decoder is shared among many words, making it less critical. On the other hand, the number of parity bits has a direct impact on the cost and must therefore be minimized. To deal with multiple errors which are physically correlated, interleaving can be used [17]. Interleaving spreads the multiple errors among different words to try to ensure that each word contains only one bit error. The use of interleaving impacts area, delay and power, and complicates the memory design. Therefore, whether it is better to use interleaving and a simple ECC, or no interleaving with a more complex ECC, depends on the specific design. The block sizes for memories are smaller than for caches, with typical values of 16, 32 and 64 bits.

The large number of parity bits for these block sizes means that OLS codes are not an attractive option for SRAM memories. Shortened versions of DS and EG codes can be a good option for data blocks of 32 bits when strong error correction is required. However, SPC for error detection and SEC-DED codes for error correction remain the default choices for SRAM protection.

4.4.5 DRAM Memories

The considerations for DRAM memories are similar to those for SRAMs. The main difference is that in DRAMs, memory modules that are composed of a number of memory ICs are commonly used. In this case, the protection against a device failure is an interesting problem and can be solved by the use of SEC combined with interleaving [19] or by the use of more advanced codes like DS or RS codes. The extension of SEC-DED codes to protect against device failures has also been studied in [67].

Table 4.7 Suitable error correction codes for different circuit types

Circuit type	Error correction codes
Registers	SPC, SEC,SEC-DED
Register file	SPC, SEC,SEC-DED,OLS
Caches	SPC, SEC,SEC-DED,OLS,EG,DS
SRAM/DRAM memories	SPC, SEC,SEC-DED,EG,DS
Content Addressable Memories (CAMs)	SPC, SEC,SEC-DED
Interconnections	SEC,OLS

4.4.6 Content Addressable Memories (CAMs)

The protection of CAMs is different than that of SRAM and DRAM memories as discussed before. A parity bit (i.e., SPC code) can be used to provide single error protection against false positives. The use of ECCs has to be combined with a modified match line to also protect against false negatives [22]. In this case, SEC codes can be used. More complex codes can be used when multiple error correction is needed. In this regard, it is important to note that decoder latency and complexity are not an issue for CAMs, as the decoder is not needed in the scheme proposed in [22]. This means that codes with large decoding complexity or latency could eventually be used if the correction of more errors is required.

4.4.7 Interconnections

The correction of single and multiple errors has been considered for interconnections. In this case the data block sizes are similar to those of memories, with common values being 16, 32 and 64. The decoding has to be simple and must also have low latency. This makes OLS codes an attractive option, while shortened DS or EG codes can be a good option for interconnections of 32 bits. When a lower degree of error protection is sufficient, SEC codes are a good option.

4.4.8 Summary

Table 4.7 summarizes the ECCs which are suitable for the different circuit types. It can be observed that while SPC and SEC/SEC-DED codes are suitable for many of these circuits, the multi-bit ECCs described above can also be useful in a wide range of circuits.

4.5 Conclusions

This chapter has presented an overview of the use of error correction codes to protect digital circuits. First, the protection and performance requirements of different circuits have been discussed. Then several error correction codes have been presented. The description of the codes has focused on their error correction capabilities and their implementation. The objective is to provide the reader with an understanding of which codes can be used to protect which circuits, as well as the tradeoffs involved. To this end, estimates of decoding complexity and latency are provided for the different codes where possible. The more theoretical aspects of these codes are not discussed as they require a strong mathematical background and our focus here is on facilitating the practitioner who wishes to select an appropriate ECC to protect a given circuit.

The discussion above has covered commonly used SEC or SEC-DED codes as well as more advanced multi-bit error correction codes. In the latter case, codes which can be implemented with low latency have been selected for the exposition, as this is an important requirement in many circuits.

References

1. J. Segura, C.F. Hawkins, *CMOS Electronics: How It Works, How It Fails* (IEEE Press/Wiley Interscience, Hoboken, 2004)
2. R.C. Baumann, Radiation-induced soft errors in advanced semiconductor technologies. IEEE Trans. Dev. Mat. Rel. **5**(3), 301–316 (2005)
3. C. Constantinescu, Intermittent faults and effects on reliability of integrated circuits. Annual Reliability and Maintainability Symposium (RAMS), (2008), pp. 370–374
4. J.-F. Li, J.-C. Yeh, R.-F. Huang, C.-W. Wu, A built-in self-repair design for RAMs with 2-D redundancies. IEEE Trans. Very Large Scale Integr. Syst. **13**(6), 742–745 (2005)
5. Z. Ji, L. Lin, J.F. Zhang, B. Kaczer, G. Groeseneken, NBTI lifetime prediction and kinetics at operation bias based on ultrafast pulse measurement. IEEE Trans. Electron Dev. **57**(3), 228–237 (2010)
6. M. Nicolaidis, Design for soft error mitigation. IEEE Trans. Device Mat. Rel. **5**(3), 405–418 (2005)
7. A. Reddy, P. Banarjee, Algorithm-based fault detection for signal processing applications. IEEE Trans. Comput. **39**(10), 1304–1308 (1990)
8. C.L. Chen, M.Y. Hsiao, Error-correcting codes for semiconductor memory applications: a state-of-the-art review. IBM J. Res. Dev. **28**(2), 124–134 (1984)
9. S. Lin, D.J. Costello, *Error Control Coding*, 2nd edn. (Prentice-Hall, Englewood Cliffs, 2004)
10. G. Dong, N. Xie, T. Zhang, On the use of soft-decision error-correction codes in NAND flash memory. IEEE Trans. Circuits Syst. I Regular Papers **58**(2), 429–439 (2011)
11. J. F. Wakerly, *Digital Design Principles and Practices*, 4th edn. (Prentice Hall, Upper Saddle River, NJ, USA, 2006)
12. M. Fazeli, S. N. Ahmadian, S. G. Miremadi, A low energy soft error-tolerant register file architecture for embedded processors. 11th IEEE High Assurance Systems Engineering Symposium, (2008), pp. 109–116
13. S. Esmaeeli, M. Hosseini, B. V. Vahdat, B. Rashidian, A multi-bit error tolerant register file for a high reliable embedded processor. 18th IEEE International Conference on Electronics, Circuits and Systems (ICECS), (2011), pp. 532–537

14. A.R. Alameldeen, Z. Chishti, C. Wilkerson, W. Wu, S.-L. Lu, Adaptive cache design to enable reliable low-voltage operation. IEEE Trans. Comput. **60**(1), 50–63 (2011)
15. H. Sun, N. Zheng, T. Zhang, Leveraging access locality for the efficient use of multibit error-correcting codes in L2 cache. IEEE Trans. Comput. **58**(10), 1297–1306 (2009)
16. E. Ibe, H. Taniguchi, Y. Yahagi, K. Shimbo, T. Toba, Impact of scaling on neutron-induced soft error rate in SRAMs from a 250 nm to a 22 nm Design Rule. IEEE Trans. Electron Dev. **57**(7), 1527–1538 (2010)
17. S. Baeg, S. Wen, R. Wong, SRAM interleaving distance selection with a soft error failure model. IEEE Trans. Nucl. Sci. **56**(4 (part 2)), 2111–2118 (2009)
18. B. Schroeder, E. Pinheiro, W. -D. Weber, DRAM Errors in the wild: a large-scale field study. ACM SIGMETRICS/Performance (2009)
19. T. J. Dell, *A White Paper on the Benefits of Chipkill-Correct ECC for PC Server Main Memory*. IBM Microelectronics, (1997)
20. K. Pagiamtzis, A. Sheikholeslami, Content-addressable memory (CAM) circuits and architectures: a tutorial and survey. IEEE J. Solid-State Circ. **41**(3), 712–727 (2006)
21. S. Mukherjee, *Architecture Design for Soft Errors* (Morgan Kaufmann, Amsterdam, 2008)
22. K. Pagiamtzis, N. Azizi, F. N. Najm, A soft-error tolerant content-addressable memory (CAM) using an error-correcting-match scheme. IEEE Custom Integrated Circuits Conference, pp. 301–304, Sept 2006
23. S. Gregori, A. Cabrini, O. Khouri, G. Torelli, On-chip error correcting techniques for new-generation flash memories. Proc. IEEE **91**(4), 602–616 (2003)
24. Y. Maeda, H. Kaneko, Error control coding for multilevel cell flash memories using nonbinary low-density parity-check codes. IEEE International Symposium on Defect and Fault Tolerance in VLSI Systems, (2009), pp. 367–275
25. A. Jiang, J. Bruck, Data representation for Flash memories, book chapter in *Data Storage*. ISBN: 978-953-307-063-6, In-Tech Publisher, (2010)
26. Q. Huang, S. Lin, K.A.S. Abdel-Ghaffar, Error correcting codes for flash coding. IEEE Trans. Inform. Theory **57**(9), 6097–6108 (2011)
27. J. Wang, T. Courtade, H. Shankar, R. D. Wesel, Soft information for LDPC decoding in flash: mutual-information optimized quantization. IEEE Global Telecommunications Conference, (2011)
28. A. Ganguly, P.P. Pande, B. Belzer, Crosstalk-aware channel coding schemes for energy-efficient and reliable NOC interconnects. IEEE Trans. Very Large Scale Integr. (VLSI) Syst. **17**(11), 1626–1639 (2009)
29. S.E. Lee, Y.S. Yang, G.S. Choi, W. Wu, R. Iyer, Low-power, resilient interconnection with Orthogonal Latin Squares. IEEE Des. Test. Comput. **28**(2), 30–39 (2011)
30. R.H. Morelos-Zaragoza, *The Art of Error Correcting Coding*, 2nd edn. (Wiley, Chichester, 2006)
31. J.J. Metzner, Convolutionally encoded memory protection. IEEE Trans. Comput. **31**(6), 547–551 (1983)
32. K. Rokas, Y. Makris, D. Gizopoulos, Low cost convolutional code based concurrent error detection in FSMs. IEEE International Symposium on Defect and Fault Tolerance in VLSI, (2003), pp. 344–351
33. E.S. Fetzer, D. Dahle, D.C. Little, K. Safford, The Parity protected, multithreaded register files on the 90-nm itanium microprocessor. IEEE J. Solid-State Circ. **41**(1), 246–255 (2006)
34. Z. Chaohuang, N. Saxena, E. J. McCluskey, Finite state machine synthesis with concurrent error detection. International Test Conference, (1999), pp. 672–679
35. M. Nicolaidis, R.O. Duarte, Fault-secure parity prediction Booth multipliers. IEEE Des. Test. Comput. **16**(3), 90–101 (1999)
36. G. C. Cardarilli, M. Ottavi, S. Pontarelli, M. Re, A. Salsano, Error detection in signed digit arithmetic circuit with parity checker. IEEE International Symposium on Defect and Fault Tolerance in VLSI Systems, (2003), pp. 401–408
37. F. Vargas, M. Nicolaidis, SEU-tolerant SRAM design based on current monitoring. Twenty-Fourth International Symposium on Fault-Tolerant Computing, (1994), pp. 106–115

38. R.W. Hamming, Error detecting and error correcting codes. Bell Syst. Tech. J. **29**, 147–160 (1950)
39. M.Y. Hsiao, A class of optimal minimum odd-weight column SEC-DED codes. IBM J. Res. Dev. **14**(4), 395–401 (1970)
40. M. Richter, K. Oberlaender, M. Goessel, New linear SEC-DED codes with reduced triple bit error miscorrection probability. IEEE On-Line Testing Symposium, (2008), pp. 37–42
41. A. Dutta, N. A. Touba, Multiple bit upset tolerant memory using a selective cycle avoidance based SEC-DED-DAEC Code. 25th IEEE VLSI Test Symposium, (2007), pp. 349–354
42. X. She, N. Li, D.W. Jensen, SEU tolerant memory using error correction code. IEEE Trans. Nucl. Sci. **59**(1), 205–210 (2012)
43. M.A. Bajura et al., Models and algorithmic limits for an ECC-based approach to hardening sub-100-nm SRAMs. IEEE Trans. Nucl. Sci. **54**(4), 935–945 (2007)
44. S. Satoh, Y. Tosaka, S.A. Wender, Geometric effect of multiple-bit soft errors induced by cosmic ray neutrons on DRAMs. IEEE Electron Dev. Lett. **21**(6), 310–312 (2000)
45. C. Wilkerson, A. R. Alameldeen, Z. Chishti, W. Wu, D. Somasekhar. S. Lu, Reducing cache power with low cost, multi-bit error-correcting codes. International Symposium on Computer Architecture, pp. 83–93, June 2010
46. P. Ankolekar, S. Rosner, R. Isaac, J. Bredow, Multi-bit error correction methods for Latency-Contrained flash memory systems. IEEE Trans. Dev. Mat. Rel. **10**(1), 33–39 (2010)
47. G.C. Cardarilli, A. Leandri, P. Marinucci, M. Ottavi, S. Pontarelli, M. Re, A. Salsano, Design of a fault tolerant solid state mass memory. IEEE Trans. Rel. **52**(4), 476–491 (2003)
48. S. Jeon, E. Hwang, K. V. Kumar, M. K. Cheng, LDPC codes for memory systems with scrubbing. IEEE Global Telecommunications Conference (GLOBECOM), (2010)
49. D. Strukov, The area and latency tradeoffs of binary bit-parallel BCH decoders for prospective nanoelectronic memories. *Proceedings of 2006 Asilomar Conference on Signals Systems and Computers*, pp. 1183–1187, Oct 2006
50. M.Y. Hsiao, D.C. Bossen, R.T. Chien, Orthogonal Latin Square codes. IBM J. Res. Dev. **14**(4), 390–394 (1970)
51. J. Dénes, A.D. Keedwell, *Latin Squares and Their Applications* (Academic, New York, 1974)
52. R. Datta, N. A. Touba, Generating burst-error correcting codes from orthogonal Latin Square codes – A Graph Theoretic Approach. *IEEE International Symposium Defect and Fault Tolerance in VLSI and Nanotechnology Systems* (DFT), (2011), pp. 367–373
53. S. Ghosh, P. D. Lincoln, Dynamic low-density parity check codes for fault-tolerant nano-scale memory. *Proceedings of Foundations of Nanoscience* (FNANO '07) (Snowbird, Utah, 2007)
54. S. Ghosh, P. D. Lincoln, Low-density parity check codes for error correction in nanoscale memory. SRI Computer Science Laboratory Technical Report, CSL-0703, (2007)
55. H. Naeimi, A. DeHon, Fault secure encoder and decoder for memory applications. *Proceedings of the IEEE International Symposium on Defect and Fault Tolerance in VLSI Systems*, (2007)
56. H. Naeimi, A. DeHon, Fault secure encoder and decoder for nanoMemory applications. IEEE Trans. Very Large Scale Integr. Syst. **17**(4), 473–486 (2009)
57. P. Reviriego, M. Flanagan, J. A. Maestro, Efficient multibit error correction for memory applications using Euclidean Geometry codes. *Proceedings of the RADECS 2011 Conference*, (2011), pp. 160–163
58. S. Liu, P. Reviriego, J.A. Maestro, Efficient majority logic fault detection with difference-set codes for memory applications. IEEE Trans. Very Large Scale Integr. (VLSI) Syst. **20**(1), 148–156 (2012)
59. P. Reviriego, M. Flanagan, J.A. Maestro, A (64, 45) triple error correction code for memory applications. IEEE Trans. Dev. Mat. Rel. **12**(1), 101–106 (2012)
60. P. Reviriego, M. Flanagan, S. Liu, J.A. Maestro, Multiple cell upset correction in memories using difference set codes. IEEE Trans. Circuits Syst. I **59**(11), 2592–2599 (2012). ISSN:1549-8328
61. P. Reviriego, C. Argyrides, J.A. Maestro, Efficient error detection in double error correction BCH codes for memory applications. Microelectron Rel. **52**(7), 1528–1530 (2012)

62. P. Reviriego, J.A. Maestro, M. Flanagan, Error detection in majority logic decoding of Euclidean geometry low density parity check (EG-LDPC) codes. IEEE Trans. Very Large Scale Integr. Syst. **21**(1), 156–159 (2013)
63. P. Reviriego, M. Flanagan, S. Liu, J.A. Maestro, Error-detection enhanced decoding of difference set codes for memory applications. IEEE Trans. Dev. Mat. Rel. **12**(2), 335–340 (2012)
64. Z. Wang, M.G. Karpovsky, K. Kulikowski, Design of memories with concurrent error detection and correction by non-linear SEC-DED codes. J. Electron. Test. **26**, 559–580 (2010)
65. C. Chen, C. Wu, An adaptive code rate EDAC scheme for random access memory Design, Automation & Test in Europe Conference & Exhibition (DATE), (2010), pp. 735–740
66. D. Yoon, M. Erez, Virtualized ECC: flexible reliability in main memory. IEEE Micro **31**(1), 11–19 (2011)
67. S. Pontarelli, G. C. Cardarilli, M. Re, A. Salsano, Error correction codes for SEU and SEFI tolerant memory systems. IEEE International Symposium on Defect and Fault Tolerance in VLSI Systems, (2009), pp. 425–430

Chapter 5
System-Level Design Methodology

Rishad A. Shafik, Bashir M. Al-Hashimi, and Krishnendu Chakrabarty

5.1 Multiprocessor System-on-Chip

Continued technology scaling has enabled the fabrication of ever more efficient and low power electronic devices for current and future generation of embedded systems. Examples of such devices include IBM's 22-nm [33] and Intel's emerging 16-nm [63] devices with promises to provide with unprecedented integration capacity and performance. However, with these technological advances, design complexity is also increasing significantly with emerging challenges related to performance, power and reliability. To address the design complexity while meeting power and performance requirements of modern applications, recently multiprocessor system-on-chip (MPSoC) has emerged as a popular embedded systems platform [44]. An MPSoC contains multiple processing elements on a single piece of silicon, each with an assigned task to define an expected application domain. The inclusion of multiple processing elements in MPSoCs has a number of benefits. These include parallel processing, low clock speed, low power consumption, etc. [72].

Depending on the type of processing elements and nature of application, MPSoCs can be either homogeneous or heterogeneous. Figure 5.1 shows example homogeneous and heterogeneous MPSoCs with processing elements and memories. As can be seen, the homogeneous MPSoC has three RISC processors (Fig. 5.1a).

R.A. Shafik (✉)
University of Bristol, Bristol, BS8 1UB, UK
e-mail: rishad.shafik@bristol.ac.uk

B.M. Al-Hashimi
School of Electronics and Computer Science, University of Southampton, Southampton, SO17 1BJ, UK
e-mail: bmah@ecs.soton.ac.uk

K. Chakrabarty
Department of Electrical and Computer Engineering, Duke University of Bristol, Durham, NC, USA
e-mail: krish@ee.duke.edu

J. Mathew et al. (eds.), *Energy-Efficient Fault-Tolerant Systems*,
DOI 10.1007/978-1-4614-4193-9_5,

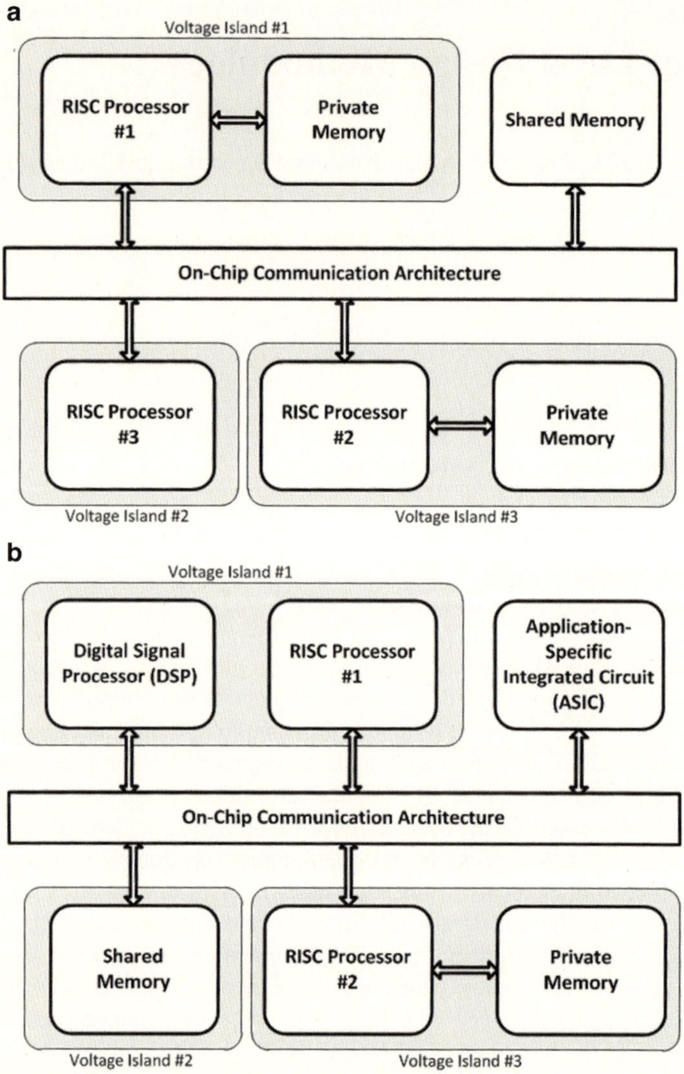

Fig. 5.1 Example MPSoCs with processing elements inter-connected by on-chip communication architectures: (**a**) a homogeneous MPSoC with similar RISC processors, (**b**) a heterogeneous MPSOC with different processing elements, both organized in different voltage islands

Due to similar instruction set architectures (ISA) homogeneous processing cores have the advantage of easier inter-core communication and mapping or migration of tasks between them. However, depending on the nature of applications often MPSoCs have different processing elements in a single piece of silicon, known as heterogeneous MPSoC. From Fig. 5.1b it can be seen that the heterogeneous MPSoC has two different kinds of processors: digital signal processor (DSP) and RISC processors. Due to different ISAs between DSP and RISC processors, communication between them can be non-trivial. Moreover, mapping and migration of tasks among the processors may require sophisticated task based modeling and virtualization techniques [51]. From Fig. 5.1 it can also be seen that depending on computational and communication requirements of the processing cores, memories can be allocated as either private (i.e. mapped and used by a particular processing core) or shared (i.e. used by multiple processing cores).

The interconnection of processings elements and memories within the MPSoC is controlled by on-chip communication architecture, which defines how inter-component communication takes place (Fig. 5.1). Since inter-component communication is crucial for system functionality and parallelism, on-chip communication architecture greatly influences the underlying performance of the system. Major MPSoC on-chip communication architectures are briefly introduced in Sect. 5.1.1.

Since MPSoCs enable modular design and layout, it is often advantageous to allow one or more processing and memory cores to share a common operating voltage, otherwise known as voltage islands. These voltage islands can be conveniently controlled in MPSoCs to reduce power consumption through dynamic voltage and frequency (DVFS) scaling. Operating at various supply voltages is very common in today's MPSoCs, which are designed specifically with low power requirements [52]. To achieve low power and high performance objectives, MPSoCs are carefully designed with higher inter-core parallelism. However, low power design of these systems can be highly challenging as reliability is seriously degraded due to significant increase in the number of hardware faults caused by electromagnetic radiations, among others [58]. MPSoC design challenges and system-level design methodologies to effectively address them are further described in Sect. 5.1.2.

5.1.1 On-Chip Communication Architectures

On-chip communication architecture facilitates communication between components within an MPSoC. Depending on how inter-component connections are laid out, on-chip communication architectures can be of three major types, as shown in Fig. 5.2: point-to-point (P2P), on-chip shared bus and network-on-chip (NoC). P2P on-chip communication architecture provides with dedicated interconnection between each communicating component within an MPSoC (Fig. 5.2a). Such interconnection gives high performance as inter-component communications are non-blocking. However, scalability of P2P architecture is poor as adding extra components can make inter-component connectivity complex, often introducing prohibitively higher layout overheads [39]. Hence, the use of P2P architectures is limited in MPSoCs [59]. On-chip bus provides with a shared communication architecture between number of MPSoC components (Fig. 5.2b). Unlike P2P architectures, on-chip shared bus provides with higher scalability and connectivity as adding more components can be easily incorporated through further bus extensions. However, due to shared and blocking communication with mutually

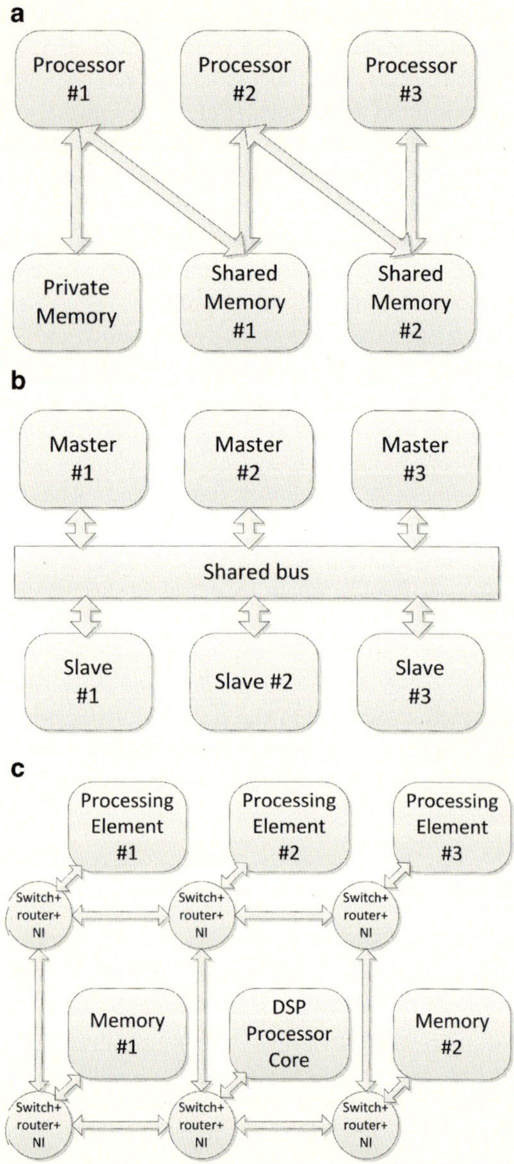

Fig. 5.2 Block diagram different MPSoC on-chip communication architectures: (**a**) point-to-point (P2P), (**b**) shared bus and (**c**) network-on-chip (NoC)

exclusive access to the interconnect, on-chip bus introduces performance overheads for inter-component communication in MPSoCs. This performance overhead can increase further as the number of communicating components in MPSoC scale higher. The Advanced Microprocessor Bus Architecture (AMBA) [5] and Advanced eXtensible Interface (AXI) [6] are dominant, industrial standard on-chip bus architectures for today's MPSoCs. To meet the demand of increased performance and scalability of current and future generations of MPSoCs, recently network-on-chip (NoC) communication architectures have been proposed. NoCs are similar to P2P except that they organize the connections through modular packet-based link-to-link interfaces with switches and routers laid in between them. Due to high level of modularity, NoCs have advantages of high scalability and performance at the expense of silicon area and complexity [20, 49]. To date a great deal of research works have already been carried out to explore efficient and reliable NoC topologies and routing techniques [38, 58, 68].

5.1.2 MPSoC Design Challenges

MPSoCs are becoming *de facto* standards for system-on-chip design due to its various advantages related to performance and modularity [34]. The performance advantage is realized through overlapped processing among the processing elements, which depends on the underlying on-chip communication architecture (see Sect. 5.1.1). These interconnected processing elements together with the memories and other peripherals are often organized in a modular design hierarchy in MPSoCs with an aim to facilitating design scalability and reducing time-to-market. However, despite these advantages, design of MPSoCs is highly challenging. In the following major challenges are discussed in further detail.

5.1.2.1 System Gap

Traditional system development introduces system gap due to separate hardware and software design and synthesis flow as shown in Fig. 5.3. As can be seen, initially the target system in partitioned into hardware and software components after systematic evaluations. The development of hardware components involves initial specifications, followed by simulations with hardware description language (HDL) modeling. The HDL model is then sufficiently validated for its functionality and then synthesized to generate actual hardware prototype. To meet the initial hardware specifications, the development steps may involve iterations between HDL based design, design validation and synthesis. Similar to hardware, software development also starts with initial specifications. Based on these specifications, initial software prototype is developed using systematic application models (e.g. task graphs and algorithmic models). The prototype is then tested and validated with sufficient constraints and portability checks with the target hardware. With both hardware and software prototypes, the system implementation proceeds with further testing and validations. This is then followed by iterative system optimizations to meet the original system constraints. Finally, when system constraints are met, market-ready system design is delivered for fabrication and customer shipment. As expected, due to separate hardware and software design and synthesis, a system gap exists in the early design stages. This system gap makes the design of MPSoCs highly time-consuming and challenging later in the design stage to optimize the design according

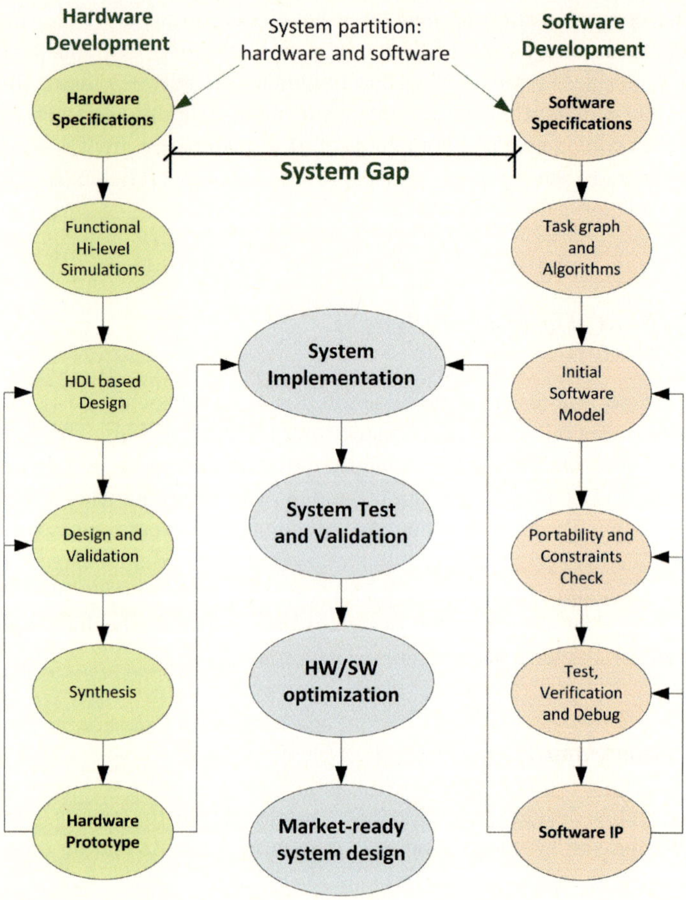

Fig. 5.3 Traditional system development steps showing separate hardware and software design and synthesis flow

to the original specifications and objectives. To make matters worse, this system gap is widening further with ever evolving complexity of hardware and software design and synthesis. Clearly, bridging this gap through system-level design methodology is a major challenge for MPSoCs, which requires extensive design space exploration early in the design phase to expedite the design optimization and integration.

5.1.2.2 Programming Model and Prototyping

The demand of performance-oriented parallel multiprocessing in MPSoCs have necessitated a design paradigm shift from computation-centric to communication-centric approaches. In communication-centric approaches, software routines and their inter-dependence need to be explicitly modeled considering the impact of underlying hardware resources (i.e. processing elements and on-chip communication architectures). Modeling such impact is critical for achieving concurrency between different software routines. However, programming such model can be tricky for the following reasons [25,44]:

- Traditional programming tools are invariably sequential in their nature. Hence, modeling concurrency between software routines can be non-trivial.
- In symmetric multiprocessing (SMP) systems, multiple processing elements share a common view of memory. Modeling memory and interconnect arbitration for such systems can be tricky.
- In asymmetric multiprocessing (AMP) systems, processing elements tend to be loosely coupled from each other with local memories. Modeling communication between processing elements for AMPs require emulating complicated interface protocols.
- To avoid all possible unwanted on-chip communication scenarios leading to asynchronizations, deadlocks, livelocks etc. programming model must reflect the actual on-chip communication behaviour. In particular, this can be highly daunting when processing elements have different on-chip interconnects and arbitration requirements between them.
- Debugging and validation in such programming model are other major challenges. To account for all possible physical scenarios, it is important that sufficient debugging and validation is carried out with the system model. Currently, there is a need for multiprocessing-enabled debugging and validation environment for MPSoCs.
- Accurate computation and communication modeling and prototyping of heterogeneous MPSoCs is highly complex. In particular, modeling computation and communication tasks with various scenarios and possibilities can be highly challenging for design space exploration.

5.1.2.3 Design Optimization

Low power consumption is a prime design requirement for MPSoCs. Dynamic voltage and frequency scaling (DVFS) is an effective power minimization technique often employed in hand-held devices to extend the battery life [2]. DVFS technique works by lowering the processor voltage and frequency according to its workload to achieve power reduction [3,57]. However, it has been reported that the reduction of supply voltage causes an exponential increase in the rate of soft errors, particularly that of single-event upsets (SEUs), leading to degradation of reliability [14,15]. This is further exacerbated by device miniaturization and continuing technology scaling, which causes hardware faults due to systematic abberations (e.g. imperfections in lithographic patterning and random effects like dopant density fluctuations) during manufacturing process [73] and electromigration during operating lifetime [18]. As a result, reliable design of MPSoCs is an emerging challenge that has been acknowledged widely [14,15,43,45,64,76].

Traditionally power-aware fault-tolerance or soft error hardening have been employed in MPSoCs using redundancy based techniques. This technique is generally effective in that the redundant resources provide a voting system from multiple resources to produce a single output in the presence of soft errors or faults [69]. Over the years, a number of redundancy-based techniques have been reported at various levels of design abstraction. The triple modular redundancy (TMR) is the most basic circuit-level redundancy technique, such as [1, 69]. The fault-tolerance is achieved through TMR technique using three hardware elements to incorporate voting a single output from multiple outputs. Due to the increase in hardware resources, large overhead in terms of power consumption and chip area is incurred using such technique [4, 8]. Another effective technique in terms of power consumption is the time redundancy, such as [10, 47]. Such technique employs multiple instances of execution to achieve fault-tolerance. Although no overhead in terms of hardware resources or area is caused, this technique has overhead in terms of performance. Information redundancy proposed in [54, 71] is also an effective fault-tolerance technique. The main idea is to append error-detection and error-correction codes along with the usual information bits to increase reliability of the system [54]. The addition of these extra codes add to the communication and computation overhead, while improving the reliability in the presence of soft errors [23]. Recently joint time and information redundancy-based technique has been proposed in a number of publications, such as [22, 35]. Using such technique has advantages of high reliability or fault-tolerance at low cost and low overhead.

Alternatives to the redundancy-based technique are the re-execution and replication of computational tasks among idle processing elements. Using this technique, no overhead in incurred in terms of extra execution time. For example, in [16, 65] low power fault-tolerance technique has been presented utilizing the idle processing elements for duplicating some of the computations. Other flexible techniques to achieve high fault-tolerance are the pre-emptive on-line scheduling, such as [75] and check-pointing during slack times between tasks, such as [36]. Using these techniques, high fault-tolerance can be achieved at the cost of increased complexity in the design of MPSoC application. However, the effectiveness of these techniques depends upon predictability of slack times, which incur large overheads and often leads to problems related to unschedulability [31]. Also, achieving fault-tolerance using check-pointing technique is limited by the schedulability and the number of check-points. The optimal number of check-points that can be inserted and scheduled in the presence of a given number faults is determined by the worst case execution time of a task [42].

Recently, researchers have shown combination of different fault-tolerance techniques to reduce system overhead for fault-tolerant and low power design. For example, fault-tolerance-based optimization of cost-constrained distributed real-time systems has been proposed in [31]. The fault-tolerance in [31] is achieved through mapping and assignment of different fault-tolerance policies to processes. Another fault-tolerant design using process re-execution and re-scheduling of low power heterogeneous MPSoC applications has been proposed in [52]. The power minimization is achieved through scheduling of voltage levels to different processes and the fault-tolerance is achieved by deciding the start times of processes and the transmission times of messages in the presence of faults. In [27] a dynamic fault-tolerance technique is presented using independent task sets scheduling with precedence relationship in MPSoC systems. Due to the use of such scheduling, the fault-tolerance technique in [27] benefits from less communicational complexity and better scheduling performance in terms of power consumption.

Traditionally power-aware fault-tolerant design techniques consider low power and reliability as two separate objectives [16, 31, 75]. For effective design optimization with low power and improved reliability as a joint objective, further studies are needed to understand reliability of applications, particularly from system- and application-level design perspective.

5.1.2.4 Runtime System Management

Operating conditions in many applications often require managing the MPSoC hardware and software resources during runtime. For example, when a particular task in an application is performance critical, higher voltage and frequency scaling may need to applied, or task may need to be mapped dynamically on a processing core with higher performance. Likewise, when a particular task requires high reliability, it may need to be mapped in multiple processing cores to ensure efficient fault detection and tolerance. With such requirements, runtime system management is of crucial importance in MPSoCs to ensure system requirements are met at all times. Various studies suggest that runtime system management has the potential to reduce the system overheads during runtime and increase system adaptability and lifetime substantially [48, 74]. However, such management often requires runtime optimization in NP-hard decision space and can potentially incur large management overheads [56].

A crucial aspect in runtime system management is the ability to accurately estimate the power, performance and reliability of the MPSoC and its different components [13]. Such estimation can be carried out by using separate sensors [74] or by using monitor modules alongside original components [46]. Runtime estimation gives an insight into comparative analysis of different components and identifies different scenarios for system adaptation and management [12]. Estimation at circuit- or device-level [46], or at architectural-level [4, 66] can be computationally extensive for complex circuits. Hence, there is increased interest in system-level estimation of different system metrics for MPSoCs [30, 62].

5.2 Hardware/Software Co-design

Over the years, various system-level design techniques and methodologies have been devised both by academia [11] and industries [17, 67]. Among them, hardware/software (HW/SW) co-design is particularly of high interest as MPSoCs

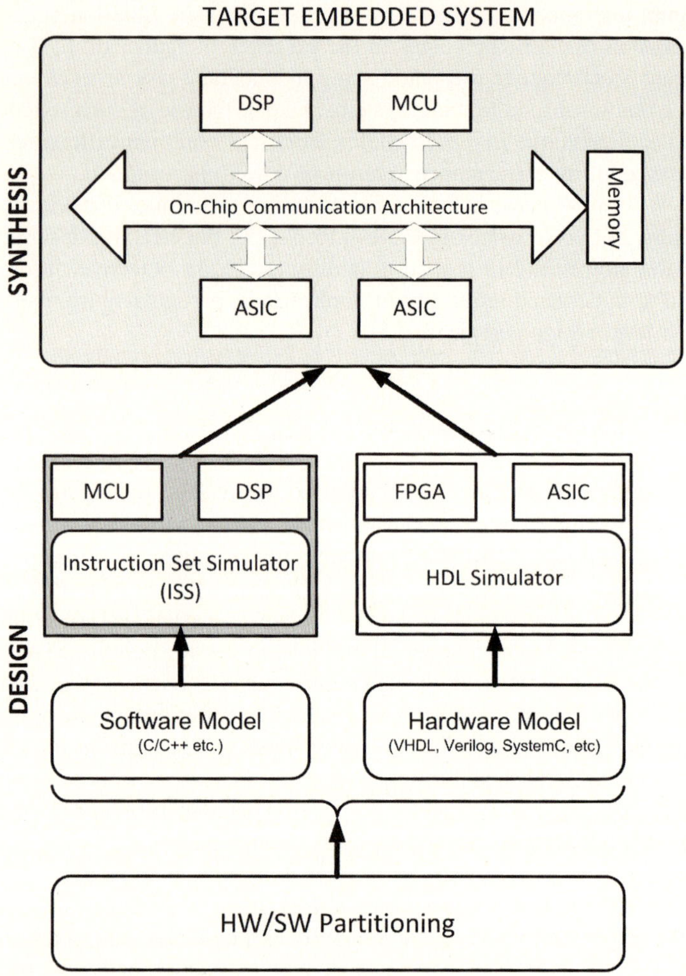

Fig. 5.4 Block diagram of hardware/software co-design and co-synthesis

are intrinsically based on software routine executing on a set of hardware resources, including processing elements and on-chip communication resources. HW/SW co-design describes the techniques for simultaneous design and synthesis with systematic combination and co-operation between hardware and software components [37]. Such design process extracts the benefits of both hardware- and software-based design and reduces time-to-market greatly [21, 26].

Figure 5.4 shows a block diagram illustrating major steps involved in HW/SW co-design. The process is initiated by partitioning of the HW/SW tasks within the embedded system (Fig. 5.4). Once a required partitioning is found, specifications for hardware (FPGA, ASIC, etc.), software (MCU, DSP, etc.) and their interfaces

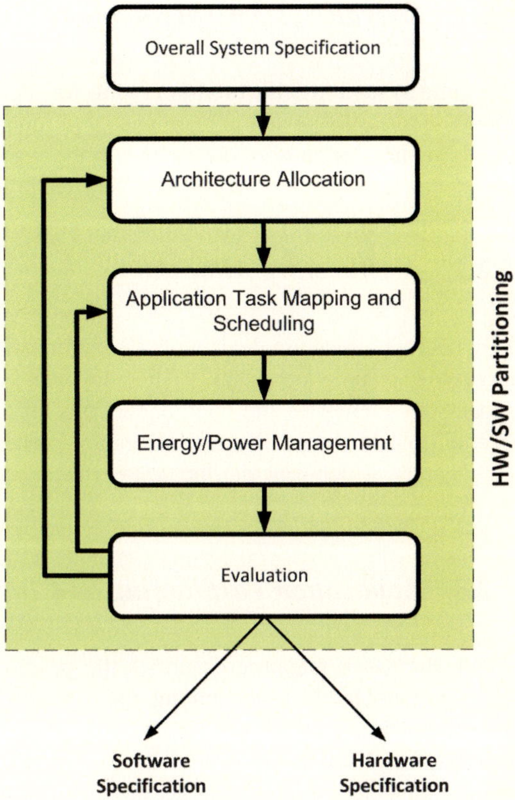

Fig. 5.5 Flowchart of hardware/software co-design steps

are defined to enable design of the prototype hardware and software components (Fig. 5.4). During design phase, software based implementations are co-simulated and interfaced with hardware based implementations. Often this is done within a real-time operating system (RTOS) environment, which emulates a multi-tasking environment for real-time applications. Finally, the target architecture is found by HW/SW co-synthesis process, which derives a mixed HW/SW implementation from the high-level hardware and software specification. To achieve optimization for target embedded system with different objectives, the whole process in Fig. 5.4 may need to be repeated several times. HW/SW co-design requires a combination of instruction set simulators and hardware emulators to effectively enable design space exploration. Over the years a number of HW/SW co-design tools have been proposed, such as COSYN [21], POLIS [9], MPARM [11].

HW/SW co-design process involves various system-level design steps to effectively allocate and partition the hardware/software communication and computation tasks and schedule them, as shown in Fig. 5.5. In the following, these steps are briefly detailed.

5.2.1 Overall System Specification

Accurate system specification is crucial for identifying the target system requirements in terms of power, performance and reliability. System specification affects the following design steps depending on the evaluated requirements.

5.2.2 Architecture Allocation

Architecture allocation deals with allocation of different design components for a target MPSoC application [57]. This may involve selection of number and types of processors, memories and interconnections for the application under concern. The overall goal of the architecture allocation process is to identify the most suitable architecture, which provides the best performance and cost under given constraints.

5.2.3 Application Partitioning and Task Mapping

Application task mapping describes the process of distributing the computation and communication tasks among the allocated processing and communicating elements [40]. The mapping considerations may include different performance and cost involved with each task. The mapping process explicitly determines the implementation related issues in hardware and software. Determining a good mapping is of crucial importance as inappropriately mapped tasks for a given allocated architecture may under-utilize its performance and increase the system costs [7, 29].

5.2.4 Task Scheduling

Task scheduling generally follows task mapping, which defines the order of execution of the different tasks, such that timing constraints are satisfied. It is generally carried out through operating system routines. Various scheduling algorithms have been proposed over the years considering various design objectives, including energy [57], performance [28] and reliability [31].

5.2.5 Energy Management and Evaluation

Architecture allocation, application task mapping and task scheduling have direct impact on performance and power consumption of MPSoC [57, 61]. Further power minimization can be achieved using DVFS technique and dynamic power management using slack time management (see Chap. 3). The power management is then followed by system evaluation, which may repeat the previous design steps to achieve optimized HW/SW partitioning and design.

5.3 Case Study: Power- and Reliability-Aware Design Optimization

To demonstrate the effectiveness of system-level design methodology in MPSoC application optimizations, a case study is investigated in detail. In particular, the impact of application task mapping, which is a crucial system-level design step of MPSoCs, is extensively studied in terms of MPSoC reliability (see Sect. 5.2.3 for details related to application task mapping). Based on this study, a system-level design optimization technique for MPSoC application through joint power minimization and reliability improvement is developed. Power minimization in the design optimization is carried out using voltage scaling technique, while reliability improvement is achieved through careful choice of application task mapping on the homogeneous MPSoC processing cores. The case study is outlined as follows. Section 5.3.1 presents MPSoC system model used, which is then followed by the study of application task mapping impact on reliability of MPSoC application in Sect. 5.3.2. Section 5.3.3 outlines a design optimization technique based on this study and Sect. 5.3.4 demonstrates the effectiveness of the technique through experimental observations.

5.3.1 System Model

In this section, the models of the system architecture, application and fault injection used in this study are introduced.

5.3.1.1 Architecture Model

In this study, an MPSoC architecture, A, based on 2D-mesh network-on-chip (NoC) with C processing cores is considered. Due to its high performance and scalability [39], NoC-based MPSoC architecture is gaining popularity and a number of academic or industrial designs have been proposed to date, such as xPIPES [19], Intel 80-core [70]. Figure 5.6 shows an example MPSoC architecture consisting of four processing cores. As shown, each block (or NoC tile) consists of a processing core and switch (Fig. 5.6). Each processing core consists of, among others, ARM7TDMI processor, an instruction cache (8 kbits), a data cache (16 kbits) and a dedicated private memory (256 kbits). Network interface attached to each

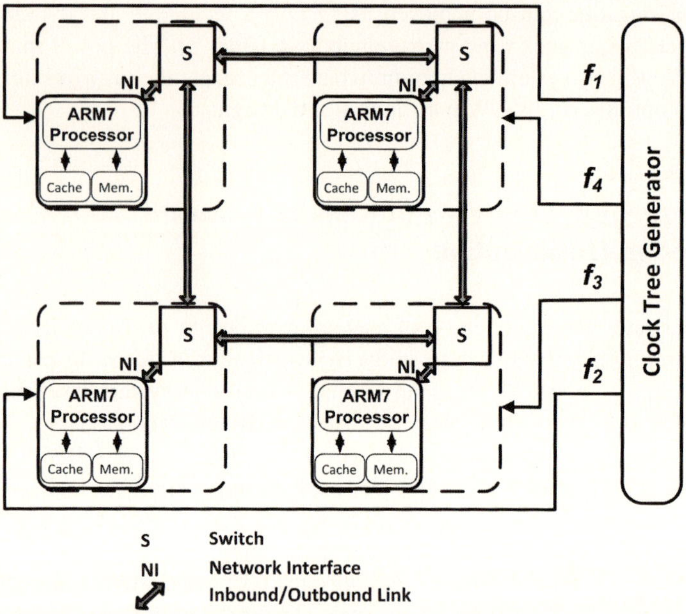

Fig. 5.6 MPSoC architecture with four processing cores and power minimization support through clock tree generator

processor (Fig. 5.6) incorporates packet-based communication with 32-bit payload and switches carry out inter-core packet-based communication with XY routing, chosen due to its performance and simplicity [39]. The cache and memory sizes have been chosen to provide high availability of data and parallelism among the processing cores. To introduce power minimization through voltage scaling in the MPSoC architecture, a clock tree generator has been included to provide different clock frequencies for the processing cores through voltage scaling (Fig. 5.6).

For a given voltage scaling, the dynamic power, P, of a processing core is given as

$$P = \alpha C_L f {V_{dd}}^2, \tag{5.1}$$

where C_L is the processor load capacitance per cycle, V_{dd} is the supply voltage, f is the operating frequency and α is the processor activity factor ($0 \leq \alpha \leq 1$). The voltage scaling technique effectively reduces the power consumption, defined by (5.1), by reducing V_{dd} and f through scaling. For ARM7TDMI processor, the relationship between V_{dd} (in volts) and f (in MHz) is given by [53] as

$$V_{dd}(f, s) = \left[0.1667 + \frac{4.1667 \times f}{10^3 \times s}\right], \tag{5.2}$$

Table 5.1 Operating frequency, f, and supply voltage, V_{dd}, for different voltage scaling of ARM7TDMI processor

Scaling, s	f, MHz	V_{dd}, V
1	200	1.00
2	100	0.58
3	66.7	0.44

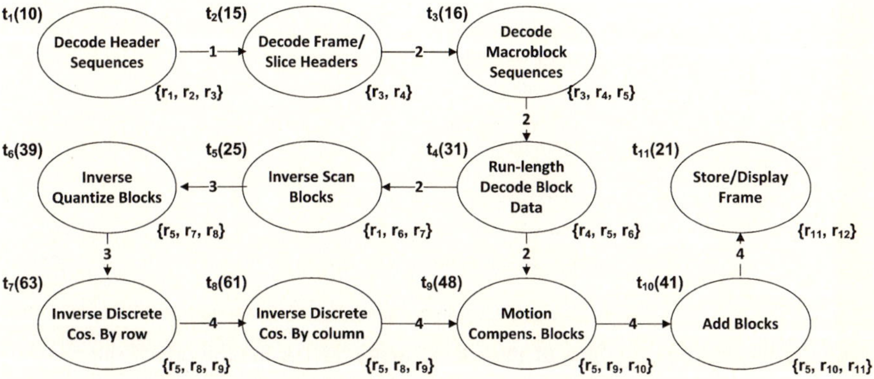

Fig. 5.7 MPEG-2 video decoder task graph with 11 tasks and associated register resources

where s is the frequency scaling constant. Table 5.1 shows the three voltage scaling options (with $s = 1, 2$ and 3 in (5.2)) used in this study. The impact of choice of voltage scaling levels on design optimization is discussed in Sect. 5.3.4.

5.3.1.2 Application Model

An application is modelled as a directed, acyclic task graph $\mathscr{G}(\mathscr{V}, \mathscr{E})$ with N nodes. Each node $t_j \in \mathscr{V}$ represents j-th computational task within the application and each edge $d_{j,k} \in \mathscr{E}$ represents inter-task communication and data dependency between j-th and k-th tasks ($j, k = 1{:}N$, where N is the number of computational tasks in the task graph). An application is realized on the MPSoC architecture by distributing the computation and communication tasks among the processing cores and their interconnects through application task mapping. Figure 5.7 shows an example task graph of MPEG-2 video decoder using 11 tasks and their associated register resources. The computational and communication costs of tasks are shown with numbers on the nodes and edges, Fig. 5.7. The computational cost represents execution time of each task and the communication cost represents the time required to transfer data between tasks (actual costs are approximate multiples of 5.5×10^6 clock cycles). The computational and communication costs are obtained using SystemC cycle-accurate simulations assuming 32-bit inter-core transfer. The computational tasks are modelled as separate task processes, while the communication between tasks is modelled as message passing queues. The communication cost

Table 5.2 Register usage of MPEG-2 video decoder tasks (Fig. 5.7) and their approximate sizes

Register set	Type	Size, bits/cycle
r_1	Scan tables and variables	4,132
r_2	Sequence types	1,124
r_3	Header sequence variables (before decoding)	640
r_4	VLC tables and variables	5,124
r_5	Header sequence variables (after decoding)	2,134
r_6	Video blocks (coded)	12,288
r_7	Scanned video blocks	13,218
r_8	Quantized video blocks	13,132
r_9	Picture ready video blocks (before motion compensation)	13,274
r_{10}	Motion compensated video data and variables	13,326
r_{11}	Decoded and motion compensated video data	13,174
r_{12}	Display/storage ready video data structure	12,288

is found by dividing the size of inter-task queue by the bandwidth of the channel (in bits per cycle). Similar evaluation of computation and communication costs has also been used in [31, 50].

Also attached with each node is a set of application registers showing the register usage by the computational tasks. Due to inter-dependent nature of the tasks of an application, the tasks share register resources among themselves. For example, the task t_1 uses the set of registers r_1, r_2 and r_3, while the task t_2 uses the set of registers r_3 and r_4. Note that between these two tasks, r_3 is shared. Table 5.2 shows the different register sets used by the MPEG-2 video decoder tasks along with their types and approximate sizes (obtained by using variable or signal type tags within the SystemC simulation environment). The actual register usage of the i-th processing core ($i = 1 : \mathscr{C}$, where $\mathscr{C}$ is the number of processing cores of an MPSoC), R_i, is found through SystemC simulation after the application tasks and their associated registers (Table 5.2) are mapped on the processing cores of an MPSoC. The R_i is given as

$$R_i = \frac{1}{T_i} \sum_{t=1}^{T_i} R_{i,t}, \qquad (5.3)$$

where $R_{i,t}$ is the register usage (in bits) at t-th clock cycle of the i-th processing cores. The R_{i_t} in (5.3) depends on the number of tasks mapped with associated resources (Table 5.2).

5.3.1.3 Fault Injection Model

In this study, fault injection is carried out using SEU-based fault model employing the fault injection simulator proposed in [60]. Using this technique the injection of SEUs is initiated through replacement of variable or signal types in the original

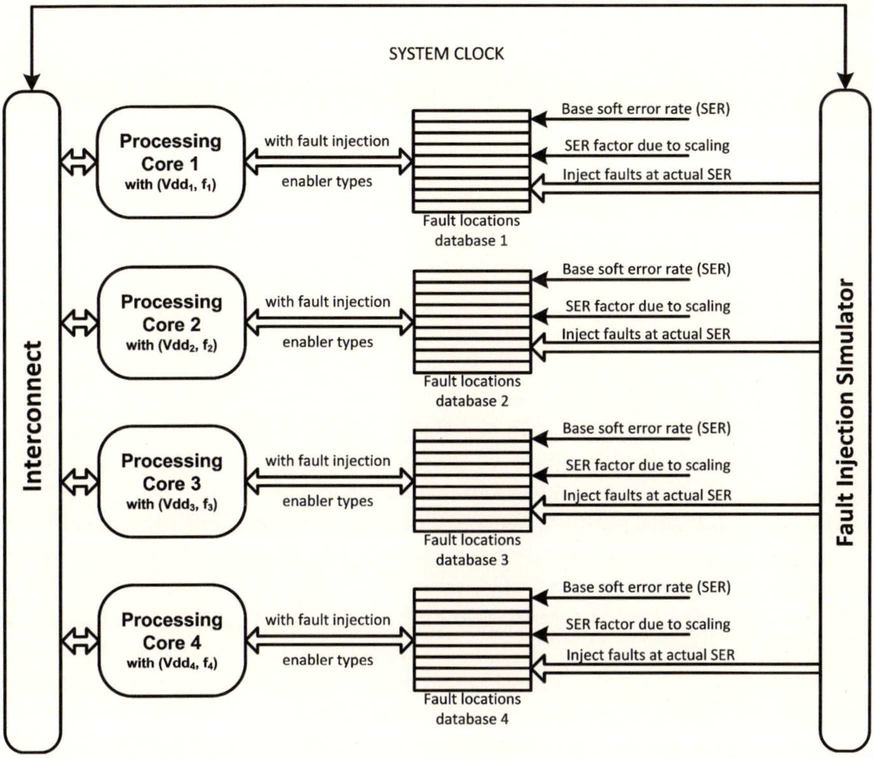

Fig. 5.8 Fault injection setup for MPSoC architecture with four processing cores

design specification to equivalent fault injection enabler types. These fault injection enabler types help form fault locations database for the device under test, which contains the target registers for SEU injection. The fault injection simulator injects SEUs in these target registers based on the specified soft error rates and probability distribution for determining fault locations. Figure 5.8 shows the fault injection setup used for the MPSoC architecture with four processing cores as an example. As can be seen, four fault locations databases are formed for four processing cores through replacement of variable/signal types to fault injection enabler types. For a given base soft error rate (SER, λ_0, in SEUs per bit per cycle considering no scaling), the actual SERs (λ, in SEUs per bit per cycle after scaling) for processing cores are found by the corresponding voltage settings used. For these actual SERs, the SEUs are injected at random locations determined by Poisson's distribution within the register space of the fault locations database. To control fault injection timings the system clock is connected to the fault injection simulator. Using SystemC monitor modules in cycle-accurate simulations, register usage and the number of SEUs experienced are found (Fig. 5.8).

5.3.2 *Impact of Task Mapping on Reliability*

Reliability of an MPSoC application in the presence of SEUs is related to the total number of SEUs experienced [41]. For a soft error rate (SER) of λ_i (SEUs per bit per clock cycle), the total number of SEUs experienced, Γ, by an MPSoC with $\mathscr{C}$ processing cores is given as

$$\Gamma = \sum_{i=1}^{\mathscr{C}} R_i T_i \lambda_i = T_M \sum_{i=1}^{\mathscr{C}} R_i \alpha_i \lambda_i, \tag{5.4}$$

where T_i is the execution time, R_i is the register usage of i-th processing core and T_M is the multiprocessor execution time (in clock cycles, $\forall_i{:}T_i = \alpha_i T_M$) of the i-th processing core. The register usage of a processing core, defined by (5.3), depends on the nature of processing being carried out by the tasks mapped, data dependency and resource sharing among them. The T_M in (5.4) affects multiprocessor performance and depends on the number of mapped tasks on a processing core and the data dependency among them. As a result, when more related tasks are mapped on a processing core, T_M increases for a given operating frequency but the register resources related to tasks are localized reducing the overall register usage ($R = \sum_i R_i$). On the other hand, when tasks are distributed among processing cores to achieve higher parallelism, T_M decreases at the expense of increased R due to higher duplication of shared register resources among tasks. Examples of this trade-off follow. In the MPEG-2 decoder (Fig. 5.7), the tasks t_5 and t_6 share about 13 kbits registers (r_7 is shared between them), while the tasks t_6, t_7 and t_8 share approximately 15 kbits registers among them (r_5 and t_8 are shared among them). To reduce register usage, for example it is possible to map tasks t_5, t_6, t_7 and t_8 on a processing core through localization of the registers. However, due to computationally intensive nature of these tasks, T_M will be high (with total computation cost of 188). To reduce T_M, an alternative option is to map tasks t_5 and t_6 on a processing core, while the tasks t_7 and t_8 can be mapped on another core. However, this gives a duplication of about 15 kbits registers (increased register usage) between the processing cores. Because of this register usage (R) and multiprocessor execution time (T_M) trade-off, the MPSoC experiences varying number of SEUs (Γ) for different task mappings given by (5.4). The Γ also depends on the voltage scaling of the MPSoC processing cores as it affects λ_i in (5.4). To demonstrate the impact of application task mapping and voltage scaling on the number of SEUs experienced (Γ), a total of 120 random task mappings were carried out using the MPEG-2 decoder (Fig. 5.7) on the MPSoC architecture (Fig. 5.6). Figure 5.9 shows the T_M, R and Γ obtained through SystemC simulation and fault injection (Sect. 5.3.1.3) using an SER of 10^{-9} SEU per bit per cycle (i.e. 1 SEU per 10 ms for 1 kbits register bank) as an example. Three key observations are made:

Observation 1: Figure 5.9a shows the trade-off between multiprocessor execution time (T_M, ms) and overall register usage (R). As can be seen, when tasks are

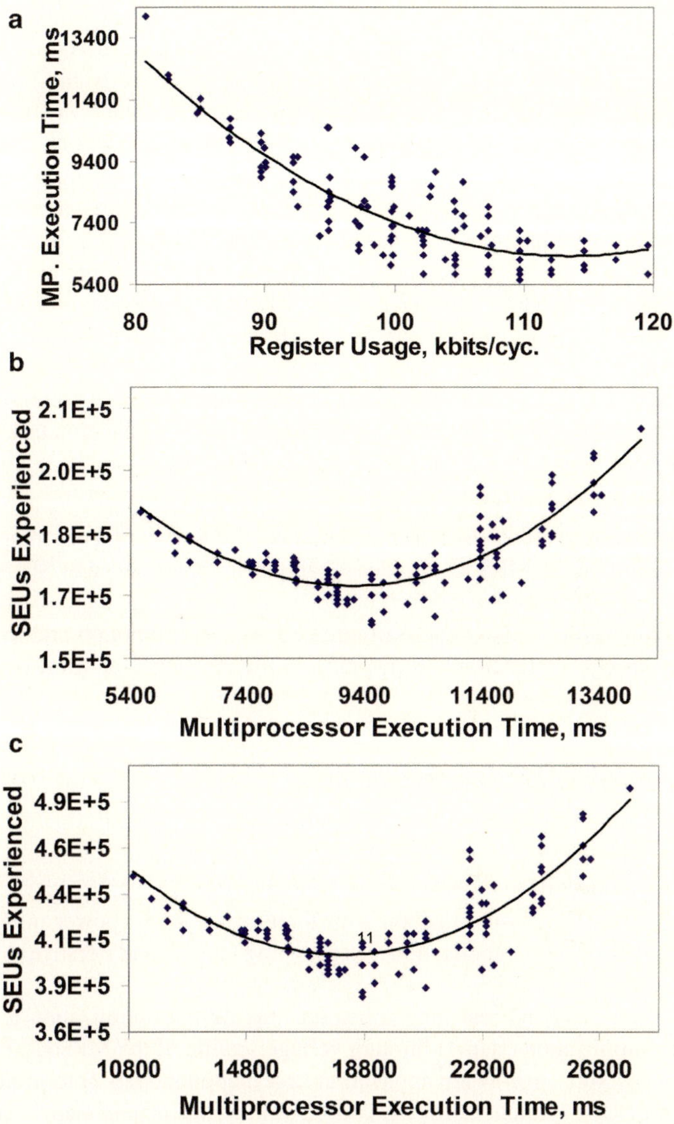

Fig. 5.9 (**a**) Trade-off between multiprocessor execution time (in ms) and register usage (in kbits/cycle), (**b**) SEUs experienced and multiprocessor execution time (in ms) when no scaling is used for MPSoC cores, and (**c**) SEUs experienced and execution time when MPSoC cores are scaled by 2; all for different task mappings of MPEG decoder with four processing cores

mapped to reduce R by localization of tasks, T_M increases. On the other hand, as tasks are mapped to reduce T_M, register resources shared among tasks are duplicated, leading to increased register usage, R.

Observation 2: Figure 5.9b shows the total number of SEUs experienced (Γ) and multiprocessor execution time, T_M (in ms), when all the decoder processing cores are scaled by 1 ($f = 200\,\text{MHz}$ and $V_{dd} = 1\,\text{V}$). It can be seen that when tasks are distributed among processing cores to reduce T_M, the decoder experiences more higher Γ given by (5.4) due to higher R (Fig. 5.9a). When tasks are localized to reduce R, the decoder also experiences higher number of SEUs due to increased T_M (Fig. 5.9a). This results in a concave curve for Γ given by (5.4), with the minimum Γ located around the middle of T_M range.

Observation 3: Figure 5.9c shows the total number of SEUs experienced (Γ) and multiprocessor execution time (T_M, in ms) when all the decoder processing cores are scaled by 2 ($f = 100\,\text{MHz}$ and $V_{dd} = 0.58\,\text{V}$). As can be seen, Γ increases by approximately 2.5 times due to V_{dd} scaling from 1 to 0.58 V (found through V_{dd} and λ relationship shown in [32]) and T_M is increased by a factor of 2 due to reduced f from 200 to 100 MHz. For example, for an application task mapping Γ increases from 1.71×10^5 to 4.12×10^5, while T_M increases from 9.5 to 18.2 s due to scaling of V_{dd} from 1 to 0.58 V.

The above observations demonstrate the impact of application task mapping and voltage scaling on the MPSoC decoder reliability in the presence of SEUs. Hence, an interesting design optimization problem is to identify suitable voltage scaling of the MPSoC processing cores to minimize power consumption and to improve reliability through application task mapping, while meeting a real-time constraint.

5.3.3 *Proposed Design Optimization*

To solve the problem of joint power minimization and reliability improvement of an application, a novel design optimization is proposed. Figure 5.10 shows flowchart of the proposed design optimization with three major steps: power minimization (step 1), soft error-aware application task mapping (step 2) and iterative assessment (step 3).

For a given SER and real-time constraint, the design optimization is initiated by power minimization (step 1) through voltage scaling of the MPSoC cores. This is followed by soft error-aware application task mapping (step 2) to minimize the number of SEUs experienced for the chosen voltage scalings in step 1. These two steps are repeated and assessed in step 3 to find a design with minimized power consumption and improved reliability in terms of SEUs experienced, while meeting the real-time constraint. The design optimization steps are discussed next.

5.3.3.1 Power Minimization

Power minimization in the proposed design optimization is performed using the voltage scaling algorithm, Fig. 5.11a. The voltage scaling algorithm, *nextScaling*, starts with the lowest voltage scaling on all identical cores and generates the next

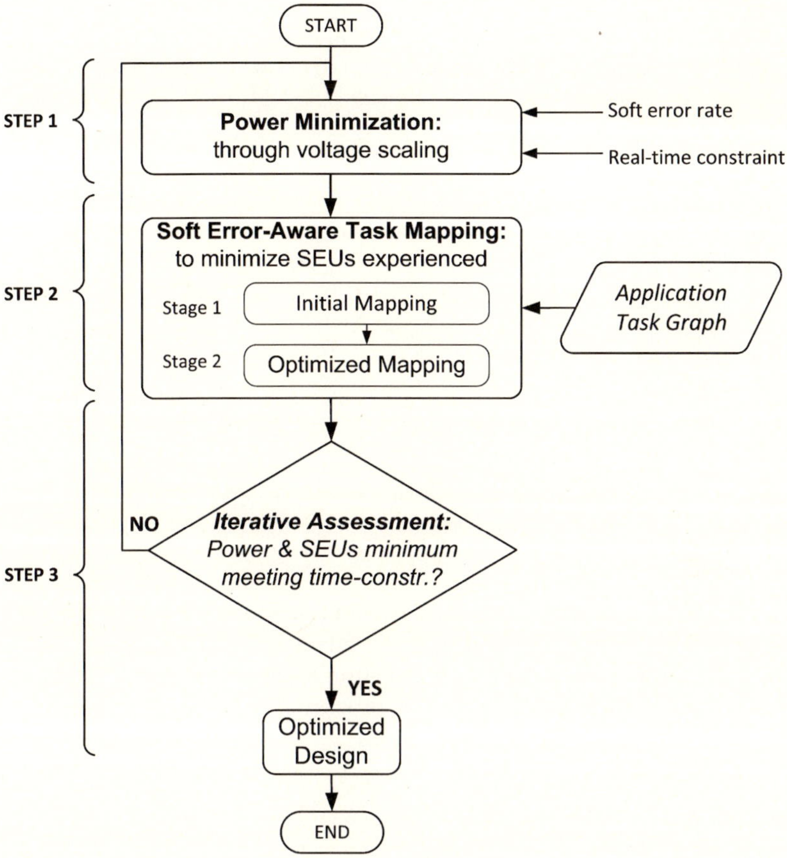

Fig. 5.10 Flowchart of the proposed design optimization

set of higher voltage scaling, *nextS*, based on the previous set of coefficients, *prevS* (Fig. 5.11a). In each iteration, *nextS* is updated as the *prevS* reduced by 1 on a processing core until the voltage scaling on the core reaches the nominal voltage scaling level ($s = 1$, lines 3–6). When the nominal level ($s = 1$) is reached, *nextS* of the core is updated by increasing voltage scaling of the core by 1 (line 9) and reducing the voltage scaling of the next processing core by 1 in steps (lines 7–11). The aim is to generate non-repetitive combinations and reduce the number of voltage scalings that need to be investigated. For example, for a homogeneous architecture with four processing cores (Fig. 5.6) and three scaling options (Table 5.1), the voltage scaling algorithm, Fig. 5.11a, generates 15 unique combinations, Fig. 5.10b, compared to a total of $3^4 = 81$ possible combinations. As can be seen, the voltage scaling starts with the highest scaling coefficient of 3 for all cores, followed by 3 for core 1, core 2, core 3, and 2 for core 4 as the next scaling combination (Fig. 5.11b). As core 4 reaches nominal value of 1, the next combination is generated by *nextScaling* algorithm as 3 for core 1, core 2, and 2 for core 3 and core 4. This

a

```
//C = no of cores, prevS = previous scaling
[nextS] = nextScaling(prevS): begin
  copy prevS into nextS
  for i := 1 to C
    if prevS[i] > 1: begin //1 is lowest scale
      nextS[i] := prevS[i]-1;
      break;
    end if
    else
      for k := i to C
        nextS[k] = prevS[k]+1;
      end for
    end else
  end for
  return nextS;
end
```

b

Iter.	Scaling Coefficients			
	Core 1,	Core 2,	Core 3,	Core 4,
#	s_1	s_2	s_3	s_4
1	3	3	3	3
2	3	3	3	2
3	3	3	3	1
4	3	3	2	2
5	3	3	2	1
6	3	3	1	1
7	3	2	2	2
8	3	2	2	1
9	3	2	1	1
10	3	1	1	1
11	2	2	2	2
12	2	2	2	1
13	2	2	1	1
14	2	1	1	1
15	1	1	1	1

Fig. 5.11 (**a**) Voltage scaling algorithm used for power minimization, (**b**) example of voltage scaling coefficients for four processing cores using voltage scaling algorithm shown in (**a**)

is followed by 3 for core 1, core 2, 2 for core 3, and 1 for core 4. The voltage scaling algorithm, thus, effectively generates all possible combinations. For a set of scaling coefficients from voltage scaling algorithm (Fig. 5.11a), the dynamic power consumption, P, of the MPSoC with $\mathscr{C}$ processing cores can be expressed as a function of voltages scaling coefficient, s_i, as

$$P = C_L \sum_{i=1}^{\mathscr{C}} \alpha_i f_i(s_i) V_{dd_i}^2(s_i), \tag{5.5}$$

where $f_i(s_i)$ and $V_{dd_i}(s_i)$ take values depending on the voltage scaling coefficient s_i (Table 5.1). For each set of voltage scaling coefficients, soft error-aware application task mapping is carried out (step 2, Fig. 5.10) to minimize the number of SEUs experienced.

5.3.3.2 Soft Error-Aware Application Task Mapping

The problem of application task mapping on MPSoC cores to minimize SEUs experienced (Γ) is an NP-complete problem [31]. In this section, a soft error-aware application task mapping is developed in two stages (step 2, Fig. 5.10): the stage 1 is the initial soft error-aware application task mapping, followed by stage 2 of search-based optimized application task mapping. Figure 5.12 shows the initial soft error-aware application task mapping algorithm (stage 1), *InitialSEAMapping*, which aims to simplify the optimization process by reducing the number of task movements. The *InitialSEAMapping* starts with mapping the task with no predecessor in task graph (G) (line 1). The dependants of the currently mapped

Fig. 5.12 Initial soft error-aware mapping algorithm, *InitialSEAMapping*

```
//C: no of cores, G: application task graph with N tasks, t is the current task
//M: mapping of all cores, A: MPSoC arch., Q: task queue, L: temporary list
[M] = InitialSEAMapping(G, C, A): begin
 1:  push G[0] into Q  // push the first task into Q
 2:  for i:= 1 to C-1 and Q is not empty
 3:    t := Q.front(); M[i].map(t); delete all mapped tasks from Q
 4:    while Ti < TMref and no. of unmapped tasks in G > (C-i)
 5:      L : = dependents of t //(sorted by minimum SEUs)
 6:      if L is empty and Q is not empty
 7:        swap last two elements in Q
 8:      else if Q is not empty
 9:        t = first element in L//task with minimum SEUs and Time
 10:       M[i].map(t); delete t from L; move tasks in L into Q and empty L
 11:     else break while; end if
 12:     t = Q.front();
 13:   end while
 14: end for
 15: return M;
end
```

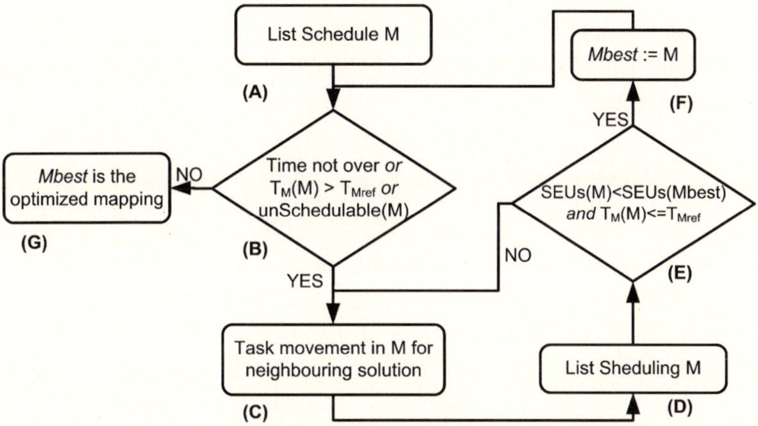

Fig. 5.13 Flowchart of optimized mapping, *OptimizedMapping*

task in G are then sorted by SEUs experienced if they are to be mapped with the current task. The sorted list of dependants is stored in L (line 5). The task that gives the minimum SEUs in L is then mapped next (lines 6–10). This is continued until the execution time of the current core does not exceed the real-time constraint ($T_{M_{ref}}$) and the number of unmapped tasks left in task graph G is higher than the number of remaining cores to ensure that tasks are mapped in all cores (lines 4–13). The unmapped tasks are stored in a queue, Q (line 10), which are then mapped gradually to the other cores using the same criteria. After all tasks are mapped, the initial mapping (M) is returned by *InitialSEAMapping* (lines 6–15).

After the initial soft error-aware task mapping (*InitialSEAMapping*, Fig. 5.12), the design optimization is continued further through optimized mapping (stage 2, step 2, Fig. 5.10). The optimization is carried out through iterative search-based mapping algorithm, *OptimizedMapping*, Fig. 5.13, employing the list scheduling for mapped tasks. The aim of such scheduling is to make an ordered grouping of tasks in processing cores to accommodate different constraints and task dependencies [31]. The *OptimizedMapping* starts with scheduling the initial task mapping, M (step A, Fig. 5.13). The mapping M is then checked to see if it violates the schedulability requirements or real-time constraints (step B). If any such violation is found within the search time, the optimization proceeds with generating neighbouring task movements to find out a possible next mapping solution (step C). With neighbouring task movement, the new mapping (M) is then list scheduled, if schedulable (step D). This is followed by comparison with the previous best solution, *Mbest*. If M is better than *Mbest* in terms of lower number of SEUs experienced and meets the given real-time constraint, it is then updated as the new *Mbest* (steps E–F). The optimization steps C–F are repeated until the specified search time is not over (step B). Once the search time is over, *Mbest* is returned as the optimized design for the chosen voltage scalings (step G).

In *OptimizedMapping*, the multiprocessor execution time (T_M, in seconds) for an application task mapping is found by the dividing the total number of execution cycles of all mapped tasks by the effective number of cycles executed by processing cores per second for chosen voltage scaling (step B, E, Fig. 5.13), i.e.

$$T_M = \left[\sum_{i=1}^{\mathscr{C}} \sum_{j=1}^{N} \left(t_j^i + \sum_{k=1}^{N} d_{j,k}^i \right)\right] / \left[\sum_{i=1}^{\mathscr{C}} \alpha_i f_i(s_i)\right], \tag{5.6}$$

where t_j^i is the execution time (in clock cycles) of the j-th task mapped on i-th processing core, $d_{j,k}^i$ is the dependency cost (in clock cycles) between j-th and k-th task ($j, k = 1 : N$) due to selection of j-th task on i-th processing core. The total number of SEUs experienced (Γ) for an application task mapping with chosen voltage scaling on MPSoC processing cores is found in *InitialSEAMapping* (line 5, Fig. 5.12) and *OptimizedMapping* (step E, Fig. 5.13) through (5.4). The per core execution time (T_i, in clock cycles) and register usage (R_i, in bits per cycle) in (5.4) are given in terms of mapped tasks as

$$\forall_i : T_i = \sum_{j=1}^{N} \left(t_j^i + \sum_{k=1}^{N} d_{j,k}^i \right), \text{ and} \tag{5.7}$$

$$\forall_i : R_i = Avg \left\{ \left| \left(\bigcup_{j=1}^{N} \bigcup_{k=1}^{N} r_{j,k}^i \right) \right| \right\}, \tag{5.8}$$

where $r_{j,k}^i$ is the set of registers shared between j-th and k-th tasks for being mapped on i-th processing core ($j = k$ defines the local register usage of j-th task). As can be seen in (5.8), R_i is given by average cardinality of the register set over the execution time arizing out of union of register usages of mapped tasks ($r_{j,k}^i$) in i-th processing core. The proposed optimization is carried out by iterative search through N tasks, with each iteration generating maximum two task movements

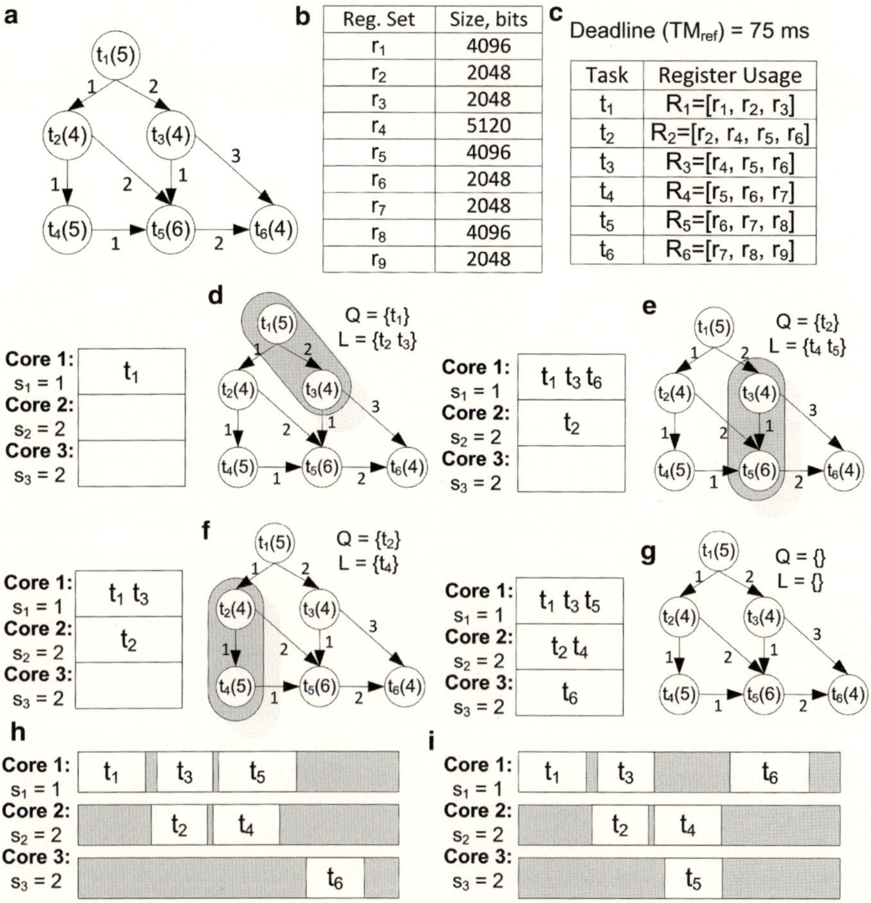

Fig. 5.14 Example illustration of the soft error-aware application task mapping (**a**) example application task graph, (**b**) sets of registers and their sizes, (**c**) register usage of different tasks of the application, (**d**) and (**f**) initial soft error-aware application task mapping (*InitialSEAMapping*, Fig. 5.12) steps, and (**g**) optimized mapping (*OptimizedMapping*, Fig. 5.13) step

out of maximum search space of $(N-1)$ dependent tasks. This is followed by second stage search through maximum $(N-1)$ tasks for minimum number of SEUs experienced. As a result, *OptimizedMapping* has worst-case complexity of $O(2N(N-1)(N-1)) \approx O(N^3)$.

An example illustrating the proposed soft error-aware application task mapping algorithm is shown in Fig. 5.14. In Fig. 5.14a, an application task graph with six tasks is shown (all costs are multiples of 60×10^4 cycles) and in Fig. 5.14b, c the application registers and their distribution for different tasks are shown. Figure 5.14d–g show the incremental task mapping using *InitialSEAMapping* algorithm, Fig. 5.12, and finally, Fig. 5.14h, i show scheduling and task movements using *OptimizedMapping*, Fig. 5.13.

The chosen voltage scaling for the processing cores are: $s_1 = 1$, $s_2 = 2$ and $s_3 = 2$ and deadline is assumed to be $T_{M_{ref}} = 75$ ms. As can be seen, after the first task, t_1, in the application task graph, Fig. 5.14a, is mapped to processing core 1, the *InitialSEAMapping* mapping algorithm selects t_3, followed by t_6 from dependency list, L. This is because task t_3 gives the least number of SEUs experienced compared to t_2 and t_5 shown in gray, Fig. 5.14d, with the $r_{j,k}$ values from Fig. 5.14c. Note that after allocating t_1, t_3 and t_5 on core 1, the deadline constraint cannot be satisfied with further allocation of tasks and the mapping algorithm carries on with mapping of tasks t_2 and t_4 in core 2, which give minimum SEUs experienced, Fig. 5.14f. Finally, the unmapped task t_6 in queue (Q) is mapped to core 3, Fig. 5.14g. After *InitialSEAMapping* (Fig. 5.12) is completed, *OptimizedMapping* list schedules the tasks, Fig. 5.14h, found through step A, Fig. 5.13. However, with the chosen voltage scalings for the architecture processing cores, this mapping cannot satisfy the real-time constraint of 75 ms. The *OptimizedMapping* swaps t_5 with t_6 in the fourth iteration as a neighbouring task mapping (step C, Fig. 5.13) and gives the minimum number of SEUs experienced for the chosen voltage scaling, while meeting $T_{M_{ref}} = 75$ ms.

5.3.3.3 Iterative Assessment

With each set of voltage scaling coefficients resulting from the voltage scaling algorithm (step 1, Fig. 5.10) soft error-aware application task mapping (step 2, Fig. 5.10) is carried out to minimize the number of SEUs experienced through application task mapping. The resulting power consumption (P) and SEUs experienced (Γ) are then iteratively assessed using a score function, Z, to produce an optimized design in terms of minimized power consumption and improved reliability, such that real-time constraints are met (similar score function for joint optimization is also used in [24]). The optimization score function, Z, is defined by a linear combination of normalized power consumption and number of SEUS experienced, given by

$$Z = 0.5\, Z_P + 0.5\, Z_\Gamma, \tag{5.9}$$

where Z_P is the score related to power consumption (P), Z_Γ is the score related to reliability improvement in terms of total number of SEUs experienced (Γ) and 0.5 is the weighting factors for Z_P and Z_Γ to give joint optimization with equal weight to power consumption and reliability improvement. The Z_P value of a design is found by normalizing power consumption (P, defined by (5.5)) due to the chosen voltage scaling (Sect. 5.3.3.1) and soft error-aware application task mapping (Sect. 5.3.3.2) by the maximum power consumption, P_{max} with the highest voltage settings, (i.e. $s_i = 1$) i.e.

$$Z_P = \frac{P}{P_{max}} = \frac{C_L \sum_{i=1}^{\mathscr{C}} \alpha_i f_i(s_i) V_{dd_i}^2(s_i)}{C_L \sum_{i=1}^{\mathscr{C}} f_{max} V_{max_i}^2}. \quad (5.10)$$

where f_{max} and V_{max} are maximum operating frequency and supply voltage (for example $V_{max} = 1\,\text{V}$, $f_{max} = 200\,\text{MHz}$ for the given scaling options in Table 5.1). The score function related to SEUs experienced, Z_Γ, in (5.9) of the same design is defined by the ratio of SEUs experienced (Γ, given by (5.4)) to the maximum number of SEUs experienced (Γ_{max}, assuming maximum activity factor $\alpha = 1$ and highest voltage scaling, $s_i = 3$, in (5.4)), i.e.

$$Z_\Gamma = \frac{\Gamma}{\Gamma_{max}} = \frac{T_M \sum_{i=1}^{\mathscr{C}} R_i \alpha_i \lambda_i}{T_M \sum_{i=1}^{\mathscr{C}} R_i \lambda_{max}}. \quad (5.11)$$

where λ_{max} is the maximum soft error rate due to highest voltage scaling (for example $V_{dd} = 0.44\,\text{V}$, $f = 66.7\,\text{MHz}$ for the given scaling options in Table 5.1). Using the optimization score function, Z, defined in (5.9), the iterative assessment (step 3, Fig. 5.10) identifies the design that gives the minimum Z value. Note that due to normalization of the power consumption of a design, (P) by the maximum power consumption (P_{max}) in (5.10) Z_P is reduced for low power consumption. However, due to normalization of SEUs experienced (Γ) by the maximum possible SEUs experienced (Γ_{max}) in (5.11) Z_Γ is increased for such design. Similarly, when power consumption increases Z_P is increased at the cost of reduced Z_Γ. As a result of this Z_P and Z_Γ relationship, the optimization score function Z gives the minimum value for a design that gives the best trade-off between P and Γ. The design that gives minimum Z value and meets the real-time constraint is chosen as the optimized design. Figure 5.15 shows an example of iterative assessment using score function, Z, to effectively find an optimized design for the MPEG-2 video decoder (Fig. 5.7) using MPSoC architecture with four processing cores (Fig. 5.6). The horizontal axis shows the subsequent voltage scaling iterations arizing from voltage scaling algorithm (Fig. 5.11a) and vertical axis gives the Z_P, Z_Γ and Z values defined by (5.10), (5.11) and (5.9). As can be seen, the voltage scaling starts with scaling by 3 on each processing core (Fig. 5.11b). For such low voltage scaling, Z_Γ is high with higher SEUs experienced (Γ) but Z_P but Z_P is low due to minimum power consumption. Higher Z_Γ results in a higher score function, Z, for this design. As the voltage scaling proceeds with next set of voltage scaling coefficients (Fig. 5.11b), Z value also varies due to the trade-offs between Z_P and Z_Γ. Note that due to the voltage settings on processing cores (with $V_{dd} = 1\,\text{V}$,

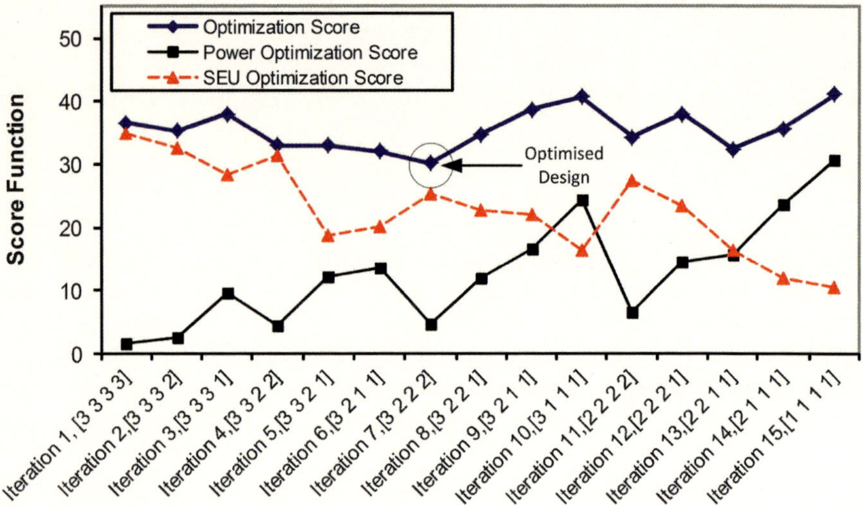

Fig. 5.15 Example iterative assessment for design optimization using MPEG-2 video decoder with four processing cores

$f = 200$ MHz on all processing core), design produced in iteration 15 results in the highest score function, Z. The design produced in iteration 7 (with $V_{dd} = 0.58$ V, $f = 100$ MHz on 3 processing cores and $V_{dd} = 0.44$ V, $f = 66.7$ MHz on a processing core) gives the best design in terms minimized Z value. Since real-time constraints are met for this design, it is returned as the optimized design for MPEG-2 video decoder (Fig. 5.7) using MPSoC architecture with four processing cores (Fig. 5.6).

5.3.4 Experimental Results

In this section, the effectiveness of the proposed soft error-aware design optimization is evaluated using four experiments, Table 5.3. The experiments are carried out using MPEG decoder implemented with the architecture of Fig. 5.6. The first three experiments, Exp:1, Exp:2 and Exp:3, are soft error-unaware optimization with different design objectives using application task mapping obtained through simulated annealing [50]. Exp:4 is the proposed design optimization. In all experiments, power minimization is obtained through iterative voltage scaling (step 1, Fig. 5.10) after application task mapping with an aim to meet the real-time constraint of decoding a *tennis* video sequence[1] of 437 frames at 29 frames per second (fps). The mapped tasks, voltage scaling on processing each core (s_i), per core execution time (T_i) and

[1] ftp://ftp.tek.com/tv/test/streams/Element/

Table 5.3 Comparison of soft error-unaware and the proposed soft error-aware optimizations using MPEG decoder MPSoC with four cores

Experiments	MPSoC core	Mapped tasks	Voltage scaling, s_i	Execution time, T_i, $\times 10^8$ cycle	Register usage, R_i, kbits/cycle	P, mW	$R = \sum_i R_i$, kbits/cycle	T_M, ($\times 10^8$) cycle	Γ, ($\times 10^5$)
Exp:1 (Optimized for reduced register usage [50])	Core 1	t_1, t_2, t_3	2	14.4	17.3	9.53	80	38.1	3.46
	Core 2	t_4, t_5	2	21.4	16.4				
	Core 3	t_{11}	2	16.7	19.1				
	Core 4	$t_6, t_7, t_8, t_9, t_{10}$	1	28.8	27.2				
Exp:2 (Optimized for parallelism [50])	Core 1	t_1, t_2, t_3, t_4, t_9	3	24.6	37.7	4.04	114	28.6	5.22
	Core 2	t_8	3	18.5	19.4				
	Core 3	t_5, t_6, t_7	2	21.2	29.3				
	Core 4	t_{10}, t_{11}	2	20.1	28				
Exp:3 (Optimized for register usage & parallelism [50])	Core 1	t_8, t_9	3	23.1	23.6	4.15	92	30.9	4.18
	Core 2	t_{10}, t_{11}	2	20.7	27.9				
	Core 3	t_6, t_7	2	21.8	22				
	Core 4	t_1, t_2, t_3, t_4, t_5	2	17.9	19.1				
Exp:4 (Proposed optimization)	Core 1	t_9	3	14.1	17	4.25	89	31.8	3.93
	Core 2	t_{10}, t_{11}	2	19.3	28				
	Core 3	t_7, t_8	2	25.8	20.4				
	Core 4	$t_1, t_2, t_3, t_4, t_5, t_6$	2	20.9	23.3				

per core register usage (R_i) are shown in column 3–6, while the power consumption (P, mW), register usage (R, kbits/cycle), the multiprocessor execution time (T_M, clock cycles) and the number of SEUs experienced (Γ) are given in columns 7–10 (Table 5.3). For each voltage scaling of the MPSoC processing cores, a time-limit of 30 min to search the design space is imposed. All experiments are carried out on an Intel(R) Core(TM)2 2 GHz CPU running RHEL5. The number of SEUs experienced (column 7, Table 5.3) is found by fault injection technique (Sect. 5.3.1.2) assuming an arbitrary SER of 10^{-9} SEUs/bit/cycle (i.e. 1 SEU per 10 ms for 1 kbits register bank). The power values are obtained by (5.1), with the α values found with the multiprocessor execution time (T_M) and execution times (T_i) of processing cores from Table 5.3 (note that switching activity factor, $\alpha = \frac{T_i}{T_M}$).

Exp:1 demonstrates the impact of design optimization with minimized register usage, R. As expected, the design produced gives the least register usage ($R = \sum_i R_i$) when compared to the other three experiments. The reduced R in Exp:1 is obtained at the expense of the highest multiprocessor execution time (T_M) of 38.1×10^8 clock cycles (as explained in Sect. 5.3.2). This makes it harder to scale down the voltages of the decoder cores. As a result, Exp:1 gives a design that has higher power consumption (9.53 mW) than the optimized design produced in Exp:4 (4.25 mW). However, the design produced in Exp:1 experiences lower SEUs than that in Exp:4 (3.46×10^5 SEUs compared 3.93×10^5 SEUs). This is because, the proposed design optimization in Exp:4 gives lower voltages of the decoder cores, and hence lower power consumption compared to the design produced in Exp:1. The design produced in Exp:2 is optimized for high parallelism. This gives reduced multiprocessor execution time (T_M) of 28.6×10^8 clock cycles, which allows the voltages of the decoder processing cores to be scaled down. As a result, Exp:2 gives lower power consumption (4.04 mW) than Exp:4 (4.25 mW). Note that this reduction in multiprocessor execution time (T_M) in Exp:2 is achieved at the expense of the highest register usage ($R = 114$ kbits per cycle). Due to lower voltage scaling of the decoder cores and higher register usage, the design optimized for high parallelism, Exp:2, experiences the highest number of SEUs ($\Gamma = 5.22 \times 10^5$) when compared to the other three experiments. In Exp:3, the design has been optimized for both register usage (R) and high parallelism. Such optimization gives a good trade-off between multiprocessor execution time and register usage, and minimizes the product: $T_M \times R$ in (5.4). However, this does not necessarily minimize of the number of SEUs experienced since optimization is carried using soft error-unaware task mapping. The design produced in Exp:4 employs soft error-aware task mapping (and minimizes Γ in (5.4) by carefully mapping the tasks to minimize the product $T_i \times R_i$ on each core) and therefore gives less number of SEUs experienced than the design produced in Exp:3 (3.93×10^5 SEUs for Exp:4 compared to 4.18×10^5 SEUs for Exp:3). Note that, although the voltage scaling of the decoder cores are similar, the proposed design optimization (Exp:4) gives about 3% higher power consumption compared to the design produced in Exp:3 due mapping of computation intensive tasks t_7, t_8 in core 3 and t_1–t_6 in core 4 of the decoder. For all design optimization approximate number of SEUs experienced can also be found by (5.4) using the T_i and R_i values shown in columns 5 and 6 (Table 5.3).

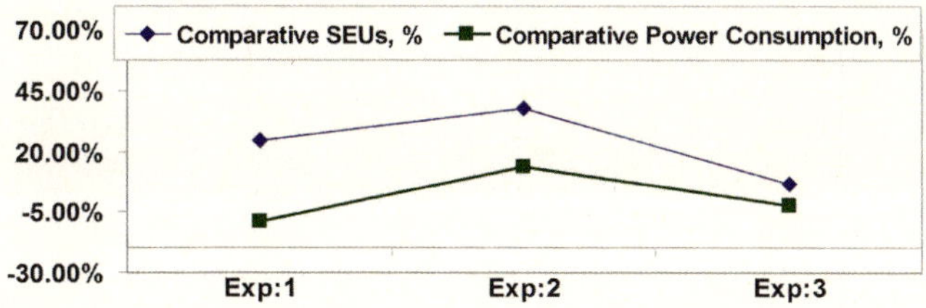

Fig. 5.16 Comparison of power consumption (P, in mW) and SEUs experienced (Γ) of Exp:1, Exp:2 and Exp:3 when compared with Exp:4

To highlight the advantages of using the proposed design optimization (Exp:4), Fig. 5.16 shows comparison of power consumption (P) and SEUs experienced (Γ) of the decoder design in Exp:1, Exp:2 and Exp:3 compared to that of Exp:4. All experiments are carried out with same voltage scaling coefficients ($s_1 = 2$, $s_2 = 2$, $s_3 = 3$ and $s_4 = 2$) for an SER of 10^{-9}. As can be seen, the design produced in Exp:4 reduces the number of SEUs experienced by upto 38% compared to the optimized design in Exp:2, while consuming 9% lower power. When compared with the design produced in Exp:1, the optimized design in Exp:4 reduces SEUs experienced by 28%, while consuming only 7% higher power.

The design optimizations in Table 5.3 were carried out using MPEG-2 decoder. To demonstrate the effectiveness of the proposed design optimization with other applications, random task graphs of 20, 40, 60, 80 and 100 tasks are also used. The random task graphs are generated using the random task and resource graph tool [55]. The cost and the number of dependants in the random task graphs are generated using uniform probability distribution with computation cost between 1 and 30, communication cost between 1 and 10 (all costs as multiples of 3.5×10^6 clock cycles), task register usage between 1 and 5 kbits and the number of dependants was found by exponential distribution between 0 and $N/2$, where N is the number of tasks. The deadline for random task graphs are set to 15, 20, 30, 40 and 50 s for random task graph with 20, 40, 60, 80 and 100 tasks, respectively. For these task graphs, the design optimization is carried out with imposed time limits of 20, 30, 40, 50 and 60 min for 20, 40, 60, 80 and 100 tasks, respectively. Table 5.4 shows the results of using the proposed design optimization (Exp:4) on the MPSoC using four processing cores (Fig. 5.6) with the random task graphs. The voltage scalings on MPSoC processing cores, per core execution time (T_i) and per core register usage (R_i) are shown in columns 3–5. The power consumption (P), overall register usage (R), multiprocessor execution time (T_M) and the total number of SEUs experienced (Γ) are shown in columns 6–9 (Table 5.4). As can be seen, depending on the application and its deadline the voltage scaling and the corresponding power consumption (P, in mW) vary (Table 5.4). However, the total register usage (R) arizing from per core register usage (R_i) increases as the number of tasks in the random task graphs increases. Also, the multiprocessor execution time (T_M, in clock cycles) and per core execution time (T_i, in clock cycles) increase as the number of tasks in the task graphs increase. Due to the

Table 5.4 Power consumption (P), register usage (R) and SEUs experienced (Γ) for different applications using Exp:4

Application	MPSoC core	Voltage scaling, s_i	Exec. time, T_i, $\times 10^8$ cycle	Reg. usage, R_i, kbits/cycle	P, mW	$R = \sum_i R_i$, kbits/cycle	T_M,cycle ($\times 10^8$)	Γ, ($\times 10^5$)
20 tasks	Core 1	3	8.3	19.1	4.34	66	12.8	2.27
	Core 2	3	8.4	14.7				
	Core 3	2	12.8	15.3				
	Core 4	2	12.4	17				
40 tasks	Core 1	3	10.9	26.9	5.2	90	23.6	2.87
	Core 2	3	9.8	20.1				
	Core 3	2	9.3	21.3				
	Core 4	1	29.8	21.2				
60 tasks	Core 1	2	13.6	25.5	5.1	107	31.2	4.82
	Core 2	2	12.4	27.9				
	Core 3	2	19.5	30.3				
	Core 4	2	30.3	23.6				
80 tasks	Core 1	3	15.5	25.2	4.4	129	41.3	6.13
	Core 2	3	15.2	38.7				
	Core 3	2	23.6	29.9				
	Core 4	1	46.4	35.8				
100 tasks	Core 1	3	15.9	32.4	4.8	149	53.7	8.25
	Core 2	3	28.7	33.6				
	Core 3	2	38.9	38.1				
	Core 4	2	49.6	44.3				

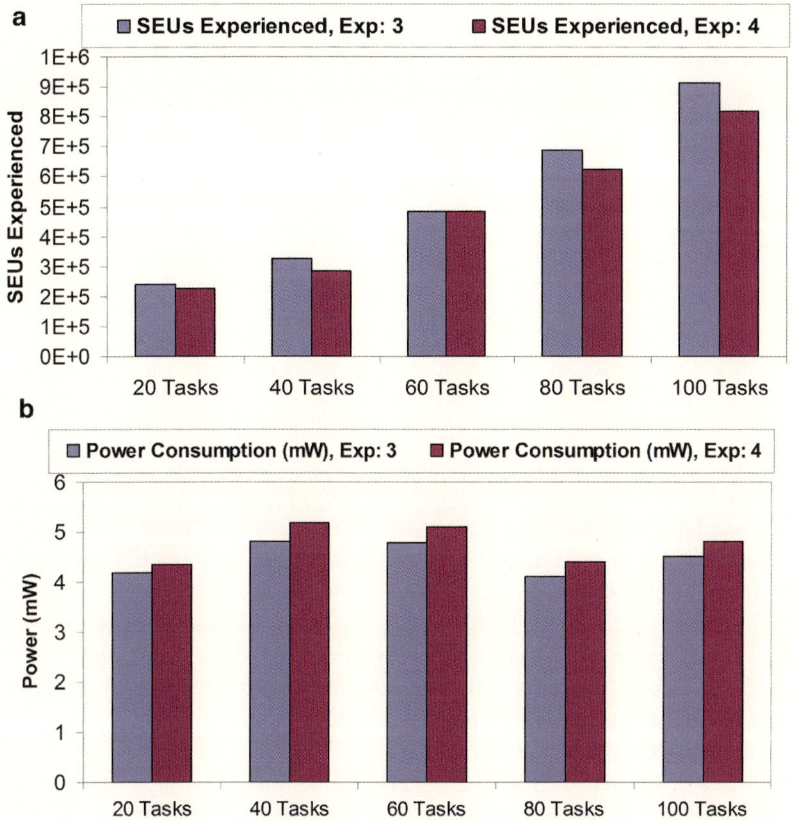

Fig. 5.17 Comparison of power consumption (P, in mW) and SEUs experienced (Γ) between Exp:3 and Exp:4 for different random task graphs

increased per core register usage (R) and per core execution time (T_i), the total number of SEUs experienced (Γ) also increases as the number of tasks in the random task graphs increases. For example, the random task graph with 100 tasks experiences the highest number of SEUs (i.e. 8.25×10^5 SEUs), while the random task graph with 20 tasks experiences the lowest number of SEUs (i.e. 2.27×10^5). Figure 5.17a, b show the comparisons of power consumption (P) and the number of SEUs experienced (Γ) using the design optimizations in Exp:3 and Exp:4 for the different random task graphs. As can be seen, the design optimization in Exp:4 consistently outperforms the design optimization in Exp:3 in terms of the number of SEUs experienced (Γ) due to soft error-aware application task mapping carried out in Exp:4 (Sect. 5.3.3.2). For example, for the random task graph with 80 tasks the proposed design optimization (Exp:4) reduces the SEUs experienced by 9.6% compared to the design optimization in Exp:3. This reduction in SEUs experienced is achieved with only 5% increase in the power consumption (P).

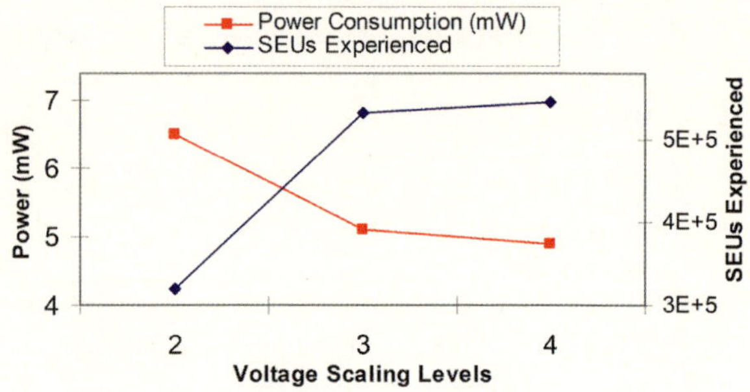

Fig. 5.18 Power consumption (P, in mW) and SEUs experienced (Γ) for different scaling levels using the proposed design optimization technique

To show the impact of choice of voltage scaling levels, Fig. 5.18 shows the power consumption (mW) and the number of SEUs experienced (Γ) by the optimized designs produced in Exp:4 with different voltage scaling levels. The design optimizations are carried out using MPSoC with 4 processing cores with random task graph of 60 tasks and employing the following voltage scaling levels: 2 levels (with 1 V–200 MHz, and 0.58 V–100 MHz), 3 levels (Table 5.1) and 4 levels (introducing 1.2 V–236 MHz in Table 5.1). As can be seen, with 4 scaling levels the proposed design optimization (Exp:4) is able to minimize power further by 4% with only 3% increase in the number of SEUs experienced compared to 3 scaling levels. This is because with more scaling options, the power minimization (step 1, Fig. 5.10) has higher flexibility with more combinations of voltage scaling generated by the voltage scaling algorithm (Fig. 5.11a). With 2 scaling levels, it is possible to reduce the number of SEUs experienced by 42% at the cost of 28% higher power consumption compared to 3 scaling levels due to limited voltage scaling options (Fig. 5.18).

5.3.5 *Architecture Allocation*

Architecture allocation is a system-level design step for MPSoCs that deals with allocation of processing elements and their interconnections into the architecture. In this study, architecture allocation is referred to as the allocation of number of processing cores in the MPSoC architecture. Table 5.5 shows the mapped tasks using the optimized mapping in Exp:4 for different allocations from two cores to six cores using MPEG-2 video decoder task graph (Fig. 5.7). The architecture allocation is shown in column 1 and per core mapped tasks of the decoder task graph (Fig. 5.7) are shown in columns 2 and 3 (Table 5.5). To demonstrate the impact of architecture allocation, Fig. 5.19 shows the multiprocessor execution time (T_M, in clock cycles) and register usage (R, in kbits per cycle) using MPEG decoder MPSoCs. The voltage scaling of processing cores is carried out using three scaling levels (Table 5.1) and application task mapping is performed with the optimized mapping algorithm, *OptimizedMapping*, of the proposed design optimization (Exp:4). The T_M and R are found while decoding a *tennis* video sequence of 437 frames at 29

Table 5.5 Task distribution of MPEG-2 video decoder (Fig. 5.7) among cores for different architecture allocations using the optimized task mapping in the proposed design optimization technique (Fig. 5.10)

Allocation	Core	Mapped tasks
2 cores	Core 1	$t_1, t_2, t_3, t_4, t_9, t_{10}, t_{11}$
	Core 2	t_5, t_6, t_7, t_8
3 cores	Core 1	t_1, t_2, t_3, t_4, t_5
	Core 2	t_6, t_7, t_8
	Core 3	t_9, t_{10}, t_{11}
4 cores	Core 1	$t_1, t_2, t_3, t_4, t_5, t_6$
	Core 2	t_7, t_8
	Core 3	t_9
	Core 4	t_{10}, t_{11}
5 cores	Core 1	t_1, t_2, t_3, t_4
	Core 2	t_5, t_6
	Core 3	t_7, t_8
	Core 4	t_9
	Core 5	t_{10}, t_{11}
6 cores	Core 1	t_1, t_2, t_3, t_4
	Core 2	t_5, t_6
	Core 3	t_7
	Core 4	t_8
	Core 5	t_9
	Core 6	t_{10}, t_{11}

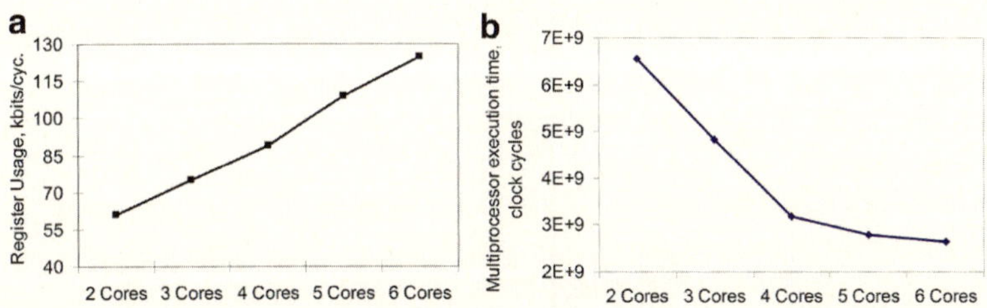

Fig. 5.19 (**a**) Register usage (R, in kbits/cycle), and (**b**) multiprocessor execution time (T_M, in clock cycles) of the MPEG-2 decoder MPSoC for different architecture allocations

frames per second. The architecture allocation is varied from two processing cores to six processing cores. As can be seen, with increase in the number of allocated cores, the register usage increases (Fig. 5.19a). This is because with increased number of allocated cores the tasks mapping or distribution causes more duplication of the shared register resources among tasks. Also, as expected with increased number of allocated cores in the MPSoC architecture, the multiprocessor execution time (T_M) decreases with higher parallelism among the mapped tasks on processing cores (Fig. 5.19b). Table 5.6 shows the impact of architecture allocation on the power consumption (P) and the number of SEUs experienced (Γ) using the optimized design produced in Exp:4. A number of applications, including MPEG decoder and random task graphs of 20, 40, 60, 80 and 100 tasks were used. The power

Table 5.6 Power consumption (P, in mW) and SEUs experienced (Γ, $\times 10^5$) for different applications and different architecture allocations

Application	2 cores		3 cores		4 cores		5 cores		6 cores	
	P, mW	$\Gamma \times 10^5$	P, mW	$\Gamma \times 10^5$	P, mW	$\Gamma \times 10^5$	P, mW	$\Gamma \times 10^5$	P, mW	$\Gamma \times 10^5$
MPEG (11 tasks)	9.1	2.13	5.9	3.17	4.25	3.93	6.34	4.95	7.24	5.36
20 tasks	10.1	0.47	4.15	1.13	4.34	2.27	5.16	2.73	6.36	3.49
40 tasks	6.2	1.07	5.1	1.78	5.2	2.87	6.16	3.46	7.11	4.35
60 tasks	7.8	1.87	4.13	3.25	5.1	4.82	4.9	5.74	5.3	7.15
80 tasks	11.2	1.95	6.1	3.76	4.4	6.13	6.14	7.24	6.69	9.13
100 tasks	10.4	2.40	5.48	4.58	4.8	8.25	5.94	8.83	6.34	11.13

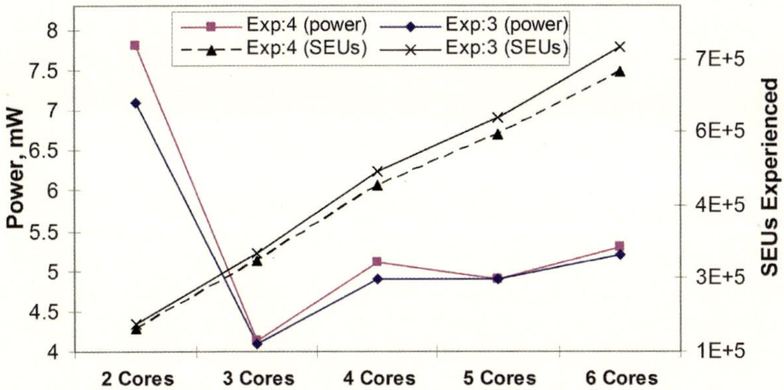

Fig. 5.20 Comparison of power consumption (P, in mW) and SEUs experienced (Γ) between Exp:3 and Exp:4 for different architecture allocations using random task graph of 60 tasks

consumption (P, in mW) and the number of SEUs experienced (Γ) for different architecture allocations are shown in columns 2–6 (Table 5.6). Two observations are made. Firstly, the architecture allocation with minimum power consumption (P) depends on the application and given real-time constraint. For example, in the case of the MPEG decoder, the least power consumption is found with four cores for the given real-time constraint of decoding *tennis* video sequence at 29 fps. Secondly, with increased number of architecture cores, the number of SEUs experienced increases. The increased number of SEUs can be explained as follows. With higher number of cores, multiprocessor execution time (T_M) reduces and the overall register usage (R) increases (Fig. 5.19). Due to reduced multiprocessor execution time, there is more opportunity for voltage scaling to reduce power consumption, which eventually increases the SER and the SEUs experienced. This is further exacerbated by the increased register usage caused by distribution of tasks with increased number of cores in MPSoC architecture (Fig. 5.19a). For example, the decoder with six processing cores experiences the highest number of SEUs, compared to the lowest for the decoder with 2 processing cores (row 2, Table 5.6). Similar observations for power consumption and the number of SEUs experienced are also observed with the random task graphs.

To compare between the soft error-aware and soft error-unaware design optimizations for different architecture allocations, Fig. 5.20 shows the power consumption (P, in mW) and the SEUs experienced (Γ) by the optimized designs produced in Exp:4 and Exp:3 using the random task graph of 60 tasks. As can be seen, the proposed optimization, Exp:4, consistently outperforms the design produced using joint optimization of reduced R and high parallelism, Exp:3, with upto 7% reduction of SEUs experienced for an SER of 10^{-9}. This reliability improvement is achieved with only 3% higher power consumption using an MPSoC with six processing cores.

5.3.6 Concluding Remarks

Increasing design complexity of current and future generations of embedded systems, particularly of MPSoCs, has necessitated design paradigm shift from traditional separate hardware and software design to electronic system-level (ESL) design methodology. Using modular and well-defined design steps early in the design phase, design space can be drastically reduced, while meeting different requirements in terms of power, performance and reliability. Moreover, using holistic system modeling, ESL design methodology can effectively address various MPSoC design challenges effectively. A case study of system-level design optimization has also been presented based on the study of impact of application task mapping on the reliability of MPSoC application (Sect. 5.3.2). Trade-off analyses of other system-level design steps have also been illustrated in detail, underlining ways to achieve joint system optimization in terms of low power consumption and high reliability.

References

1. J.A. Abraham, D.P. Siewiorek, An algorithm for the accurate reliability evaluation of triple modular redundancy networks. IEEE Trans. Comput. **23**(7), 682–692 (1974)
2. B.M. Al-Hashimi (ed.), *System-on-Chip: Next Generation Electronics*, chap. 17 (IEE Circuits, Devices and Systems, London, 2006)
3. S. Aminzadeh, A. Ejlali, A comparative study of system-level energy management methods for Fault-Tolerant hard real-time systems. IEEE Trans. Comput. **60**(9), 1288–1299 (2011)
4. L. Anghel, D. Alexandrescu, M. Nicolaidis, Evaluation of a soft error tolerance technique based on time and/or space redundancy, in *Proceedings of the International Symposium on Integrated Circuit Design and System Design*, Manaus (IEEE Computer Society, Los Alamitos, 2000), p. 237
5. ARM, Advanced microprocessor bus architecture (AMBA) specification, v2.0 (1999), http://www.arm.com
6. ARM, Advanced eXtensible Interface (AXI) specification, AMBA v3.0 (2007), http://www.arm.com
7. G. Ascia, V. Catania, M. Palesi, An evolutionary approach to network-on-chip mapping problem. in *IEEE Congress on Evolutionary Computation*, Edinburgh, vol. 1, 2–5 Sept 2005, pp. 112–119
8. J.R. Azambuja, F. Sousa, L. Rosa, F.L. Kastensmidt, Evaluating large grain TMR and selective partial reconfiguration for soft error mitigation in SRAM-based FPGAs, in *Proceedings of the 15th IEEE International On-Line Testing Symposium (IOLTS)*, Sesimbra-Lisbon, 24–26 June 2009, pp 101–106
9. F. Balarin, P.D. Giusto, A. Jurecska, M. Chiodo, C. Passerone, A.S.-V. Harry Hsieh, E. Sentovich, B. Tabbara, L. Lavagno, K. Suzuk, *Hardware-Software Co-design of Embedded Systems: The POLIS Approach* (Springer, USA, 1997)
10. H. Beitollahi, S.G. Miremadi, G. Deconinck, Fault-tolerant earliest-deadline-first scheduling algorithm, in *Proceedings of the IEEE International Parallel and Distributed Processing Symposium IPDPS 2007*, Long Beach, 26–30 Mar 2007, pp. 1–6
11. L. Benini, D. Bertozzi, A. Bogliolo, F. Menichelli, M. Olivieri, MPARM: exploring the multiprocessor SoC design space with SystemC. J. VLSI Signal Process. **41**(2), 169–182 (2005)

12. J.B. Bernstein et al., Electronic circuit reliability modeling. Microelectron. Reliab. **46**(12), 1957–1979 (2006)
13. D. Brooks et al., Power, thermal, reliability modeling in nanometer-scale microprocessors. IEEE Micro **27**, 49–62 (2007)
14. V. Chandra, R. Aitken, Impact of technology and voltage scaling on the soft error susceptibility in nanoscale CMOS, in *Proceedings of the IEEE International Symposium on Defect and Fault Tolerance of VLSI Systems*, Boston, 2008, pp. 114–122
15. V. Chandra, R. Aitken, Impact of voltage scaling on nanoscale SRAM reliability, in *Proceedings of the DATE'09. Design, Automation and Test in Europe Conference and Exhibition*, Nice, 20–24 Apr 2009, pp. 387–392
16. G. Chen, M. Kandemir, E. Li, Energy-aware computation duplication for improving reliability in embedded chip microprocessors, in *Proceedings of the Asian and South Pacific Design Automation Conference (ASPDAC)*, Yokohama, Japan, 2006, pp. 134–139
17. CoCentric, Designware System-level Library (2008), http://www.synopsys.com/Tools/SLD/VirtualPlatforms/Pages/SLLibrary.aspx
18. F. Dabiri, N. Amini, M. Rofouei, M. Sarrafzadeh, Reliability-aware optimization for DVS-enabled real-time embedded systems, in *Proceedings of the 9th International Symposium on Quality Electronic Design (ISQED)*, San Jose, 17–19 Mar 2008, pp. 780–783
19. M. Dall'Osso, G. Biccari, L. Giovannini, D. Bertozzi, L. Benini, xPIPES: a latency insensitive parameterized network-on-chip architecture for multi-processor SoCs, in *ICCD'03: Proceedings of the 21st International Conference on Computer Design*, San Jose (IEEE Computer Society, Washington, DC, 2003), p. 536
20. W. Dally, Performance analysis of k-ary n-cube interconnection networks. IEEE Trans. Comput. **39**(6), 775–785 (1990)
21. B.P. Dave, G. Lakshminarayana, N.K. Jha, COSYN: hardware-software co-synthesis of embedded systems, in *DAC'97: Proceedings of the 34th Annual Design Automation Conference*, Anaheim (ACM, New York, 1997), pp. 703–708
22. A. Ejlali, B.M. Al-Hashimi, M.T. Schmitz, P. Rosinger, S.G. Miremadi, Combined time and information redundancy for SEU-tolerance in energy-efficient real-time systems. IEEE Trans. Very Large Scale Integr. Syst. **14**(4), 323–335 (2006)
23. A. Ejlali, B.M. Al-Hashimi, P. Rosinger, S.G. Miremadi, Joint consideration of fault-tolerance, energy-efficiency and performance in on-chip network, in *Proceedings of the Conference on Design, Automation and Test in Europe (DATE)*, Nice, 2007, pp. 1647–1652
24. S. Ghosh, S. Basu, N.A. Touba, Joint minimization of power and area in scan testing by scan cell reordering, in *Proceedings of the Annual Symposium on VLSI (ISVLSI)*, Tampa, 2003, p. 246
25. E. Gutiérrez, O. Plata, E.L. Zapata, Optimization techniques for parallel irregular reductions. J. Syst. Archit. **49**(3), 63–74 (2003)
26. R. Gupta, G. De Micheli, Hardware-software cosynthesis for digital systems. IEEE Des. Test Comput. **10**(3), 29–41 (1993)
27. J. Han, Q. Li, Dynamic power-aware scheduling algorithms for real-time task sets with fault tolerance in parallel and distributed computing environment, in *International Parallel and Distributed Processing Symposium*, Denver, 2005, pp. 6–16
28. C.-C. Han, K.G. Shin, J. Wu, A fault-tolerant scheduling algorithm for real-time periodic tasks with possible software faults. IEEE Trans. Comput. **52**(3), 362–372 (2003)
29. J. Hu, R. Marculescu, Energy- and performance-aware mapping for regular NoC architectures. IEEE Trans. Comput Aided Des. Integr. Circuits Syst. **24**(4), 551–562 (2005)
30. A. Israr, S.A. Huss, Specification and design considerations for reliable embedded systems, in *Proceedings of the DATE'08*, Munich, 2008, pp. 1111–1116
31. V. Izosimov, P. Pop, P. Eles, Z. Peng, Design optimization of time-and cost-constrained fault-tolerant distributed embedded systems, in *Proceedings of the Conference on Design, Automation and Test in Europe (DATE)*, Munich, 7–11 Mar 2005, pp. 864–869
32. T. Karnik, P. Hazucha, Characterization of soft errors caused by single event upsets in CMOS processes. IEEE Trans. Dependable Secur. Comput. **1**(2), 128–143 (2004)

33. R.H. Kim et al., 22 nm technology node active layer patterning for planar transistor devices, in *Proceedings of the Optical Microlithography XXII*, San Jose, vol. 7274, ed. by H.J. Levinson, M.V. Dusa (SPIE, Bellingham, 2009) pp. 72742X–72742X-6
34. T. Kogel, R. Leupers, H. Meyr (eds.), System level design principles, in *Integrated System-Level Modeling of Network-on-Chip Enabled Multi-processor Platforms* (Springer, Dordrecht, 2006), pp. 33–42
35. D. Kwai, B. Parhami, Fault-tolerant processor arrays using space and time redundancy, in *Proceedings of the IEEE Second International Conference on Algorithms and Architectures for Parallel Processing, ICAPP*, Singapore, 11–13 June 1996, pp. 303–310
36. D. Kwai, B. Parhami, A unified approach for fault tolerance and dynamic power management in fixed-priority real-time embedded systems. IEEE Trans. Comput Aided Des. Integr. Circuits Syst. **25**(1), 111–125 (2006)
37. M. Lajolo, A. Raghunathan, S. Dey, L. Lavagno, A. Sangiovanni-Vincentell, A case study on modeling shared memory access effects during performance analysis of HW/SW systems, in *Proceedings of the International Workshop on Hardware-Software Codesign*, San Diego, March 1998, pp. 117–121
38. H. Lee, U.Y. Ogras, R. Marculescu, N. Chang, Design space exploration and prototyping for on-chip multimedia applications, in *Proceedings of the Design Automation Conference (DAC)*, San Francisco, 24–28 July 2006
39. H. Lee, N. Chang, U. Ogras, R. Marculescu, On-chip communication architecture exploration: a quantitative evaluation of point-to-point, bus, and network-on-chip approaches. ACM Trans. Des. Autom. Electron. Syst. **12**(3), 1–20 (2007)
40. C. Lee, H. Kim, H.-W. Park, S. Kim, H. Oh, S. Ha, A task remapping technique for reliable multi-core embedded systems, in *Proceedings of the International Conference on Hardware/Software Codesign and System Synthesis (CODES+ISSS)*, Scottsdale, 24–29 Oct 2010, pp. 307–316
41. X. Li, D. Yeung, Exploiting application-level correctness for low-cost fault tolerance. J Instrum. Level Parallel. **10**, 1–28 (2008)
42. G. Li, F. Hu, L. Yuan, An energy-efficient fault-tolerant scheduling scheme for aperiodic tasks in embedded real-time systems, in *Proceedings of the Third International Conference on Multimedia and Ubiquitous Engineering MUE'09*, Qingdao, 4–6 June 2009, pp. 369–376
43. A. Maheshwari, W. Burleson, R. Tessier, Trading off transient fault tolerance and power consumption in deep submicron (DSM) VLSI circuits. IEEE Trans. Very Large Scale Integr. Syst. **12**(3), 299–311 (2004)
44. G. Martin, Overview of the MPSoC design challenge, in *Proceedings of the 43rd Annual Conference on Design Automation*, San Francisco, 2006, pp. 274–279
45. R. Melhem, D. Mosse, E. Elnozahy, The interplay of power management and fault recovery in real-time systems. IEEE Trans. Comput. **53**(2), 217–231 (2004)
46. N. Miskov-Zivanov, D. Marculescu, Circuit reliability analysis using symbolic techniques. IEEE Trans. CAD **25**(12), 2638–2649 (2006)
47. M. Nicolaidis, Time redundancy based soft-error tolerance to rescue nanometer technologies, in *Proceedings of the 17th IEEE VLSI Test Symposium*, San Diego, 25–29 Apr 1999, pp. 86–94
48. V. Nollet, D. Verkestt, A quick Safari through the MPSoC run-time management Jungle, in *IEEE/ACM/IFIP Workshop on Embedded Systems for Real-Time Multimedia, ESTIMedia*, Salzburg, 2007, pp. 41–46
49. U. Ogras, J. Hu, R. Marculescu, Key research problems in NoC design: a holistic perspective, in *Proceedings of the CODES+ISSS*, Jersey City, 2005, pp. 69–74
50. H. Orsilla, T. Kangas, E. Salminen, T. Hämäläinen, D. Timo, M. Hännikäinen, Automated memory-aware application distribution for multi-processor system-on-chips. J. Syst. Archit. Euromicro J. **53**(11), 795–815 (2007)
51. L. Ost, Luciano, S. Varyani, L.S. Indrusiak, M. Mandelli, G.M. Almeida, E. Wachter, F. Moraesm, G. Sassatelli, Enabling adaptive techniques in Heterogeneous MPSoCs based on virtualization. ACM Trans. Reconfigurable Technol. Syst. **5**(3), 1936–7406 (2012)

52. P. Pop, K. Poulsen, V. Izosimov, P. Eles, Scheduling and voltage scaling for energy/reliability trade-offs in fault-tolerant time-triggered embedded systems, in *Proceedings of the International Conference on Hardware/Software Codesign and System Synthesis (CODES+ISSS)*, Salzburg, 2007, pp. 233–238
53. J. Pouwelse, K. Langendoen, H. Sips, Dynamic voltage scaling on a low-power microprocessor, in *Proceedings of the 7th Annual International Conference on Mobile Computing and Networking*, Rome, July 2001, pp. 251–259
54. I. Profeta, J.A., N.P. Andrianos, B. Yu, B. W. Johnson, T.A. DeLong, D. Guaspart, D. Jamsck, Safety-critical systems built with COTS. Computer **29**(11), 54–60 (1996)
55. RTRG, Random task and resource graph tool (2010), http://www.zepler.net/~ras06r/rtrg. Accessed 23 Apr 2010
56. M. Ruggiero, Dynamic power management techniques for system-on-chip. Ph.D. thesis, University of Bologna, 2008
57. M.T. Schmitz, B.M. Al-Hashimi, P. Eles, *System-Level Design Techniques for Energy-Efficient Embedded Systems* (Kluwer, Dordrecht, 2004)
58. R.A. Shafik, B.M. Al-Hashimi, Reliability analysis of on-chip communication architectures: an MPEG-2 video decoder case study. Microprocess. Microsyst. **35**(2), 285–296 (2011)
59. R. A. Shafik, P. Rosinger, B. M. Al-Hashimi, MPEG-based performance comparison between network-on-chip and AMBA MPSoC, in *IEEE Design and Diagnostics of Electronic Circuits and Systems (DDECS)*, Bratislava, Slovakia, Apr 2008, pp. 98–103
60. R.A. Shafik, P. Rosinger, B.M. Al-Hashimi, SystemC-based minimum intrusive fault injection technique with improved fault representation, in *Proceedings of the International On-Line Testing Symposium (IOLTS)*, Rhodes, July 2008, pp. 99–104
61. R.A. Shafik, B.M. Al-Hashimi, S. Kundu, A. Ejlali, Soft error-aware voltage scaling technique for power minimization in application-specific MPSoC. J. Low Power Electronics. **5**(2), 145–156 (2009)
62. R.A. Shafik, B.M. Al-Hashimi, J. Mathew, D.K. Pradhan, S.P. Mohanty, RAEF: a power normalized system-level reliability analysis and estimation framework, in *IEEE Computer Society Annual Symposium on VLSI (ISVLSI)*, Amherst, 2012, pp. 189–194
63. S.R. Shenoy, A. Daniel, Intel architecture and silicon cadence: the catalyst for industry innovation. Technical report, Intel Corp., 17 Apr 2009
64. N. Soundararajan, N. Vijaykrishnan, A. Sivasubramaniam, Impact of dynamic voltage and frequency scaling on the architectural vulnerability of GALS architectures, in *ISLPED'08: Proceedings of the 13th International Symposium on Low Power Electronics and Design*, Bangalore (ACM, New York, 2008), pp. 351–356
65. D.J. Soudris, P. Poechmueller, E.D. Kyriakis-Bitzaros, M.K. Birbas, C.E. Goutis, M. Glesner, Design methodology for systematic derivation of fault-tolerant array processors, in *Proceedings of the CompEuro'92 Computer Systems and Software Engineering*, The Hague, Netherlands, 4–8 May, 1992, pp. 562–567
66. J. Srinivasan et al., The case for lifetime reliability-aware microprocessors, in *Proceedings of the 31st International Symposium on Computer Architecture, ISCA'04*, Washington, DC, 2004, p. 276
67. Synposys, Primetime (2008), http://www.synopsys.com/Tools/Implementation/SignOff/Pages/PrimeTime.aspx
68. R.R. Tamhankar, S. Murali, G.D. Micheli, Performance driven reliable link design for networks on chips. in *Proceedings of the Conference on Asia South Pacific Design Automation*, Shanghai, 2005, pp. 749–754
69. W.J. Van Gils, A triple modular redundancy technique providing multiple-bit error protection without using extra redundancy. IEEE Trans. Comput. **C-35**(7), 623–631 (1986)
70. S. Vangal, J. Howard, G. Ruhl, S. Dighe, H. Wilson, J. Schanz, D. Finan, P. Iyer, A. Singh, T. Jacob, S. Jairr, S. Venkataramarr, Y. Hoskote, N. Borkar, An 80 tile 1.28 tflops network-on-chıp ın 65 nm CMOS, in *Proceedings of the International Solid State Circuit Conference (ISSCC)*, San Francisco, CA, USA, 2007, pp. 98–100

71. S. Wang, J. Hu, S.G. Ziavras, On the characterization and optimization of on-chip cache reliability against soft errors. IEEE Trans. Comput. **58**(9), 1171–1184 (2009)
72. W. Wolf, The future of multiprocessor systems-on-chips, in *Design and Automation Conference (DAC)*, San Diego, 2004, pp. 681–685
73. B. Wong, F. Zach, V. Moroz, A. Mittal, G. Starr, A. Kahng, *Nano-CMOS Design for Manufacturability* (Wiley, Hoboken, 2009)
74. F. Zanini, D. Atienza, C.N. Jones, G. De-Micheli, Temperature sensor placement in thermal management systems for MPSoCs, in *Proceedings of the IEEE International Symposium on Circuits and Systems (ISCAS)*, Bangkok, Jun 2003, pp. 1065–1068
75. Y. Zhang, K. Chakrabarty, Energy-aware adaptive checkpointing in embedded real-time systems, in *Proceedings of the International Conference on Design, Automation and Test in Europe (DATE)*, Munich, 2003, p. 10918
76. D. Zhu, R. Melhem, D. Mosse, The effects of energy management on reliability in real-time embedded systems, in *IEEE/ACM International Conference on Computer Aided Design (ICCAD)*, San Jose, 2004, pp. 35–40

Chapter 6
Fault-Tolerant Reconfigurable On-Chip-Network

Mohammad Hosseinabady and Jose L. Nunez-Yanez

Fault-tolerant reconfigurable on-chip networks are infrastructure communication architectures for the future computing platforms that can run many applications with dynamic work-load in the presence of different types of faults in the system. This chapter introduces the architecture of such platforms and studies their fault-tolerant features.

6.1 Introduction

With ever increasing number of transistors available on a single die, designers are able to embed different types of computational and communicational resources in a single chip to cope with insatiable demand for running different complex applications with different constraints. Therefore, designers propose Multi-Processor Systems-on-Chip (MPSoC) frameworks [1, 2] which contain many different types of components such as processors, memories, communication modules, accelerators and even Field Programmable Gate Arrays (FPGAs).

According to the International Technology Roadmap for Semiconductors (ITRS) predictions [3], shown in Fig. 6.1, the number of processing elements in Systems-on-Chip (SoC) consumer portable products will grow rapidly in the future. Moreover, the amount of memory will also increase proportionally with the number of processing elements.

M. Hosseinabady (✉)
Institute of Electronics, Communications and Information Technology (ECIT)
Queens University Belfast, Belfast, Northern Ireland, UK
e-mail: m.hosseinabady@qub.ac.uk

J.L. Nunez-Yanez
Department of Electrical and Electronic Engineering, University of Bristol, UK
e-mail: j.l.nunez-yanez@bristol.ac.uk

J. Mathew et al. (eds.), *Energy-Efficient Fault-Tolerant Systems*,
DOI 10.1007/978-1-4614-4193-9_6,

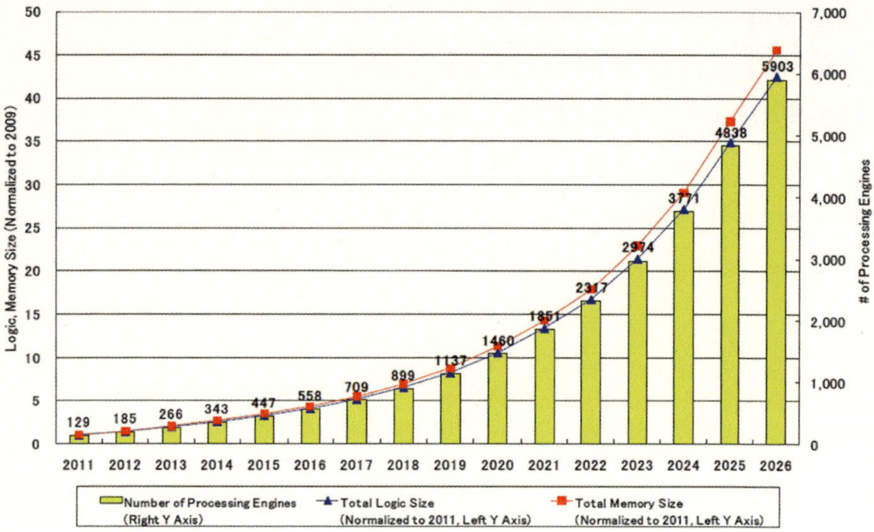

Fig. 6.1 SoC consumer portable design complexity trends [1]

As the number of components increases in an MPSoC, the communication architecture among them plays the main role in determining the speed of the system. The traditional bus-based architectures as the simplest way to connect components become the main bottleneck because of its poor scalability and high amount of power consumption. As the number of components connected to a bus increases, the length of the bus and the number of interfaces between the bus and components increase which in turn increase the capacitance of the wire constituting the bus. This capacitance is the main parameter in determining the bus speed [4]. In addition, using buses on a chip to connect many components results in layout issues in manufacturing the whole chip [5]. Coping with these issues, researchers have augmented the traditional bus architecture with new techniques such as bus segments [6], crossbar switches [7], arbiters [7, 8], buffer [9], and asynchronous transactions [10].

However, these new bus architectures cannot fully cope with the scalability problem. For example, although, using multi-level or hierarchical bus structures can handle the communication among a few cores, still a regular mechanism is required to scale up the communication structure towards multi- and many-core MPSoCs.

Addressing the scalability problem and alleviate the power consumption in the communication structure, researchers have proposed Network-on-Chip (NoC) [11, 12]. This new on-chip interconnection scheme utilizes identical routers to connect components through a network topology such as Mesh. In addition, a packet based communication performs the transaction among components. The main benefit of the NoC which makes that scalable is the regularity in topology, routing and router architecture.

Static NoC-based MPSoCs, which their parameters and structures are determined at design time, are the first group of these computational platforms. However, a general static NoC-based MPSoC cannot be efficient because of dynamic behavior of applications and high-probability of occurring faults in the chip during its life time. While such an MPSoC can efficiently execute a group of applications, it may not show high performance in running other types of applications because of their different functionality constraints or different types of traffic that they may inject into the NoC. In addition, a statically designed NoC-based MPSoC may be unable to run applications or lose its performance in running application in the event of permanent, transient or intermittent faults which are very common in new advanced nanotechnologies.

Adding dynamic reconfigurable features in terms of software or hardware reconfigurability can provide a versatile NoC-based SoC which can be used in different scenarios by adapting itself to different constraints required by applications. Employing software program running on processor components in an MPSoC is a well-known technique to add software versatility enhancement to MPSoCs. However, this kind of programmability is not enough to adapt the communication network to different traffics. Therefore, hardware reconfigurability can be considered as a solution to provide an adaptable NoC-based MPSoC which can be used for efficiently executing different applications with wide verity of computational and communicational constraints.

This chapter studies different approaches for dynamic reconfigurable NoC-based MPSoCs and explains the benefits and drawbacks of each approach by focusing on their fault-tolerant feature. The rest of this chapter is organized as follows. The next section, explains the details of NoC-based MPSoC and introduces some terminologies that will be used in the rest of the chapter. Sect. 6.3 explains some of the key issues and problems in designing a NoC. The reconfigurable NoCs are discussed in Sect. 6.4. Finally, Sect. 6.5 explains a fault-tolerant stochastic task mapping technique on a Network on Reconfigurable Chip (NoRC).

6.2 NoC-Based MPSoC

The network-on-chip has been proposed as an alternative to on-chip bus and point-to-point communication architectures to cope with ever increasing demand for bus bandwidth and power consumption in new SoCs with many heterogeneous cores [11, 12].

Figure 6.2 shows a typical NoC-based MPSoC platform in which cores communicate together through a network of routers. This MPSoC comprises of two main components: cores and Network-on-Chip (NoC). These components are explained in the sequel.

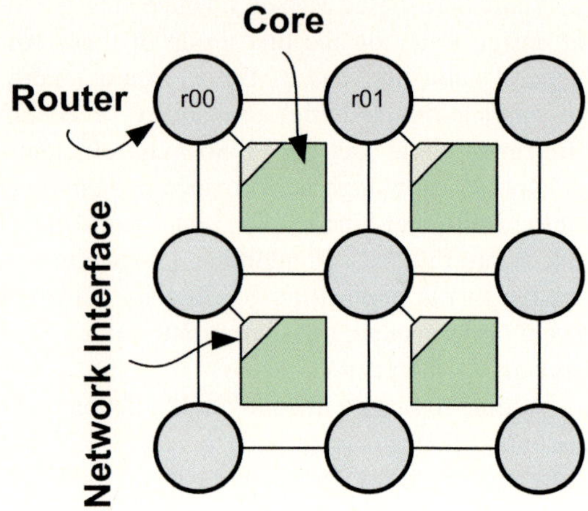

Fig. 6.2 A typical NoC-based MPSoC platform

6.2.1 Cores

Cores consist of processors such as general-purpose processor, Graphics Processing Units (GPUs) or Digital Signal Processors (DSPs), accelerators, memories, and controllers and so on which provide storage and computational components in the platform. Different architectures have been proposed for cores in an MPSoC that can be categorized as homogeneous and heterogeneous cores.

In a homogeneous MPSoC, cores are of the same type which usually consists of at least a processing element (PE) and a local memory which may be connected via a local bus. TeraFLOPS is a homogeneous NoC-based MPSoC manufactured by Intel [13] consisting of 80 identical processors. Figure 6.3 depicts the processor in this MPSoC which is called *processor engine*. The processor engine consists of two independent fully-pipelined single-precision floating-point multiply-accumulator units, 3 KB single-cycle instruction memory, and 2 KB of data memory.

Cores in a heterogeneous MPSoC are not identical and each of which can efficiently executes different types of tasks. Xpipes [14] is a homogeneous NoC-based MPSoC comprises of general-purpose processors, DSPs, Reduced Instruction Set Computing (RISC) processors, memories and other modules as cores. Figure 6.4 depicts a heterogeneous NoC-based MPSoC for the MPEG4 decoder which consists of different types of cores connected together through a NoC.

A comparison between homogeneous and heterogeneous MPSoC has been done in [15] and [16]. Using Spidergon NoC topologies, Saponara and Fanucci [15] compare two heterogeneous and homogenous architectures for some real applications including H.264/MPEG AVC video codec and a low-distortion digital audio amplifier. Based on their results, the heterogeneous architecture shows less power consumption and lower area occupation. However, homogeneous architecture is more flexible to run different applications and standards. Taking 4G baseband modem applications, which require high computing power with real-time constraints

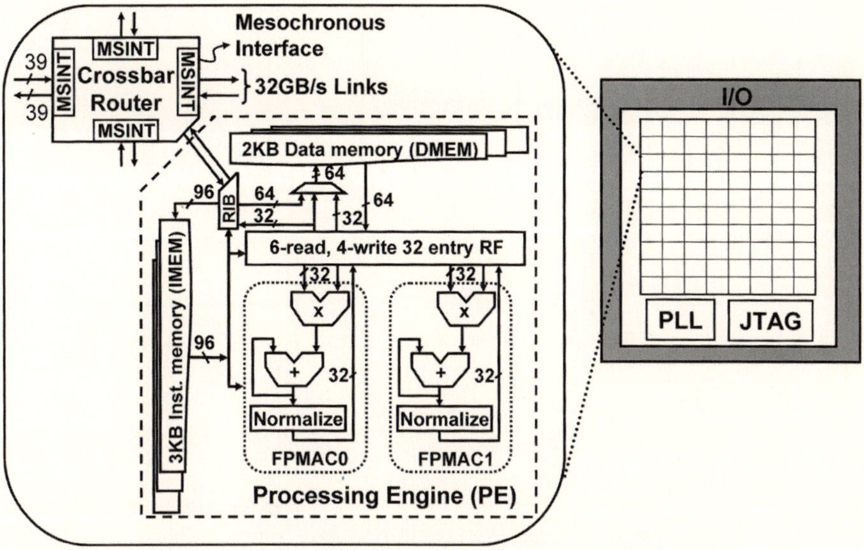

Fig. 6.3 Core architecture in TeraFLOPS NoC-based SoC [13]

and low-power consumption as well as their rapid evolution of different standards, Jalier et al. [16] show that the homogeneous technique can be more efficient and flexible than the heterogeneous technique.

In summary, since heterogeneous MPSoCs use different types of cores (i.e., processing elements) and usually application specific cores, they can provide a high power efficiency and smaller area occupation for specific applications. However, using identical cores, homogeneous MPSoCs provide high flexibility and regularity which make them more scalable and reconfigurable at run-time to cope with faults and dynamic workloads.

6.2.2 NoC

A packet-based communication scheme is used to transfer data from a source core to a destination core. In this scheme, the source core encapsulates the data into some packets and sends them to the underlying communication platform to be delivered to the destination core. A packet comprises of data, routing information, error detection or correction codes, quality of service information or even some configuration data. Some of the information in the packets is used by the destination core and some of them are used by the communication platform. To provide a communication infrastructure to be used in different scenarios and applications a communication protocol such as Open Systems Interconnection (OSI) model [17], a multi-level communication protocol used in computer network, should be

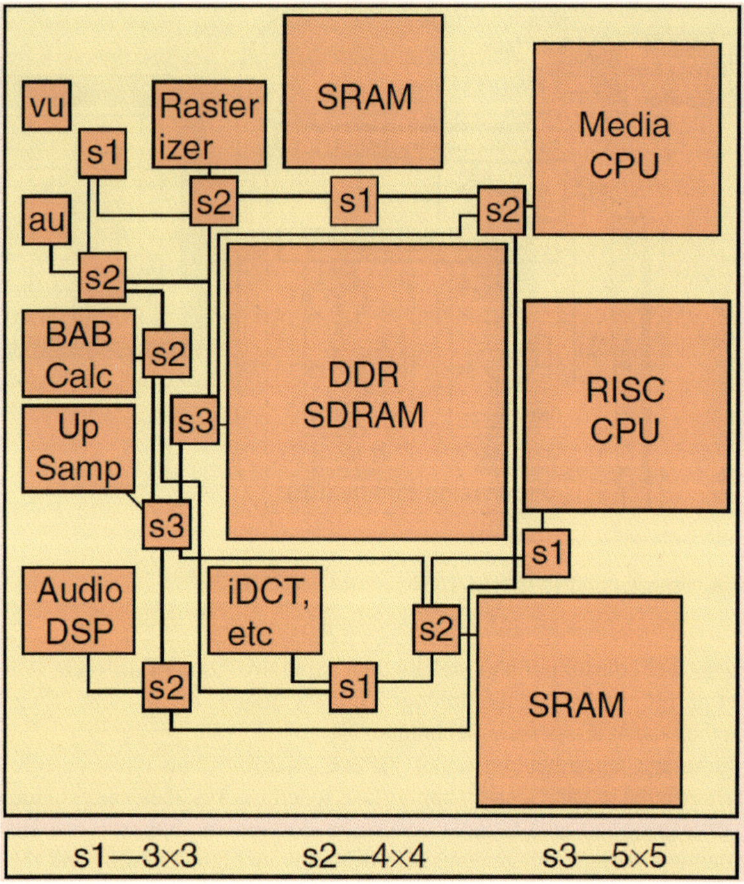

Fig. 6.4 A heterogeneous NoC configurations for the MPEG4 decoder [14]

used to formally define the communication protocol in a NoC. Therefore, some researchers have proposed and used such protocols [18–21]. In a common protocol, cores send/receive the data in the form of *messages*. Then, the messages transmitted over NoC divided into *packets* which in turn partitioned into flow control units called *flits* [22]. In addition, a switching technique determines how packets traverse from one point to another. Many different switching techniques are proposed in NoC including wormhole, store and forward, and Virtual-Cut-Through (VCT) [21]. The *wormhole* switching is the most promising technique used in NoCs due to the limited availability of buffering resources and low latency requirements.

A typical NoC consists of Network Interface (NI) and routers connected together through links which are explained in the sequel.

Network Interfaces (NIs): An NI is used to connect a core to the NoC. The main role of an NI is to form packets from data and retrieve data from packets. It also should split the generated packet to flits to be sent to the attached router and merge the received flits from router to create the packet. Despite of the data, NI should encapsulate the routing information to the packets. Based on the addressing technique used in the NoC, NI can have a significant area contribution in the NoC. For example, the NIs in a TV companion chip redesigned with a NoC as the interconnect fabric [23] shows 78% of increase in chip area.

Routers: Routers are the main components in NoC which have been the focus of most of the research in NoC area. A router consists of a few input and output ports with some amount of buffers, a switch fabric to create the path between input ports and output ports, and a routing algorithm which determines the output port for the received packets (or flits). A router usually performs mechanisms for quality of service, fault-detection and fault correction. Based on the switching techniques and rooting algorithm different structure for a router have been proposed in the literature [20, 21].

Network topology: Network topology defines how to connect routers in a NoC. This topology determines all the potential paths between source and destination cores which will be established by routers between the cores for data transfer. If the topology cannot provide enough paths between cores in a NoC to support the bandwidth requirement of cores, then routers and other parts of NoC cannot efficiently transfer data between cores. Therefore, topology has a key role to provide the required bandwidth in an NoC which has been studied by researchers [20, 21].

The next section explains the NoC design techniques and some of the issues and problems in designing a NoC.

6.3 NoC-Based MPSoC Design

Based on the NoC structure and types of cores and applications, different scenarios have been proposed for designing a NoC-based MPSoC. One of the common design flows [24, 25] is shown in Fig. 6.5 which consists of three stages: *task mapping*, *task scheduling* and *core mapping*. This design flow starts with an application (described by a task graph) and finishes by mapping the application on NoC-based MPSoC.

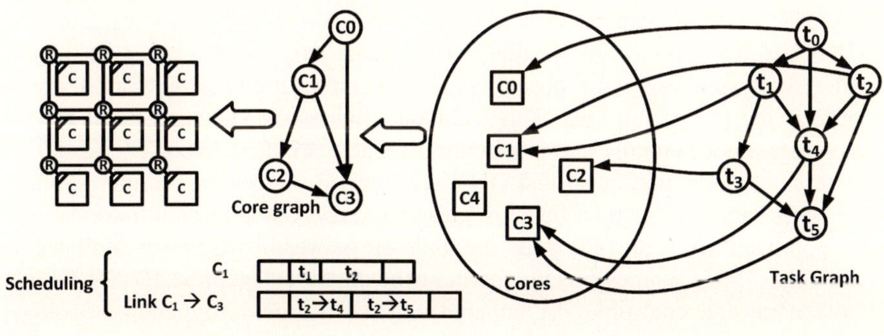

Fig. 6.5 A typical NoC design flow

Task-graph is a traditional technique to model concurrency in an application. In this technique, each application is divided into tasks which usually represent the computation parts of the application. These tasks are connected through links which denote the data or control dependency among tasks. Therefore, vertices in a task-graph show the tasks in the application and links between two tasks represent the dependency between them. A task-graph is able to model different constraints in an application such bandwidth, throughput, power consumption, execution time and so on by augmenting tasks and links by numbers and weights.

Choosing suitable cores to run tasks is the first step in the design-flow shown in Fig. 6.5. Different factors such as instruction set in cores, area of core; amount of work required in a task can be used to make this decision. The chosen cores can be modeled by a graph called *core-graph* which its nodes represent the cores and its links represent the dependency among cores. It is possible to map more than one task to a core during the task mapping step. In this case, a scheduling algorithm should define the order of task execution in the core which hosts more than one core as well as the order of data transaction on the links used by more than one task to send their data. After, mapping and scheduling the tasks in an application, the resulted core graph should be mapped on NoC architecture. Each core should be connected to a router such that all constraints in the application graph are met. Different objectives such as communication power consumption, packet latency, and throughput are considered to efficiently map cores in this step [26].

6.3.1 Design-Time Synthesis

Scalability and modularity of NoC structure can be used to synthesis and adapt the communication architecture to performance, bandwidth, power consumption and area constraints of specific use cases. These adaptations can be performed at design time by utilizing static design techniques. These techniques mostly try to synthesis the NoC architecture by finding a suitable topology or by finding proper values for different parameters in the NoC, such as the amount of buffers in the routers or NIs, the number and size of flits or packets and so on. A design exploration of application-specific MPSoCs in order to find a NoC with optimal cost-performance trade-offs design constrains, such as power, area, and wire-length is explained in [27]. Focusing on application specific NoCs, a linear programming based techniques for topology synthesis is proposed in [28] generating irregular NoC topology to minimize the power and area. This technique divides the synthesis problem into two stages: floor planning and interconnection network generation. The results of this research show that mesh and QNoC [29] based topologies consume 2.3 and 1.75 times more power than their irregular topology. Using the multi-commodity flow approach, Hu et al. [30] study the trade-off between NoC power and latency. The results of this synthesis technique show the improvement of 52.1%, 29.4% and 35.6% in power latency product compared with mesh, torus and hypercube topology, respectively.

6.3.2 *Major Problems in Statically Designed NoC*

Since, statically designed NoCs are synthesized and configured for specific use-cases and platform conditions, as long as these conditions and use cases are not changed the designs show a good performance. However, faults occurring in the system may change the platform architecture or emerging new standards or algorithms in the applications can drastically reduce the efficiency of these platforms. Three main sources of issues in statically designed NoC-based SoCs are as follows:

- Dynamism in applications
- Platform efficiency degradation after using the system for a while
- High probability of faults in the system

Each of these issues is explained in the sequel.

6.3.2.1 Dynamism in Applications

The number and types of applications running on a SoC platform are increasing rapidly which makes the platform difficult to satisfy all their requirements and constraints in terms of bandwidth and latency. For example, TI OMAP™ 4 platform [31] is a general purpose SoC targeted for future mobile phones and mobile-Internet-devices (MIDs) to support different types of applications such as web-browsing, HD video, location-based services and social networking with diversity of bandwidth and quality of service requirements. This platform consists of ARM Cortex-A9 processors, graphics accelerator, image signal processor and a few controllers of other modules. The Philips Nexperia™ digital video platform [32] is another SoC for vide variety of image, video and signal processing applications.

The dynamism in applications can be categorized into two groups: *inter-application* dynamism and *intra-application* dynamism. Example of applications with inter-application dynamism are MPEG-4 [33] and Reconfigurable Video Coding (RVC) [34] which consist of many features and encoding/decoding techniques that some of them do not have a clear definition and not proposed yet. Any implementation of these applications should be compatible with different number of standards even with those which are not developed yet. In addition, running multiple applications with different constraints and behavior on a NoC-based MPSoC results in an inter-application dynamism. Considering inter- and intra-application dynamisms, a design exploration technique is explained in [35].

6.3.2.2 Platform Efficiency Degradation

NoC-based platforms lose their efficiency and system utilization (up to 60% [36]) mainly because of task dispersal and resource fragmentations issue after mapping and releasing applications [37, 38].

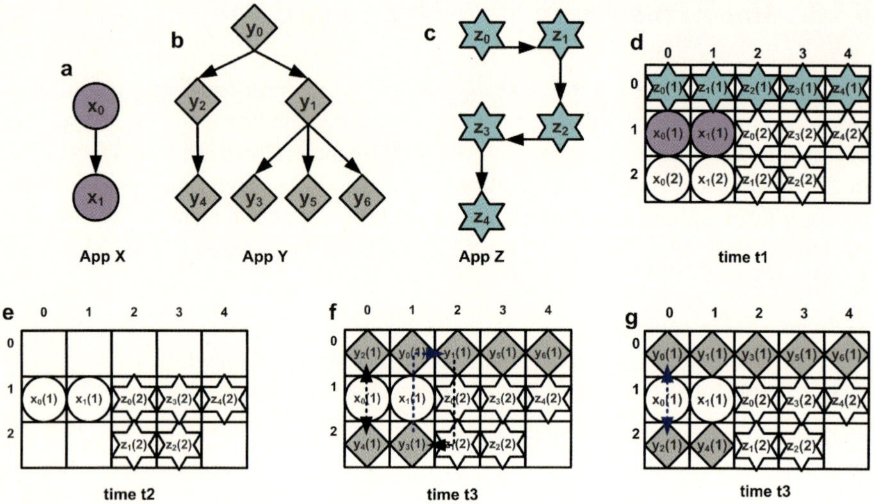

Fig. 6.6 Task dispersal [37]

Figure 6.6 illustrates an example which shows the task-dispersal in a NoC which results in external contention (contention between the traffics of two different applications) in a non-contiguous task mapping scheme which in turn reduces the system efficiency [37]. Three simple applications, *App X*, *App Y* and *App Z*, described by task graphs are shown in Fig. 6.6a–c, respectively. Figure 6.6d shows a 3×5 mesh-based NoC at time *t1* which runs two instances of *App X* and two instances of *App Z*. Let us assume, at time *t2* the instance one of *App Z* and instance two of *App X* finish their execution and release the allocated resources and a new request for *App Y* arrives. As it can be seen, the available resources spread in three disjoint convex regions which are {(0,0), (0,1), (0,2),(0,3),(0,4)}, {(2,0),(2,1)} and {(2,4)}. Figure 6.6(f) and (g) show two different tile allocations for the requested application. Two communicating tasks mapped into two different regions cause external contention. For example, Fig. 6.6f shows the mapped *App Y* in which tasks y_2 and y_1 communicate with tasks y_4 and y_3, respectively. Because these communications should cross the regions, they cause external contention with the traffic of other mapped applications. Figure 6.6g illustrates the other tile allocation which shows less external contention because only the communication between tasks y_0 and y_4 cross the regions [37].

6.3.2.3 Fault

As the critical dimensions in technology of manufacturing NoC shrink and the number of devices on a chip increases the reliability of NoCs decreases drastically. Note that, faults in a NoC can bring the entire system to a complete halt. Therefore, techniques are required to keep the system alive even in the event of faults. The highly scaled NoC designs are very vulnerable to different types of faults including *permanent*, *intermittent* and *transient* faults. Permanent faults occurring due to transistor wear-out [33] and electro-migration [39] of a conductor can make a NoC erroneous during its life time. Intermittent fault can at random moments exhibit its erroneous effect, or not. Transient faults [40] including caused by neutron radiations from cosmic rays or alpha particles from packaging materials pose a major challenge to the design of memories and logic circuits in nanometer technologies.

To cope with aforementioned issues, researchers propose reconfigurable NoC-based MPSoC in which dynamic techniques can be used to adapt the system to the changes in the application and architecture. The next section explains dynamic NoCs and their design techniques.

6.4 Reconfigurable NoC

This section studies the reconfigurable NoCs [41, 42] with a specific focus on fault-tolerant techniques. NoC reconfiguration techniques are the schemes which tune the initial design of a NoC to cope with the dynamic behavior of applications or any changes in the NoC structure at runtime. A typical reconfigurable technique changes the state of the NoC in terms of its structure, components, parameters, or software. These reconfigurable techniques can be done in two ways with and without stalling the system [43]. The former techniques apply a stall in the system before applying the reconfiguration techniques. Therefore, it guaranteed that packets do not deliver to a wrong destination during reconfiguration process. However, the idea with the latter techniques is to allow applications and traffic continue their tasks without interruption during reconfiguration process. In this case, it is possible that some packets are delivered to the wrong destination. Therefore, the reconfiguration techniques also should guarantee that packet will be delivered to their destinations.

Reconfigurable techniques can be applied on different components in a NoC which can be categorized as *application-level*, *protocol-level*, and *hardware-level* reconfigurable schemes. The details of these techniques are explained in the sequel.

6.4.1 Application-Level

Application level also known as system-level reconfigurable techniques usually change the state of a NoC-based MPSoC in terms of the tasks mapped on processors. *Dynamic task mapping*, *task scheduling* and *task migration* are some of the techniques in this category [44, 45]. Usually a dedicated core in the NoC as the *manager* controls the status of the entire system via a specific controller [46] or operating system [47] and perform the application-level reconfiguration techniques.

The fault-tolerant NoC based on application-level reconfiguration techniques primarily consist of three parts: a dynamic fault detection technique [48, 49], a group of system assessment metrics [50] and a fault-aware resource management [50].

Considering redundant and spare resources in a system to be used as replacements for faulty resources is a well-known technique to provide fault-tolerant feature. Shamshiri and Cheng [51] have shown that using spare cores and wires in a NoC-based SoC can significantly improves reliability of the system. This idea has been the motivation of some application-level fault-tolerant techniques for NoCs.

Considering permanent, transient and intermittent faults in NoC-based SoCs, an application-level fault-aware resource management technique is proposed in [50] in order to optimize the entire system performance and communication energy consumption. This technique is considers a network of routers connected through a mesh topology with deadlock-free, minimal-path routing and virtual-channels. Attached to each router is a core which can be a processor or memory. A specific core called manager runs a platform OS to detect fault and to manage applications and resources in the NoC. It also assumes that applications have been analyzed statically and the core graph (Sect. 3) for each application is available. This research studies three different distributions for spare cores which are side assignment (in which spare core are assigned to the boundary cores), random assignment and uniform assignment. If there is a faulty core in the system, optimizing Manhattan distance among cores [52], core fragmentation and link contentions this work finds a proper spare core to run tasks in the faulty core.

Stochastic task mapping [38] is another group of application-level fault-tolerant techniques in NoC [53, 54]. A stochastic fault-tolerant technique based on [38] will be discussed as a case study in Sect. 5.

6.4.2 Protocol-Level Reconfigurability

Using adaptable routing algorithms, flow control techniques and dynamic quality of service (QoS) schemes, protocol-level reconfigurable techniques provide a reconfigurable NoC to cope with the dynamic behavior of application and occurring fault in the system [55]. A fault-tolerant routing scheme is proposed in [56] for 2D mesh NoCs which combines the North-last and South-last turn models to create a robust hybrid NoC routing scheme.

6.4.3 Hardware-Level Reconfigurability

Using the concept of partial reconfiguration available in new FPGAs [57], can be used to reconfigure hardware components in a NoC-based MPSoC without interrupting the tasks in the rest of the system.

An FPGA-based NoC called Programmable NoC (PNoC) has been proposed by [58] in which a lightweight router, requiring a few FPGA resources, has been used as its key feature. The PNoC is reconfigurable at design time and runtime. At design time, different network architecture can be constructed, whereas, some parameters including communication path width, flow control and timeout handling can be determined for tuning the entire system. At runtime, reprogramming the routing tables, it is possible to add and remove nodes dynamically with support of partial reconfiguration in FPGAs. Using dynamically reconfigurable FPGAs, Jovanovic et al. [59] have proposed a scalable dynamic NoC. Nollet et al. [60] proposes a heterogeneous NoC-based MPSoC comprising of general-purpose processors (GPPs), specialized processors (DSPs and accelerators), and reconfigurable hardware tiles with different sizes. Using partial dynamic reconfigurable FPGAs, a run-time reconfigurable NoC framework has been proposed in [61]. This framework dynamically can change the number of links between cores in the NoC as well as the routing table in routers in order to reduce the latency in the system.

6.5 Case Study: Network on Reconfigurable Chip

This section studies the stochastic fault-tolerant task mapping technique on Network-on-Reconfigurable-Chip (NoRC) [38]. First, a brief explanation of the NoRC architecture and then the stochastic task mapping technique will be discussed.

6.5.1 NoRC Architecture

Figure 6.7a shows an overview of the NoRC architecture model consisting of eight tiles connected through a 2×4 mesh topology. Each tile consists of a router, a run-time reconfigurable region (such as the regions that can be created in Virtex-5 or

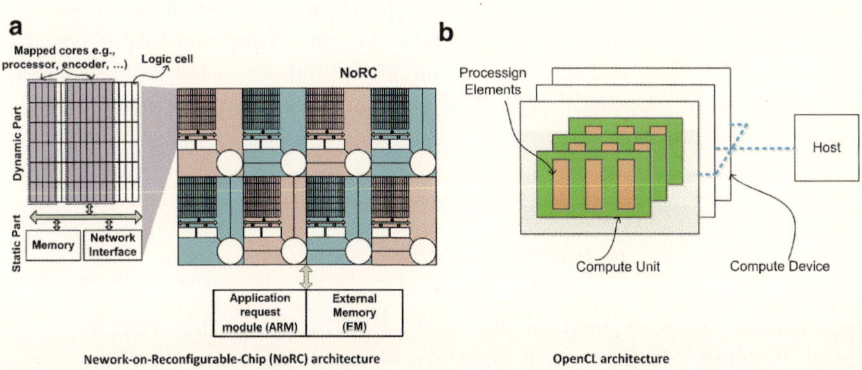

Fig. 6.7 Network-on-Reconfigurable-Chip (NoRC) architecture

Table 6.1 Analogy between NoRC and OpenCL framework

NoRC	OpenCL
ARM + EM	Host
Tile	Compute device
Dynamic part	Compute unit
Mapped core	Processing elements

Virtex-6 FPGAs using the partial reconfiguration design flow) that can implement different types of cores (e.g. processors, communication cores etc.), a small local memory and a network interface. The small local memory in a tile is used to save the portion of the code and data needed for the task in that tile. Because of the structure of the tiles in the proposed NoRC architecture, the mesh interconnection topology is selected because of its simplicity, regularity and suitability for VLSI implementation. The routers provide an XY routing algorithm with a wormhole switching technique. An external module called application request module (ARM), which can access the tiles using a simple control network, sends application requests to the platform while an external memory (EM) keeps all configuration data for the applications that need to be mapped and run. In the general case, the ARM could be mapped in a host processor and the NoRC platform used as a reconfigurable coprocessor resource.

The proposed architecture is conceptually similar to OpenCL [62] platform model shown in Fig. 6.7b. The model consists of a host connected to one or more OpenCL devices. An OpenCL device is divided into one or more compute units (CUs) which are further divided into one or more processing elements (PEs). Computations on a device occur within the PEs. Table 6.1 summarises the analogy between NoRC platform components and their OpenCL equivalents. However, cores in the proposed platform utilise message passing to communicate together. In this analogy, ARM and EM play the role of host processor which controls the task mapping. A tile is a compute device in the OpenCL model that can consists of a few dynamic parts. The dynamic part corresponds to the compute unit of OpenCL that can host a few core. Finally, a core is the processing element in the OpenCL. The main advantage of the proposed architecture in compared with the OpenCL model is the communication among cores. Whereas, global and local memories are used as a shared communication medial among kernels (i.e., processing elements) in OpenCL which is not scalable and restricts the communication due to the limited memory bandwidth, in the proposed NoRC, cores are communicate through a NoC which is scalable and distributed.

6.5.2 Stochastic Routing Algorithm

Applications to be mapped on the NoRC architecture are described by a task graph in which nodes represent the computational tasks and a link between two tasks denotes the data communication among them. Because of its inherent and

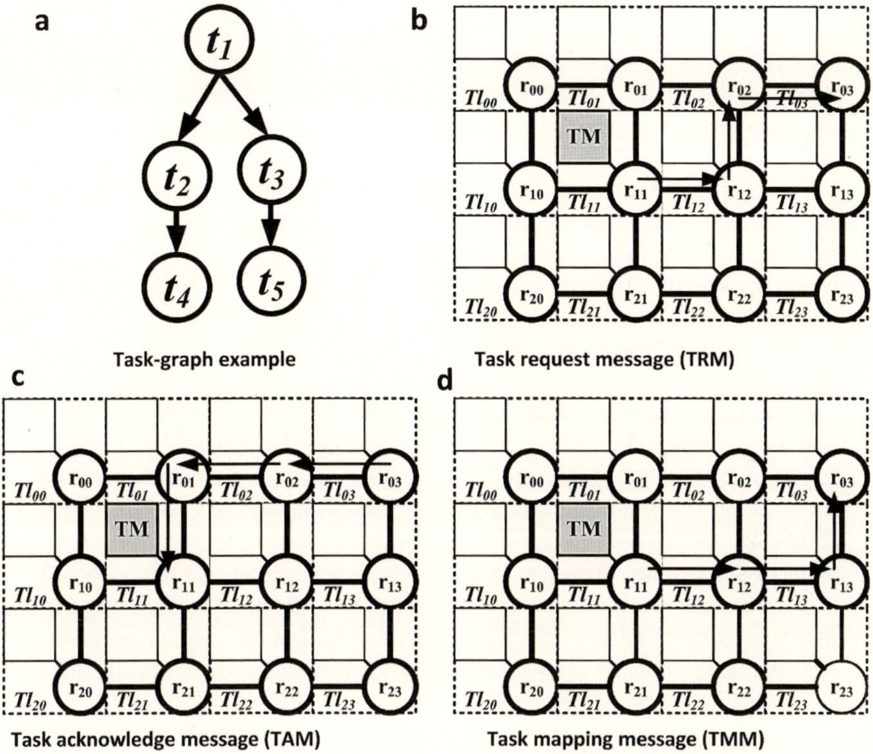

Fig. 6.8 Task mapping scheme

high capability of fault-tolerance, a stochastic routing algorithm is used to map applications (which consist of tasks) on the NoRC platform. In this scheme, for each application to be mapped on the platform there is a special task called task manager that is responsible for mapping and monitoring the tasks of that application. The task manager is mainly performs the task mapping, fault-detection, task migration and energy management.

The task mapping algorithm consists of three parts each of which run on ARM, task manager, and network interfaces. In the request of mapping an application, first, the ARM randomly finds a free tile to map a task manager. Then, using a stochastic routing, the task manager will be responsible to map the tasks in the given application, sequentially. For this purpose, each network interface in a tile has a clear contribution in this stochastic task mapping. Taking a simple example, we explain the stochastic task mapping algorithm.

Figure 6.8a shows an application represented by its task graph which comprises of five tasks. When ARM receives a request to map this application, it randomly selects a tile to map a task manager. For example tile Tl_{11} is selected as shown in Fig. 6.8b. Then the task manager maps the tasks sequentially using a three-phase random walk search scheme.

- **Send task request message:** In this step, the task manager creates a task request message (TRM) consisting of task information and a message life time. The message life time is a counter which determines the number of routers as the length of the random walk. Then, it sends that to the tile's network interface. The network interface randomly selects one of its neighbouring tiles and sends it the message (e.g., TL_{12} in Fig. 6.6(b)). The network interface in that tile (i.e., Tl_{12}) checks the capability of hosting the task and modifies the message by this information and decrements the life time counter. Then it sends the modified message to a randomly selected tile around itself (e.g., Tl_{02}), except the tile which has sent the TRM. This mechanism is repeated by network interface while the life time is not zero.
- **Receive task acknowledge message**: The network interface that receives a TRM with life time of zero, modifies the message and changes its type to task acknowledge message (TAM) and sends it to the task manager using the XY routing algorithm (Fig. 6.8c). Task manager examines the received TAM, and selects a proper tile from the randomly visited ones that can host the task. In one scenario, the tile which is close to the previously mapped task can be selected to reduce the task dispersion.
- **Send task mapping message**: When the task manager selects the proper tile to host the task, it sends a task mapping message (TMM) to that tile using the XY routing algorithm (Fig. 6.8d). Then, the network interface which receives this message communicates with ARM for other information such as bitstream, code and data required realizing the task.

Algorithm 1 ARM routine

```
Data: app an application
Data: NoRC platform
Result: mapped app on NoRC
while app_retry < APP_RETRY do
   while manager_retry < MANAGER_RETRY do
      tl = randomly select a tile on the NoRC
      if tl can host a task manager then
         Map manager in tl
         manager_map_success = true
         Break
      else
         manager_map_success = false
         manager_retry++
      end
   end
   if manager_map_success == false then
      application_retry++;
      app_map_success = false;
      Continue;
```

```
        end
        wait for response from manager;
        if app is rejected by task manager then
            application_retry++
            app_map_success = false;
        else
            app_map_sucess == true;
            Break;
        end
    end
    if app_map_success == false then
        Reject the app
    end
```

Algorithm 2 **Task manager routine**

```
Data: app an application
Result: mapped app on NoRC
while app! = NULL do
    tsk = selectT ask(app)
    app = app − {tsk}
    bool task_map_sucess = false
    while task_retry < TASK_RETRY do
        msg = TRM,tsk,LIFE_TIME
        p = randomly selected an output port
        Send out msg through p;
        wait for response
        mdg = received message
        if rmsg is TAM then
            Tl = findT ile(rmdg)
            if Tl! = NULL then
                msg = TMM, tsk
                Send out msg to tile Tl
                task_map_success = true
                Break;
            else
                task_retry++;
            end
        end
        if task_map_success == false then
            Reject the application
        end
    end
end
```

Algorithm 3 **Network interface routine**

```
Data: msg received message
Result: modify msg or map a task
if msg is a task request message then
   if the tile can host the requested task then
      msg = msg + (tile_id)
   end
   if msg.life_time! = 0 then
      msg.life_time−−;
      p = a randomly selected output port
        Send out msg though p
     else
        Change the msg type to TAM
        Send back the msg to manager by using XY routing
     end
  end
```

Algorithm 1-Algorithm 3 show the ARM, task manager and network interface contributions in the stochastic task mapping techniques. In order to increase the probability of mapping tasks and applications, ARM and task manager use a retry mechanism in the event of any failure to map a task or an application. Three parameters restrict the number of retries in ARM and task manager. The ARM retries $MANAGER\_RETRY$ times (Line 5 in Algorithm 1) if the randomly selected tile cannot host the task manager. In addition, the ARM retries APP_RETRY times (Line 4 in Algorithm 1) if mapping the application fails. In addition, the mask manager retries $TASK\_RETRY$ times (Line 7 in Algorithm 2) if it fails to find a proper tile to host a task.

6.5.3 Dynamically Fault-Tolerance in NoRC

Different types of permanent or temporary dynamic faults can occur at run-time during the system's lifetime due to component wear-out (caused by electromigration) [63] or cosmetic radiations [64]. In this section, we consider faults that may occur on links; routers and tiles (consist of network interfaces and reconfigurable areas). The major causes of faults on links are electromigration and stress-induce defects [65] which may occur at manufacturing time or at runtime.

FPGAs usually implement the combinational and sequential logic in programmable complex logic blocks (CLBs), which are customized by loading configuration data (bitstream) in the SDRAM memory cells. These memories are vulnerable to charged particle strikes to the cells which can modify the function of the task mapped on the reconfigurable part.

6.5.3.1 Fault Detection

Fault detection is an important step in a fault-tolerant platform. There are two mechanisms in detecting a fault in the platform.

- **Regular fault detection**: This mechanism is run by routers in which a router regularly checks its connections to its neighboring routers, and sends the results to the ARM, regularly. For this purpose, each router sends a test-packet to its neighboring routers. The router that receives this test-packet sends back a test-acknowledge-packet and informs the sender of its status. If the sender receives an incorrect acknowledge massage or does not receive the acknowledge massage then it reports the ARM of the probability of fault in the target router or communicating link. Otherwise, it reports the ARM that the target router and the link are non-faulty. The ARM has a database to keep track of all these reports. Algorithm 4 shows that how ARM manage these reports.

Algorithm 4 Updating fault database in ARM

```
Data: repi received reports from rith router
Result: updated fault database
if repi received then
   if repi reports router rj is non-faulty then
      Mark router rj and communication link between routers ri and rj as non-faulty
   else
      if repi reports a faulty router rj then
         Mark the router rj as faulty router
      else
         Mark the communication link between routers ri and rj as faulty
      end
   end
else
   Mark router ri as faulty
End
forall the ri do
   if all its communication links are faulty then
      Mark ri as a faulty router
   end
end
```

The ARM checks if all router connected to a specific router report a link fault then that router marked as faulty. For example, let's consider the faulty platform shown in Fig. 6.9 in which router r_{32} is faulty. In this case, routers r_{31}, r_{42}, r_{33}, and r_{22} report to the ARM that their link to the router r_{32} is faulty. Then, ARM mark

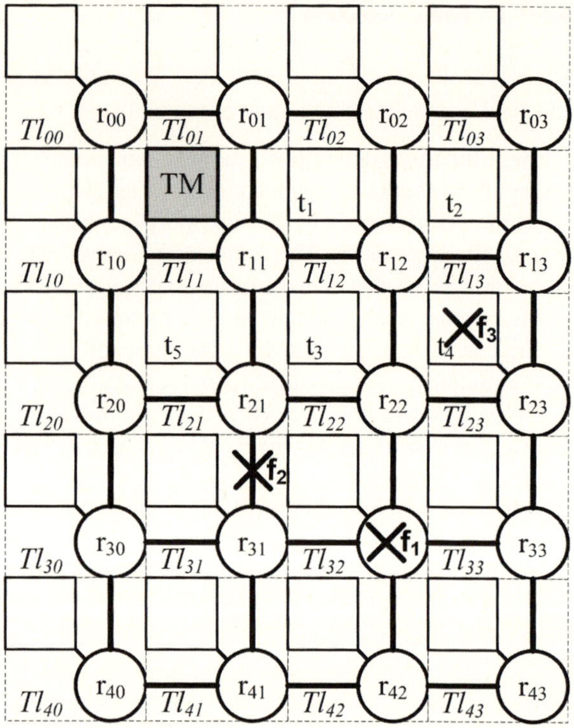

Fig. 6.9 A faulty platform

the router as a faulty router. This platform also contains a faulty link between tiles Tl_{21} and Tl_{31}. In this case, two routers r_{21} and r_{31} reports that link as a faulty link.

- **Run time fault detection**: The second mechanism is performed by the mapped tasks in an application. Since tasks are communicating together through an acknowledgement scheme, if there is any problem in this scheme the destination tiles will be reported to the task manager as the faulty tile. Then the task manager migrates the task in the faulty tile to another tile and reports the ARM about the fault in the tile. For example, let's consider the faulty platform in Fig. 6.9 in which the application shown in Fig. 6.8a is mapped and tile Tl_{23} becomes faulty during running the application. In this case, tasks t_2 and t_4 are connected together and mapped on tiles Tl_{13} and Tl_{23}. As the tile Tl_{23} is faulty, the task t_2 does not receive any acknowledgement from task t_4, therefore, it informs the task manager of this fault. Then the task manager tries to remap the task t_4 on another non-faulty tile.

The ARM information about the faulty routers, tiles and links has two impacts on the task mapping algorithm. First, the ARM probability to find a tile for mapping a task manager will be decreased. Second, TRM life-time will be increased in order to compensate the reduction in the number of tiles and paths available to map a task. The next section explains and calculates these effects.

6.5.4 *Fault Effects*

This section explains the effect of faults on the mapping algorithm explained in Sect. 5.3. A fault adds some overhead to the application mapping problem and reduces the probability of mapping the applications. This overhead will be analyzed in this section.

6.5.4.1 Mapping a Task Manager

Faulty tiles reduce the probability of mapping a task manager by ARM. Considering a uniform distribution for the occupied and faulty tiles, the probability of finding a free tile to map a task manager is $q = 1-(u+u_f)$, in which u is the tiles utilization factor (i.e., percentage of the occupied tiles) and u_f is percentage of faulty tiles. Therefore, ARM will find a free tile after r_{mgr} retries with the probability of

$$P_{r\,mgr}(M) = 1-(1-q)^{r_{mgr}} = 1-\left(u+u_f\right)^{r_{mgr}} \tag{6.1}$$

6.5.4.2 Mapping a Task

Let's assume p is the probability of mapping a task after a random-walk with *l*-steps, then the probability of mapping a task after r_{tsk} retry is as Eq. 6.2

$$P_{r_{tsk}}(t) = 1-(1-p)^{r_{tsk}} \tag{6.2}$$

Note that, the value of p depends on the platform utilisation, the TRM life-time and percentage of the faulty tiles (i.e., u_f).

The probability of mapping an application, shown by Eq. 6.3 is equal to probability of mapping the task manager multiply by the probability of mapping tasks in the application provided that the task manager has been mapped. Assuming that the probability of mapping tasks in an application are independent and the same and the probability of mapping a task after r_{tsk} -retry is $P_{r_{tsk}}(t)$ then Eq. 6.5 shows the probability of mapping an application consists of $|A|$ tasks. Equation 6.6 shows this probability after substituting Eqs. 6.1 and 6.2.

$$\mathrm{P_l(A) = P(M)\,P_l(A|M)} \tag{6.3}$$

$$= \mathrm{P(M)}\left(\prod_{\forall t\in A} \mathrm{P_{r_{tsk}}(t)}\right) \tag{6.4}$$

$$= \mathrm{P(M)(P_{r_{tsk}}(t))^{|A|}} \tag{6.5}$$

$$= \mathrm{(1-(u+u_f)^{r_{mgr}})(1-(1-p)^{r_{tsk}})^{|A|}} \tag{6.6}$$

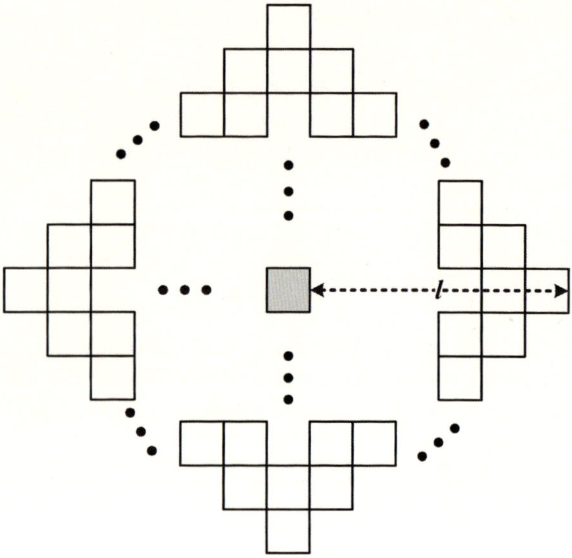

Fig. 6.10 Search space with lifetime l around a tile ($N = 2l(l+1)$)

Therefore, the probability of mapping an application with r_{App} -retry is as Eq. 6.7

$$\begin{aligned} P_{r_{App}}(A) &= 1 - P_1(A)^{r_{App}} \\ &= 1 - (1 - u^{r_{mgr}})\left(1 - (1-p)^{r_{tsk}}\right)^{|A|^{r_{App}}}. \end{aligned} \tag{6.7}$$

Note that increasing the three retry factors (i.e., r_{mgr}, r_{tsk} and r_{App}) can improve the probability of mapping an application in the event of fault. However, it drastically increases the consumed energy to map an application. Life-time which determines the probability p in Eq. 6.7 is another factor to increase this probability and decrease the retry factors. To find the minimum life-time to get the best probability of mapping an application, the search space of a TRM with life-time of l should be investigated. Figure 6.10 shows the search space for life-time l around a tile. The number of tiles in this space is $2\times[1+3+5+\ldots+(2l-1)+2l = 2l(l+1)$. Therefore, an application with maximum number of task $|A| = 2l(l+1)$ can be mapped in this area. Solving this, Eq. 6.8 shows the minimum life-time.

$$l \geq \frac{\sqrt{2 \times |A| + 1} - 1}{2} \tag{6.8}$$

This lower bound is correct for a blank platform (i.e., with not previously mapped application or faulty tile). Considering the uniform distribution for occupied and faulty tiles, the number of free tile in the search space of Fig. 6.10 reduces to $(1 - u - u_f)(2l(l+1))$ which means that the application size should be less than $|A| = (1 - u - u_f)(2l(l+1))$. Therefore, Eq. 6.8 will be changed to

$$l \geq \frac{\sqrt{2 \times \frac{|A|}{1-(u+u_f)} + 1} - 1}{2} \tag{6.9}$$

Therefore, the lower bound life time is directly related to $\sqrt{|A|}$ and $\frac{1}{\sqrt{1-(u+u_f)}}$. Reference [66] has shown that the probability of reaching tills which are away from the task manager, using the proposed random walk is very low. It has also shown that the random walk contains all lattice points of the ball of radious $(1 - \epsilon)\sqrt{l}$ where $0 < l < 1$ and l is the number of steps or lifetime. Therefore, we need $l = d^2$ steps to travel a distance of d in a mesh topology. Hence, we increase the lifetime lower bound to Eq. 6.10 meaningfully increase the chance of finding a free tile to map a task.

$$l \geq \left(\frac{|A|}{\sqrt{1 - \left(u + u_f\right)}} \right)^2 \geq \frac{|A|}{2} \tag{6.10}$$

This shows that lower bound lifetime is proportional to $|A|$ and we explained also it is proportional to $\frac{1}{\sqrt{1-(u+u_f)}}$ therefore, Eq. 6.11 show the lower bound lifetime in which β is a constant and has shown in [38] that it is about 1.

$$l \geq \beta \frac{|A|}{\sqrt{1 - (u + u_f)}} \tag{6.11}$$

6.5.4.3 Energy Model

We use bit energy model described in [52] to evaluate the energy overhead caused by faulty tiles. In this model, the energy is determined by the number of bits transferring between a source core and a destination core. According to this definition, Eq. 6.12 shows the energy consumption for sending a bit from a source to the destination.

$$E_{bit} = E_{srcNI_{Bit}} + n E_{Router_{Bit}} + (n-1)\, E_{Wire_{Bit}} + E_{desNI_{Bit}} \tag{6.12}$$

where $E_{srcNI_{Bit}}$, $E_{Router_{Bit}}$, $E_{Router_{Bit}}$ and $E_{desNI_{Bit}}$ denote the bit energy dissipated in the source network interface, a router, a communication link and the destination network interface, respectively.

6.5.5 Experimental Results

A SystemC simulation has been used to model the NoRC platform and evaluate its fault-tolerant features. The SystemC NoRC model is based on the technique proposed in [67], a fast SystemC-based transaction level modeling simulator for large NoCs.

Fig. 6.11 263 decoder mp3 decoder task graph

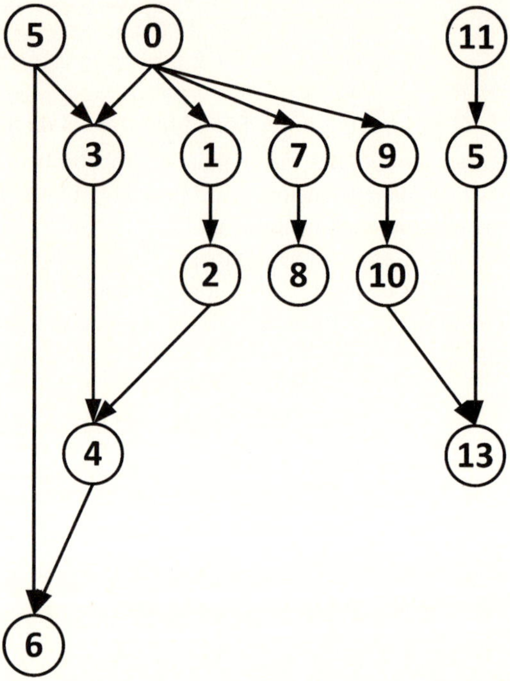

For the evaluation process, five groups of applications have been considered as the benchmarks. Each of the first four benchmarks which generated using the task graph for free (TGFF) package [68] contains 50 task graphs. The last one contains two task graphs of real applications including 263 decoder mp3 decoder (Fig. 6.11) [28] and a multi-window display (MWD) (Fig. 6.12) [69].

Table 6.2 shows the statistics of the synthetic benchmarks. Benchmark_1, Benchmark_2, and Benchmark_3 contain small, medium, large size task graphs. Benchmark_4 contains a mixture of small and large task graphs.

We have sent 2,000 requests to map and release randomly selected task graph from each benchmark to a NoRC platform consisting of 64×64 tiles. In addition, 50 faulty tiles have been considered during the simulation. Table 6.3 shows the number of rejected applications in the non-faulty and faulty NoRC. As it can be seen, the faulty routers have a very small impact on the stochastic task mapping.

Table 6.4 shows the number of steps required to map the requested applications. As it can be seen, the faulty NoRC required more steps than that of the non-faulty NoRC to map applications. According to the bit energy model, the number of steps can directly determine the energy consumption during task mapping process. The last three columns in this table show the overhead of task mapping in the faulty NoRC for three different value of β.

We have sent 2,000 requests to randomly map and release selected applications from 50 instances of real task graphs of Figs. 6.11 and 6.12 to the 64×64 NoRC

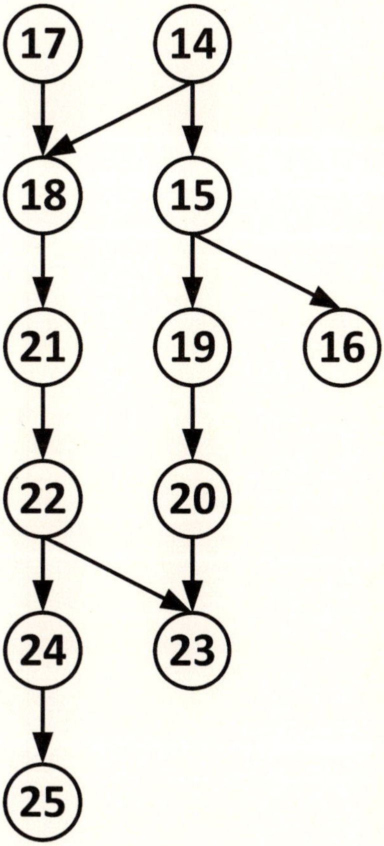

Fig. 6.12 Multi-window display (MWD) task graph

Table 6.2 Synthetic benchmarks statistics

	Min # of task	Max # of task	Ave # of task
Benchmark_1	2	6	2.32
Benchmark_2	9	46	23.35
Benchmark_3	23	114	60.71
Benchmark_4	2	119	23.56

Table 6.3 Number of rejected applications in synthetic benchmarks application mapping

	Non-faulty NoRC			Faulty NoRC		
	$\beta = 0.5$	$\beta = 1.0$	$\beta = 1.5$	$\beta = 0.5$	$\beta = 1.0$	$\beta = 1.5$
Benchmark_1	5	0	0	6	0	0
Benchmark_2	459	440	425	461	441	429
Benchmark_3	1,203	1,199	1,188	1,213	1,234	1,348
Benchmark_4	36	12	10	37	13	11

Table 6.4 Total number of steps in synthetic benchmarks task mapping

	Non-faulty NoRC			Faulty NoRC			Overhead %		
	$\beta = 0.5$	$\beta = 1.0$	$\beta = 1.5$	$\beta = 0.5$	$\beta = 1.0$	$\beta = 1.5$	$\beta = 0.5$	$\beta = 1.0$	$\beta = 1.5$
Benchmark_1	11,938	19,579	26,674	24,425	24,179	31,634	51.12	19.0	15.68
Benchmark_2	1,519,949	2,872,456	4,396,379	3,538,732	3,872,456	5,386,305	57.04	25.8	18.38
Benchmark_3	14,936,289	27,945,438	42,802,813	34,597,327	35,145,438	53,739,420	56.83	20.5	20.35
Benchmark_4	916,716	1,764,645	2,559,103	2,126,382	2,364,645	3,097,265	56.89	25.4	17.38

Table 6.5 Real task graph simulation results

Non-faulty NoRC		Faulty NoRC		
# of rejected Apps	# of total steps	# of rejected Apps	# of total steps	# of total steps overhead %
15	1,123,457	18	1,357,834	17.26

Table 6.6 Average MDpL for mapped applications

	Non-faulty NoRC	Faulty NoRC	Overhead %
Benchmark_1	4.36	4.47	2.46
Benchmark_2	4.72	4.93	4.25
Benchmark_3	6.34	7.21	12.06
Benchmark_4	4.10	4.33	5.31
Real task graphs	3.94	4.1	3.91

platform. Also, we have injected 50 faulty tiles in the platform. Table 6.5 shows the number of rejected applications and total number of steps for non-faulty and fault NoRC.

As it can be seen, the proposed stochastic task mapping is almost successful to map the request applications with 17.26% overhead in the total number of steps which is a representative for the overhead of extra consumed energy.

To evaluate the effects of faulty tiles on the mapped application average Manhattan Distance per Link (MDpL) as defined in [38] for each mapped application is computed for non-faulty and faulty NoRC. MDpL is defined by the sum of all Manhattan distances between two connected tasks in the application divided by the total number of links in the application.

Table 6.6 shows the results. To get a sense of this number, let's consider the real application results, as the average distance between two tasks is 3.94 and 4.1 in the non-faulty and faulty NoRC, respectively and average distance between tasks in the ideal case is $(15 + 12)/2 = 13.5$ (average number of links in the two task graphs of Figs. 6.11 and 6.12), then the total average distance overhead between tasks is $(4.1 * 13.5 - 3.94 * 13.5)/(4.1 * 13.5) = 3.91\%$ which is acceptable with the assumption of 50 faulty tiles.

Note that, based on the proposed bit energy model in which the consumed energy is proportional to the Manhattan distance, the overheads represented in the last column of Table 6.1 also show the average overheads of consumed energy in the presence of faults in the NoRC.

References

1. A. Jerraya, W.Wolf, *Multiprocessor Systems-on-Chips.* (Morgan Kaufmann Publishers, San Francisco, CA 94111, 2004)
2. W. Wolf, A.A. Jerraya, G. Martin, Multiprocessor System-on-Chip (MPSoC) technology. IEEE Trans. Comput-Aided Des. Integr. Circuits Syst. **27**(10), 1701–1713 (2008)

3. ITWG, *International Technology Roadmap for Semiconductors: System Drivers.* (Semiconductor Industry Association, 2011) [online]. http://www.itrs.net/Links/2011ITRS/2011Chapters/2011SysDrivers.pdf
4. R. Mehra, L.M. Guerra, J.M. Rabaey, A partitioning scheme for optimizing interconnect power. IEEE J. Solid-State Circ. **32**(3), 433–443 (1997)
5. P. H. Wu, T. Y. Ho, Thermal-aware bus-driven floorplanning, in *Proceedings of International Symposium on Low Power Electronics and Design* (ISLPED), (2011), pp. 205–210
6. J.Y. Chen, J. Wen-Ben, W. Jinn-Shyan, L. Hsueh-I, T.F. Chen, Segmented bus design for low-power systems. IEEE Trans. VLSI Syst **7**(1), 25–29 (1999)
7. ARM Ltd., International technology roadmap for semiconductors: system drivers, ARM Limited, ARM IHI 0011A, (1999)
8. R. Lu, Cheng-kok Koh, AMBA-Bus, a high performance bus architecture for system-on-chips, in *Proceedings of International Conference Computer-Aided Design*, (2003), pp. 8–12
9. STMicroelectronics, STBus communication system concepts and definitions, STMicroelectronics, 14178 Rev 2, (2012)
10. A.J. Martin, M. Nystrom, Asynchronous techniques for system-on-chip design. Proc. IEEE **94**(6), 1089–1120 (2006)
11. L. Benini, G. De Micheli, *Networks on Chips: Technology and Tools* (Morgan Kaufmann Publishers, Amsterdam, 2006)
12. W. Dally, B. Towles, Route packets, not wires: on-chip interconnection networks, DAC'01, (2001), pp. 684–689
13. S.R. Vangal, J. Howard, G. Ruhl, S. Dighe, H. Wilson, J. Tschanz, D. Finan, A. Singh, T. Jacob, S. Jain, V. Erraguntla, C. Roberts, Y. Hoskote, N. Borkar, S. Borkar, An 80-tile sub-100-W TeraFLOPS processor in 65-nm CMOS. IEEE J. Solid-State Circ. **43**(1), 29–41 (2008)
14. D. Bertozzi, L. Benini, Xpipes: a network on chip architecture for gigascale systems-on-chip. IEEE Circ. Syst. Mag. **4**(2), 18–31 (2004)
15. S. Saponara, L. Fanucci, Homogeneous and heterogeneous MPSoC architectures with Network-On-Chip connectivity for low-power and real-time multimedia signal processing, *Hindawi Publishing Corporation, VLSI Design*, vol. 2012, Article ID 450302
16. C. Jalier, D. Lattard, A. A. Jerraya, G. Sassatelli, P. Benoit, L. Torres, Heterogeneous vs homogeneous MPSoC approaches for a mobile LTE modem, *IEEE Design, Automation and Test in Europe* (DATE'10), (2010), pp. 184–189
17. "Information technology – open systems interconnection – basic reference model: the basic model," ISO/IEC 7498-1:1994(E)
18. L. Benini, D. Bertozzi, Network-on-chip architectures and design methods. IEE Proc.-Comput. Digit. Tech. **152**(2), 261–271 (2005)
19. M. Coppola, S. Curaba, M. D. Grammatikakis, G. Maruccia, F. Papariello, OCCN: A Network-On-Chip modeling and simulation framework, in *Proceedings of the Design, Automation and Test in Europe Conference and Exhibition Designers'* Forum (DATE'04), (2004), pp. 174–179
20. A. Agarwal, C. Iskander, R. Shankar, Survey of network on chip (NoC) architectures & contributions. J. Eng. Comput. Archit. **3**(1) (2009) [online]. http://www.scientificjournals.org/journals2009/articles/1423.pdf
21. T. Bjerregaard, S. Mahadevan, A survey of research and practices of network-on-chip. ACM Comput. Surv. **38**(1), 1–51 (2006)
22. S. Pasricha, N. Dutt, *On-Chip Communication Architectures: System on Chip Interconnect* (Morgan Kaufmann Publishers, Amsterdam, 2008)
23. F. Steenhof, H. Duque, B. Nilsson, K. Goossens, R. P. Llopis, Networks on Chips for high-end consumer electronics TV system architectures, *IEEE Design, Automation and Test in Europe* (DATE'06), (2006), pp. 148–153
24. J. Hu, R. Marculescu, Communication and task scheduling of application-specific networks-on-chip. IEE Proc. Comput. Digit. Tech. **152**(5), 643–651 (2005)
25. R. Marculescu, U.Y. Ogras, L.S. Peh, N.E. Jerger, Y. Hoskote, Outstanding research problems in NoC design: system, microarchitecture, and circuit perspectives. IEEE Trans. Comput.-Aided Des. Integr. Circ. Syst. **28**(1), 3–21 (2009)

26. S. Murali, G. De Micheli, Bandwidth-constrained mapping of cores onto NoC architectures, in *Proceedings of Design, Automation and Test in Europe Conference*, pp. 896–901, Feb 2004
27. C. L. Chou, R. Marculescu, U. Ogras, S. Chatterjee, M. Kishinevsky, D. Loukianov, System interconnect design exploration for embedded MPSoCs, in *Proceedings of International Workshop on System Level Interconnect Prediction* (SLIP'11), (2011), pp. 1–8
28. K. Srinivasan, K. S. Chatha, G. Konjevod, Linear programming based techniques for synthesis of network-on-chip architectures, in *Proceedings of IEEE International Conference on Computer Design: VLSI in Computers and Processors*, (ICCD'04), (2004), pp. 422–429
29. E. Bolotin, I. Cidon, R. Ginosar, A. Kolodny, Cost considerations in network on chip. Integ. VLSI J. **38**(1), 19–42 (2004)
30. Y. Hu, Y. Zhu, H. Chen, R. Graham, C. Cheng, Communication latency aware low power NoC synthesis, in *Proceedings of Design Automation Conference* (DAC'06), (2006), pp. 574–579
31. Texas Instruments Inc., OMAP 4: mobile applications platform, (2011) [Online]. http://www.ti.com/lit/ml/swpt034b/swpt034b.pdf
32. J. A. de Oliveira, Nexperia computing architecture for connected consumer applications, in *Proceedings of the 20th International Conference on VLSI Design,* (2007)
33. O. Avaro, A. Eleftheriadis, C. Herpel, G. Rajan, L. Ward, MPEG-4 systems: overview. Signal Process. Image Commun. **15**, 281–298 (2000)
34. S.S. Bhattacharyya, J. Eker, J.W. Janneck, C. Lucarz, M. Mattaveli, M. Raulet, Overview of the MPEG reconfigurable video coding framework. J. Signal Process. Syst. **63**(2), 251–263 (2009)
35. P. v. Stralen, A. Pimentel, Scenario-based design space exploration of MPSoC, in *Proceedings of IEEE International Conference on Computer Design* (ICCD), (2010), pp. 305–312
36. Y. Zhu, Efficient processor allocation strategies for mesh connected parallel computers. J. Parallel Distrib. Comput. **16**(4), 328–337 (1992)
37. M. Hosseinabady, J. L. Nunez-Yanez, and A. M. Coppola, Task dispersal measurement in dynamic reconfigurable NoCs. *IEEE Annual Symposium on VLSI,* (2010), pp. 167–172
38. M. Hosseinabady, J.L. Nunez-Yanez, Run-time stochastic task mapping on a large scale network-on-chip with dynamically reconfigurable tiles. IET Comput. Dig. Tech. **6**(1), 1–11 (2012)
39. C. Constantinescu, Trends and challenges in VLSI circuit reliability. IEEE Micro **23**(4), 10–19 (2003)
40. M. Hosseinabady, P. Lotfi-Kamran, J. Mathew, S. Mohanty, D. Pradhan, single-event transient analysis in high speed circuits, in *Proceedings of International Symposium on Electronic System Design* (ISED'11), (2011), pp. 112–117
41. R. Dafali, J-Ph. Diguet, M. Sevaux, Key research issues for reconfigurable network-on-chip, in *Proceedings of International Conference on Reconfigurable Computing and FPGAs*, (2008), pp. 181–186
42. C. Killian, C. Tanougast, F. Monteiro, A. Dandanche, A new efficient and reliable dynamically reconfigurable network-on-chip. J. Electrical Comput. Eng. Article ID 843239 (2012)
43. T. Pionteck, R. Koch, C. Albrecht, Applying partial reconfiguration to networks-on-chips, in *Proceedings of Field Programmable Logic and Applications* (FPL'06), (2006), pp. 1–6
44. E.L.S. de Carvalho, N.L.V. Calazans, F.G. Moraes, Dynamic task mapping for MPSoCs. IEEE Des. Test Comput. **27**(5), 26–35 (2010)
45. S. Manolache, P. Eles, Z. Peng, Fault and energy-aware communication mapping with guaranteed latency for applications implemented on NoC, in *Proceedings of 42nd Annual Design Automation Conference* (DAC 05), (2005), pp. 266–269
46. C. Chou, U.Y. Ogras, R. Marculescu, Energy and performance aware incremental mapping for network on chip with multiple voltage levels. IEEE Trans. Comput.-Aided Des. Integr. Circ. Syst. **27**(10), 1866–1879 (2008)
47. V. Nollet, T. Marescaux, D. Verkest, Operating-system controlled Network-on-Chip, in *Proceedings of Design Automation Conference* (DAC'04), (2004), pp. 256–259
48. C. Grecu, A. Ivanov, R. Saleh, E. S. Sogomonyan, P. P. Pande, On-line fault detection and location for NoC interconnects, in *Proceedings of IEEE International On-Line Testing Symposium*, (2006), pp. 145–150

49. C. Killian, C. Tanougast, F. Monteiro, A. Dandanche, Online routing fault detection for reconfigurable NoC, in *Proceedings of Field Programmable Logic and Applications* (FPL'06), (2010), pp. 183–186
50. C. Chou, R. Marculescu, FARM: fault-aware resource management in NoC-based multiprocssor platforms, in *Proceedings Design, Automation & Test in Europe Conference & Exhibition* (DATE'11), (2011), pp.1–6
51. S. Shamshiri, K.-T. Cheng, Modeling yield, cost, and quality of a spare-enhanced multicore chip. IEEE Trans. Comput. **60**(9), 1246–1259 (2011)
52. T. T. Ye, L. Benini, G. De Micheli, Analysis of power consumption on switch fabrics in network routers, in *Proceedings of Design Automation Conference* (DAC'02), (2002), pp. 524–529
53. P. Bogdan, T. Dumitras, R. Marculescu, Stochastic communication: a new paradigm for fault-tolerant networks-on-chip, Hindawi Publishing Corporation VLSI Design Volume (2007), Article ID 95348.
54. W. Song, D. Edwards, J. L. Nunez-Yanez, S. Dasgupta, Adaptive stochastic routing in fault-tolerant on-chip netwroks, in *Proceedings of ACM/IEEE International Netwrok-on-Chip* (NoCS'9), (2009), pp. 32–37
55. S. Murali, D. Atienza, L. Benini, G. De Micheli, A multi-path routing strategy with guaranteed in-order packet delivery and fault-tolerance for networks on chip, in *Proceedings of 42nd Annual Design Automation Conference* (DAC 06), (2006), pp. 845–848
56. S. P. Yong Zou, Ns-ftr: a fault tolerant routing scheme for networks on chip with permanent and runtime intermittent faults in Design Automation Conference (ASP-DAC), pp. 443–448. IEEE, Jan 2011
57. Xilinx, Partial reconfiguration user guide, UG702 (v12.3) October 5, (2010)
58. C. Hilton, B. Nelson, PNoC: a flexible circuit-switched NoC for FPGA-based systems. IEE Proc. Comput. Dig. Tech. **153**(3), 181–188 (2006)
59. S. Jovanovic, C. Tanougast, S. Weber, C. Bobda, *CuNoC:* A scalable dynamic NoC for dynamically reconfigurable FPGAs, in *Proceedings of Field Programmable Logic and Applications* (FPL'07), (2007), pp. 753–756
60. V. Nollet, P. Avasare, H. Eeckhaut, D. Verkest, H. Corporaal, Run-time management of a MPSoC containing FPGA fabric tiles. IEEE Trans. Very Large Scale Integr. (VLSI) Syst. **16**(1), 24–33 (2008)
61. V. Rana, D. Atienza, M. D. Santambrogio, D. Sciuto, G. De Micheli, A reconfigurable Network-on-Chip architecture for optimal multi-processor SoC communication, *16th IFIP/IEEE International Conference on Very Large Scale Integration*, (2008), pp.321–326
62. "The OpenCL Specification," Khronos OpenCL Working Group, Version 1.1, (2011)
63. S. Borkar, Designing reliable systems from unreliable components: the challenges of transistor variability and degradation. IEEE Micro **25**(6), 10–16 (2005)
64. M. Hosseinabady, P. Lotfi-Kamran, Pejman, J. Mathew, S. Mohanty, D. Pradhan, Single-event transient analysis in high speed circuits, in *Proceedings of the 2011 International Symposium on Electronic System Design* (ISED '11), (2011), pp. 112–117
65. T. Tonegawa, M. Hiroi, K. Motoyama, K. Fujii, H. Miyamoto, Suppression of bimodal stress-induced voiding using high-diffusive dopant from Cu-alloy seed layer, in *Proceedings of the International Interconnect Technology Conference*, (2003), pp. 216–218
66. M.B.G. Lawler, D. Griffeath, Internal diffusion limited aggregation. Ann. Probab. **20**(4), 216–218 (1992)
67. M. Hosseinabady, J.L. Nunez-Yanez, Fast and low overhead architectural transaction level modeling for large-scale network-on-chip simulation. IET Comput. Dig. Tech. **6**(6), 384–395 (2012)
68. K. Vallerio, Task graph for free (2003) [Online], http://ziyang.eecs.umich.edu/dickrp/tgff/
69. K. Srinivasan, K. S. Chatha, A low complexity heuristic for design of custom network-on-chip architectures, in *Proceedings of the conference on Design, automation and test in Europe* (DATE'06), (2006) pp. 130–135

Chapter 7
Bio-Inspired Online Fault Detection in NoC Interconnect

Malachy McElholm, Jim Harkin, Liam McDaid, and Snaider Carrillo

7.1 Introduction

Guaranteeing fault-free design of modern electronic systems is becoming increasingly difficult as we scale to high density chips and many-core systems, where variations in the silicon manufacturing process means living with failure is a reality as faulty conditions can occur post deployment. The presence of unreliability requires modern systems to be adaptive to faulty conditions post deployment. To address this reliability challenge researchers have looked to building brain-inspired computing architectures, based on Spiking Neural Networks, which aim to mimic the efficient and self-adaptive information processing capabilities of the human brain; i.e. provide greater levels of fault-tolerance through flexible self-adaption to identified faults.

The high level of fine-grained parallel processing in the brain via synapses and neurons is one aspect that enables fault-tolerance. Several brain inspired computing approaches have explored the use of Network-on-Chip (NoC) technology as a mechanism to facilitate the parallelism between neurons, such as EMBRACE [15]. The NoC is similar to a computer network and uses an array of routers to time-multiplex discrete packets of data across shared routers on the chip. However, ultimately such NoC structures are realised, like any other circuit, using existing silicon manufacturing processes and therefore are susceptible to permanent faults from wear-out effects, but also from exponentially increasing numbers of temporary faults caused by radiation. A key fundamental task in any fault-tolerant system is the detection of faults, in particular the ability to sense temporary as well as permanent faults.

M. McElholm (✉) • J. Harkin • L. McDaid • S. Carrillo
Intelligent Systems Research Centre, University of Ulster, Magee Campus,
Derry, North Ireland, UK
e-mail: m.mcelholm@ulster.ac.uk; jg.harkin@ulster.ac.uk; lj.mcdaid@ulster.ac.uk;
Carrillo_Lindado-S@email.ulster.ac.uk

J. Mathew et al. (eds.), *Energy-Efficient Fault-Tolerant Systems*,
DOI 10.1007/978-1-4614-4193-9_7, © Springer Science+Business Media New York 2014

To advance current NoC-based systems including brain inspired computing paradigms requires the capability to initially detect such fault types within the NoC interconnect. A major challenge is the development of a fault detection architecture that can offer adequate fault coverage during real-time operation (i.e. online) without incurring substantial cost in terms of area and power consumption; the latter are the limiting factors in achieving scalable interconnect. This chapter presents SMART a low-cost, online strategy for detecting hardware faults in NoC interconnect using biological synapses and neurons. The chapter is structured as follows with Sect. 7.2 providing a review of existing work in the area of fault detection and approaches used in NoC. Section 7.3 proposes the SMART strategy with Sect. 7.4 providing analysis of the strategy through circuit simulation. Section 7.5 details a Hardware implementation of the Fault Detection Unit with results and area/power comparison to existing Fault detection approaches.

7.2 Related Work

This section outlines related work in fault detection techniques and approaches used in NoC, with a summary on the key remaining challenges. In addition, background on the function of synapses and neurons is also provided.

Research into fault-tolerant architectures has attempted to categorize faults as either data or control flow errors. Data errors occur when the content of a variable stored in memory or controller is altered. Errors occurring in the data transmission between processing elements, i.e. in NoCs, can also be classified as data errors. Control flow errors occur when the content of a memory cell or register storing an instruction is altered. For a processor this results in executing an incorrect sequence of instructions with the potential to induce fatal program errors. Within this context, both data and control faults can be classified at a circuit level into either permanent or transient faults.

Examples of permanent faults include open faults, which occur when the connection between two test points no longer exists or high resistance is experienced. Short circuit faults display the opposing characteristics where a low resistance connection between two isolated test points is encountered. Concatto et al. describe a test methodology for both AND-short and Or-short faults [6]. Stuck at faults can be characterised as nets or wires that are permanently connected to, or stuck at a particular value or voltage. At sub-micron device level non-permanent or transient faults are becoming increasingly common. For example, crosstalk is a transient fault condition that can occur across several wires and can have a detrimental effect on an individual signal path. These effects may include signal glitches as well as rise and fall time delays or speed-ups that can induce system errors [12]. Similarly [26] identify other transient fault conditions such as Single Event Upset (SEU) that can be caused by electromagnetic interference or cosmic radiation, and has been shown to lead to eventual failure. Many approaches have been investigated in exploring the detection of such faulty conditions in electronic computing systems.

7.2.1 *Fault Detection Techniques*

Common Built-In Self Test (BIST) programs can detect hard failures in processors, memories and interconnect. Such programs are usually run at power up or on demand and involve sequencing a known set of test vectors into the circuit and testing the results for fault indicators. Built-In Self Test (BIST) is a diagnostic technique that has been applied to NoC architectures [12]. Such diagnostic programs can detect permanent faults however they are unable to detect transient and intermittent faults during run-time operation.

7.2.1.1 Circuit Level Approaches

Transient and intermittent faults have been traditionally protected against by using error detection and error correction mechanisms. Parity codes is a popular coding techniques that operates by adding a bit to the data packet to indicate if the number of 1 bits in the packet is odd or even. A Cyclic Redundancy Check (CRC) enabled device calculates a fixed length sequence based on the input code and attaches this CRC code to the transmitted block. Upon receipt, the device repeats the calculation and if the CRC do not match corrective action is taken. This error detection code usually results in a request for the source to resend the data, but is unable to determine the fault type or location.

Hamming Codes can be use as a Single Error Correction code where any given bit is represented by a unique set of parity bits. Therefore, if any bit has an error, the bit position can be determined by the parity bits. If an additional overall Parity bit is added, the hamming code can be used to detect, but not correct, any 2 bit error. This is known as a Single Error Correction- Double Error Detection code. Dual Rail Code is another technique that effectively doubles the physical lines required as both the signal and its complement are transmitted.

Using Delayed Sampling registers is another method of detecting transient faults. In this scenario three sampling registers are added to the input stage and incoming data is sampled three times at different moments, all derived from the system clock edge. A majority voter defines the correct data that will pass through the router based on the results from the three sampling registers [7]. However, the major drawback to these approaches is both additional hardware resources and computing time.

7.2.1.2 System Level Approaches

Whilst circuit level solutions employ techniques at a local level or at individual sections of a system, several higher level system solutions utilise additional hardware/software to provide the overall system with a level of fault protection. Some of the popular techniques employed in this area are outlined below.

- Redundancy

Hardware redundancy uses multiple instances of the system hardware, or part of it, to as a back up in the event of failure or detected faults. By replicating the hardware element the system can function correctly by effectively ignoring the faulty element. For example, the Triple Mode Redundancy (TMR) technique replicates the same computation in three identical modules with the output of each passed to a voter mechanism where the majority output of the three identifies the module which is faulty [22]. The obvious disadvantage of hardware redundancy is the excessive hardware costs associated with replicating hardware elements to facilitate fault-tolerance. N-version programming is another variation of redundancy aimed at detecting errors in the design by employing parallel or sequential executions of the same program and comparing the results [10]. This approach is restricted to software based designs with requirement for additional program area and time for sequential program executions.

- Control Flow Checking (CFC)

Control flow checking monitors for control flow errors that may cause a software program to deviate from its normal instruction flow. Nodoushan et al. [27] describes the basic idea of most CFC techniques as partitioning a program into basic blocks. Each block contains a set of instructions in which the program flow is not changed by a branch instruction. Therefore the program should enter the first instruction in the block and exit the last in a fault free situation. Branch instructions appear between these blocks. There are numerous different CFC methods that adopt slightly different strategies such as using signatures associated with blocks and the relationship between signatures is used to test the program flow [11]. Although this approach can detect both permanent and temporary faults, this method of Control Flow Checking only targets the detection of faults in software program flow, whilst this research is concerned with detecting and diagnosing faults in hardware systems.

- Watchdog Processor

Although CFC may be implemented in software, it is more common to use dedicated hardware such as Watchdog Processors to implement the CFC tasks. Benso et al. [2] refer to a Watchdog processor as a simple coprocessor used to perform concurrent system level error detection by monitoring the main processor behaviour. The Watchdog processor is an extension of the basic watchdog timer used in microprocessor devices to provide a method of reverting to a known good state if the program becomes unstable. Error detection is done by comparing concurrent information with that provided during setup. The disadvantages of the Watchdog processor is the significant area overhead required in realising the additional co-processor in the system.

7.2.2 Fault Detection in Networks-on-Chip (NoC)

In the context of NoCs, fault types can be localised as link errors (physical lines between routers) errors and router errors. Faults are mostly attributed to open connections, shorts, stuck-at, bridging, Single Event Upset (SEU), and crosstalk, where the duration of the fault identifies its classification as either permanent or temporary [16]. The link errors align closely with data errors [25] and occur during packet transfer from router to router. Router errors occur within the router architecture and can be classified under control flow errors. Hosseinabady et al. [18] propose a method of self testing focused on the router. It passes the same vectors to all NoC routers and detects hard faults by comparison of router outputs with each other. More recent comparison based approaches have explored the use of traditional artificial neural networks in testing outputs of routers for faults [23]. These methods only targets faults in the router and cannot be used to diagnose interconnect link errors, in addition they do not lend themselves to online testing as time is incurred due to the propagation of test vectors.

Grecu et al. [12] suggest a BIST for NoC interconnect that targets the potential for cross talk effects due to inter wire coupling. Monitoring a single wire (victim) whilst manipulating all other wires in the link (Aggressors), allows for detection of crosstalk induced errors as well as short or open connections. Herve et al. [17] propose a method to diagnose pair wise short faults within a defined neighbourhood of a mesh NoC. Each processing element interface is equipped with a Test Data Generator (TDG) and Test Error Detector block, which are used to implement a Walking-One Sequence to step test patterns through the network. The main limitation of such self test mechanisms in their inability to detect in-line faults during run-time operation due to the reliance of specific test data patterns in the system. Extensions to these approaches include token-based techniques for online diagnostics however routers are tested sequentially which incurs a large overhead in time and area due to TDGs [19].

Other work by Murali et al. [24] considers the use of error detection codes to implement self checking NoC routers. They define two categories for NoC applications which are switch-to-switch and end-to-end. In switch-to-switch all switches located in the transmission path between source node and destination node check the packet to detect probable faults. In the end-to-end case just the send and receive nodes check the validity of the packet. Obviously the switch-to-switch implementation gives a more accurate representation of potential fault location, whereas end-to-end cannot determine the fault location but can detect the presence of a fault. However the main disadvantage of this method is the area overhead required, with switch to switch method adding 70% to the NoC area thus making this technique unsuitable for scalable NoC architectures [1] extends this work and employ varied techniques to determine the best online switch fault detection, targeting the control elements of NoC as opposed to the more common link errors. Alaghi et al. confined their fault detection strategy to stuck at direction faults (stuck-at-north, stuck-at-processor) for an x-y routing mesh and proposed a number of self testable switch architectures to help fault detection under this model. Fault coverage under this strategy is limited because of the narrow fault model targeting only control errors.

Work by Grecu et al. [12] proposes a scheme for detection and location of faults based on the concept of code-disjoint error detection. Data paths are used to implement code disjoint technique where the incoming parity bit is compared to the calculated parity at input stage. This scheme claims to improve the results obtained by the switch-to-switch strategy outlined by Murali et al. [24], because it can differentiate between faults occurring in the network routers from faults in the communication channels between switches. A similar approach to the error detecting unit at each router input and output is proposed by Kohler et al. [20]. Although Fault coverage and area/power consumption are improved using these techniques the area overhead encountered limit the suitability for highly scalable NoC architectures.

Benso et al. [2] propose a watchdog monitor with strategies to address data, control flow and bus errors. Data protection is achieved by storing a local copy of some processor critical variables in the Watchdog, and checking the write and read operations to this variable. Control flow checking is implemented by splitting the program code into branch free blocks with single entry using a signature based scheme as the CFC method. To avoid modifying the application, bus protection is implemented using Automatic Repeat Requests (ARR). In this scenario the watchdog requests multiple transmissions of data until two transmissions are the same. The ARR however introduces a very high time overhead and does not identify the fault location. Moreover, the dedicated watchdog processor uses supplementary hardware resources. The approach by Giaconia et al. [28] suggests an alternative use for the watchdog with on-the-fly acquisition of data leaving the watchdog processor completely independent of the rest of the hardware/software. They propose a reasonableness check to determine if significant variables are within set constraints and program flow check controlling if some segments of code are executed within predefined time intervals. The Watchdog is configured to detect read or writes to these addresses and interrupt the main processor accordingly. As with many watchdog based approaches, this technique provides a degree of fault-tolerance but does not provide any method of fault diagnosis and necessitates a dedicated hardware resource to implement the watchdog function. Harkin et al. [13] suggests a method of fault detection targeted at self repairing NoC systems where two types of Watchdog, a Master and Slave/Router Watchdog (RW) are explored. The approach provides good detection capabilities however the logic RW module exhibits a 13% area overhead in comparison to the existing Router and standard processing element, which adversely affects its suitability for scalable NoC architectures. More recently Dai et al. [29] proposed a NoC monitoring circuit using the rate of successful ACK (Acknowledge) and NACK (negative ACK) signal from the link of a NoC router to detect and diagnose fault types. The frequency of ACK versus NACK events over a defined period is used to classify faults as either transient or non-transient/permanent faults. The proposed system uses an accumulator controlled by a timer mechanism to add the number of events until a threshold is reached, or resets the accumulated events after a certain period. This is predominantly a check for the link interconnects but storing a cumulative result for the link failures attached to each router. If a pattern emerges surrounding the link errors associated with a router that router can be assumed faulty. This work provides a useful model for distinguishing between transient and permanent faults but is once again targeting a parallel data path comprising a number of individual wires. The major drawback with this approach is the need to include a large timer for each router on hardware which is expensive in terms of area, and therefore not scalable for large NoC-based system implementations.

In summary, many approaches to fault detection have been proposed with good fault coverage however the key requirements for NoC-based systems which have not been fully addressed to date include; online capabilities (i.e. detect in real-time and in a non-intrusive manner with NoC traffic) and maintaining NoC scalability.

7.2.3 Spiking Neural Networks and EMBRACE

7.2.3.1 Spiking Neural Networks

Research has shown that we can learn a lot from biology, in particular from neuroscience, as to how synapses and neurons in the brain process and communicate information in a robust and power-efficient manner. The robust information processing capability in the brain comes from the vast parallel density of synapses and neurons, where the brain contains an estimated 10^{10} neuron cells and almost 10^{15} synapses [8]. Various abstract models of the brain have been investigated over the last 60 years with Spiking Neural Networks (SNNs) being the most promising neural network model to date. SNNs aim to emulate information processing and communication capabilities of the brain by mimicking the efficient function of excitatory and inhibitory synapses and threshold neurons. SNN have been used in many application areas including pattern classification, dynamic control and signal processing where, unlike traditional neural networks, the temporal nature of input data is exploited.

SNNs communicate through pulses (spikes), the timing of which is used to transmit information and perform computations. SNNs model the biological neuron which consists of a cell body (i.e. soma) that possesses many input branches called dendrites that carry information from other neuron cells, see Fig. 7.1a. The output of the neuron is called an axon, which communicates information to other neurons in form of action potential or spikes (pulses). Figure 7.2 illustrates some examples of pulsing (spiking) patterns. Spikes are transmitted between neurons via weighted synaptic connections called synapses, see Fig. 7.1b. Excitatory and inhibitory are two basic types of synapses used in the current SNN paradigm which closely model biology. The excitatory and inhibitory weighted synapses have the basic operation of facilitating (strengthening the spike input activity) and depressing (suppressing the activity), respectively. Adjusting the weights of the synapse enables particular

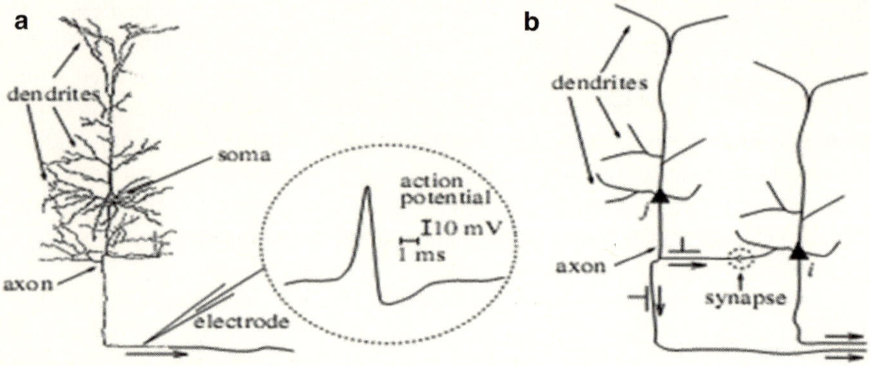

Fig. 7.1 Biological neuron cell (**a**) shows the soma, axon and dendrites; (**b**) an action potential transmission between neurons j and i [8]

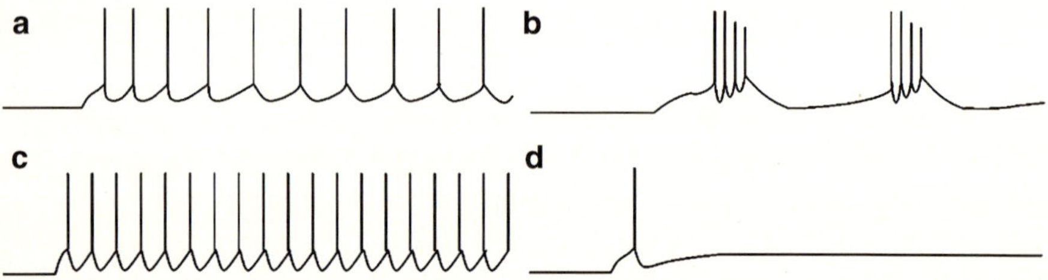

Fig. 7.2 Example spiking patterns (**a**) Regular-spiking neurons (**b**) bursting-spiking neurons (**c**) fast-spiking neurons and (**d**) rebound-spiking neurons [8]

sequences of spikes (pulse frequencies) to be effectively filtered, where the spikes are either communicated to a neuron or suppressed. The combined effect from the facilitating and depressing actions is reflected in the neuron output where a neuron will generate an output spike when the sum of its weighted input spikes (from the synapses) exceeds a firing threshold value. A neuron will maintain the generation of spike outputs (pulse train) while its input is maintained above this threshold. It is the integration of the synapse outputs with neurons which provides the abstract basis for information processing [3].

7.2.3.2 Embrace

To achieve real-time processing of SNNs, hardware implementations have been explored to fully exploit the high levels of parallelism between synapses and neurons. One recent hardware architecture, EMBRACE (EMulating Biologically-inspiRed ArChitectures in hardwarE), which achieves real-time acceleration, uses Network-on-Chip (NoC) and analogue spiking neuron cell structures as building

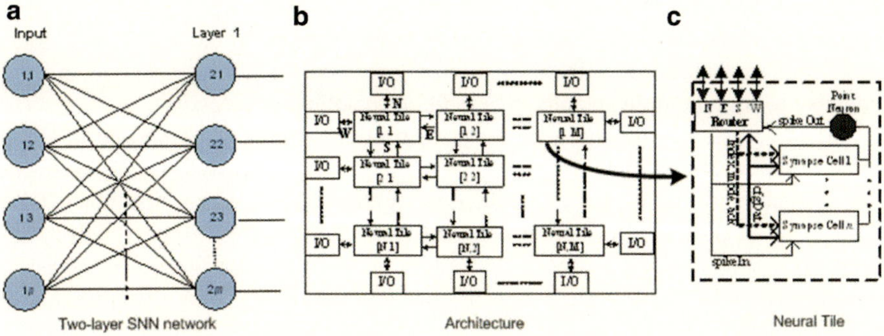

Fig. 7.3 EMBRACE architecture (**a**) shows a regular N × M SNN (**b**) the corresponding 2D NoC implementation using EMBRACE, and (**c**) shows the neural tile composed of a digital router and an analogue synapse cell and neuron

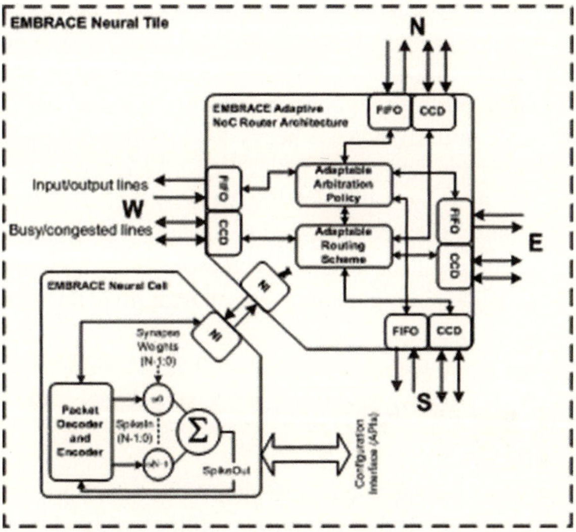

Fig. 7.4 EMBRACE Tile with the neural cell and adaptive NoC router

blocks [4, 14] to provide a scalable implementation solution for SNNs. The EMBRACE architecture is composed of a 2D array of Neural Tiles where each tile contains the neural cell and NoC router. Figure 7.3 illustrates the architecture and method for realising SNNs across the array of neural tile.

Novel hardware blocks have been developed including an adaptive NoC router [5] and efficient excitatory and inhibitory hardware synapses [21]. Figure 7.4 illustrates the adaptive NoC router of EMBRACE's tile structure which facilitates the interconnect between neurons. The router has four channel links including North, East, South and West directions. Neurons and synapses are connected to the NoC router via a dedicated network interface (NI), i.e. fifth link. These routers are then interconnected via point to point data links that enables time-multiplexing of discrete packets of data (spike events) from source to destination across several links. The EMBRACE NoC router is a key building block in facilitating the large-scale implementation of SNNs in hardware and has two key aspects, namely the adaptive routing module and adaptive arbitration policy.

Adaptive routing module: The adaptive routing scheme is composed of three main components; (1) an XY routing algorithm which receives the packet from the adaptable arbitration policy module, (2) a channel congestion detector (CCD) which, based on the information received from neighbouring routers, selects an output port direction and passes this information to (3) the adaptive routing decision module. The CCD provides a means of detecting the current state of traffic in any given direction.

Adaptive arbitration policy: As the traffic pattern of a spiking neuron is highly non-uniform and asynchronous, the adaptive router uses a hybrid arbitration policy which combines the strong fairness policy of the round-robin arbiter and the priority scheme of a first-come first-serve approach. The arbitration policy uses an event register to store information regarding each spike event for each router input buffer, and five distributed control units, i.e. one for each port. This allows the scheduler to manage thread communication without incurring task-switching overhead. Only the input buffers that contain information are serviced, thus avoiding wasted clock cycles. Similarly when a heavy load traffic scenario occurs, all ports are serviced, based on the same approach as the round-robin.

The long term goal of EMBRACE is to support the exploitation of SNNs in providing future bio-inspired processing paradigms which can perform self-repair under the presence of faults. Therefore, it is important that faults can be detected in the dominant component of EMBRACE, the NoC interconnect.

7.2.4 Key Challenges

The key challenges for NoC fault detection is the ability to detect temporary and permanent faults during runtime (online), without impacting on the NoC throughput while under the constraints of low area and power implementations. The area/power constraint is imposed due to the scalability criteria for NoCs, and is critical for large scale implementations as we approach the target of +250 processing cores per chip in 2015 (+1,000 within the decade). Considering the discrete event based nature of digital systems and that the discerning difference between temporary and permanent faults is the frequency of occurrence over time, it is possible to use the pulse filtering properties of excitatory and inhibitory synapses and neurons to detect temporary and permanent faults. Therefore, SMART aims to address these challenges by exploiting the temporal and spatial capability of biological synapses and neurons, which exhibit low area/power, in detecting faults. No work to date has explored the concept of using SNN synapses and neurons in this manner before, and given the availability of hardware synapses and neurons [9] with low area and power requirements, it is now possible to realise such a strategy in hardware.

7.3 SMART Strategy

Information in SNNs is communicated through sequences of pulses or frequencies where synapses act as filters to allow selected frequencies to be transmitted between neurons. We can view this process of encoding information in the brain as temporal/spatial filtering and if we look into digital multiprocessing systems, a similar analogy can be applied to data communication and processing between interconnected components in the NoC. For example, as digital data passes along NoC channel wires certain frequencies are exhibited. When faults occur the communication frequencies change as data communication is perturbed by the faults. Using synapses at either end of a NoC channel to detect when data arrives (or fails to arrive), enables a neuron to generate a corresponding output frequency and signify the occurrence of a fault.

The SMART strategy has the basic principle of using biologically inspired neural network based models for online fault detection in NoC based multiprocessor systems. The strategy is based on the novel Fault Detection Unit (FDU) which uses available hardware synapse models with excitatory (facilitation) and inhibitory (depression) responses as a method for fault detection. In effect the FDU detects faults and draws a correlation between fault types and their temporal behaviour. For example, Fig. 7.5 illustrates the FDU where two points A and B on a NoC channel wire are connected to a single excitatory (ES) and inhibitory (IS) synapse, respectively, and their outputs connected to a summing neuron. The excitatory synapse is activated to represent a signal event (logic 0-1 transition), and a paired inhibitory synapse is also stimulated on receipt of the signal event at another location in the signal path.

Under non-faulty conditions, as shown in Fig. 7.5, the wire can communicate digital pulses where a pulse experienced at point A initiates the ES to produce an excitatory (facilitation) response, while at point B the pulse initiates the IS to produce an inhibitory (depressing) response. The right of Fig. 7.5 illustrates the output current of facilitation and depression from stimulus at points A and B, respectively. The ES and IS have reciprocal responses and therefore when both are activated they have mirrored outputs. The role of the neuron, *D*, is to sum both ES and IS currents and produce a pulsing, digital output frequency if the sum, *S*, exceeds a defined detection threshold, Nth ($S > N$th is the basis for detection of a fault).

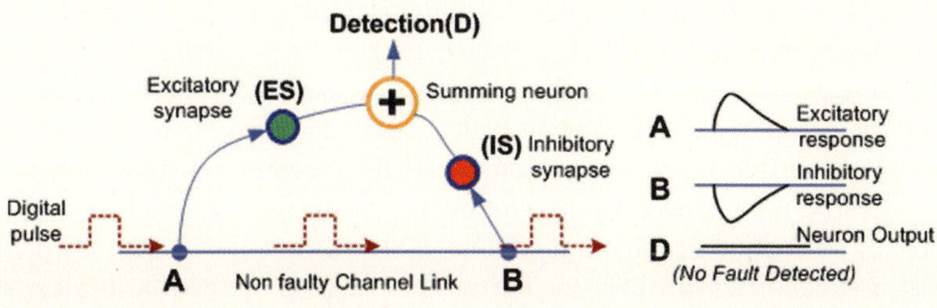

Fig. 7.5 FDU applied to a single channel

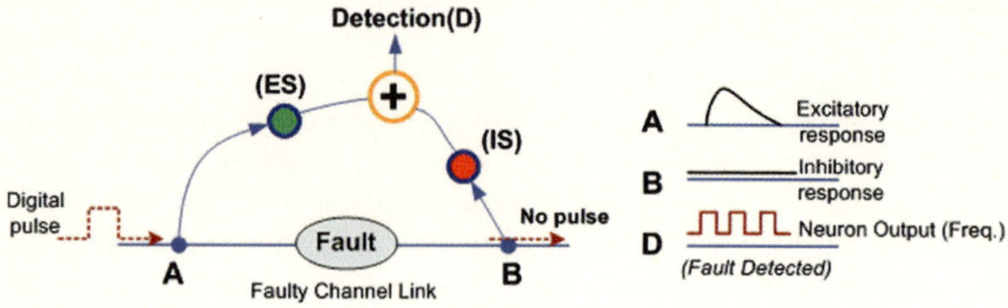

Fig. 7.6 FDU applied to a single faulty channel

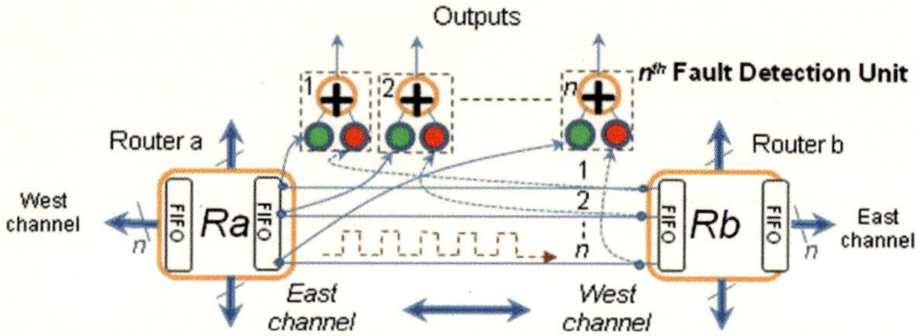

Fig. 7.7 SMART applied to example NoC channel

In Fig. 7.5 where no fault is present, the summing of the ES and IS output currents cancel each other producing a net sum which is below the threshold ($S > N\text{th}$), and therefore no fault is detected as the neuron output remains at zero. However, in Fig. 7.6 a fault is illustrated which stops the signal from propagating between points A and B (due to a temporary/permanent fault, e.g. stuck-at, transient, open-circuit etc.), only the ES synapse is activated as the IS synapse at point B does not receive the pulse stimulus. This results in the neuron accumulating the full current from the facilitating (ES) synapse which causes the sum to exceed the threshold ($S > N\text{th}$), and therefore identifies a fault by producing a pulsing output from the neuron.

Figure 7.7 illustrates how the SMART strategy can be applied to detect faults on the example East-to-West channel between two NoC routers, *Ra* and *Rb*. To facilitate online detection each of the *n* wires in the channel is connected to a single FDU and NoC traffic on the wires stimulate the FDUs. As traffic passes from *Ra* to *Rb* on the channel, a fault on wire 1 causes FDU 1 to generate a frequency output for a period of time. Figure 7.8 illustrates four example outputs that the FDU produces when the NoC traffic pattern on the wire is alternating (e.g. walking ones $= 0 - 1 - 0 - 1$). Using the walking one example pattern we can see in Fig. 7.8 that temporary and permanent faults can be identified by the temporal nature (duration) of the frequency output from the FDU. Number (1) shows a zero FDU output when no fault is detected and numbers (2)–(4) show examples for detected permanent, transient and temporary faults. This information can be used further to assist in diagnostics however the current information on the nth location of the FDU indicates which actual line in the channel is faulty. This can be applied to each channel of the router.

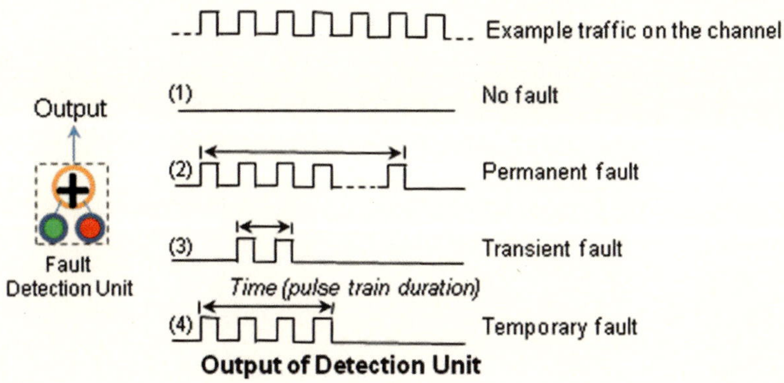

Fig. 7.8 Example outputs from FDU

The key advantage of SMART is the ability to use the existing traffic communicating across a channel to detect faults, thereby providing online detection which is non-intrusive to the throughput of the channel. Therefore, SMART does not require the generation or injection of test patterns which can incur a time penalty or, the management thereof to perform tests during non-busy periods of the channel. These are key to maintaining NoC scalability cross the time and area domains.

7.4 Analysis

The online fault detection capability of the FDU is analysed in this section. Simulink models have been developed that apply the FDU to a number of circuit configurations to evaluate faulty and non-fault conditions. For example, the FDU is applied to a single channel wire model and this is extrapolated to detect faults on a multiple wire channel structures such as a parallel bus. The FDU is also modelled to detect faults across a synchronous logic block such as a Flip-flop. Finally the FDU is applied across a number of lines in a channel to determine if a fault detected is a result of crosstalk between adjacent lines. The diagrams provided have been generated by Matlab simulink models of the FDU in the described scenarios.

7.4.1 Single Channel Wires

Figure 7.9 illustrates the Simulink simulation of the single wire Fault Detection Unit model shown in Fig. 7.5 where no fault is present or detected. Plots (a) and (b) in Fig. 7.9 represent the digital pulse detected at points A and B respectively. A digital pulse at point A initiates an excitatory synapse (ES) response shown in subplot (c). A corresponding pulse at point B produces an inhibitory synapse

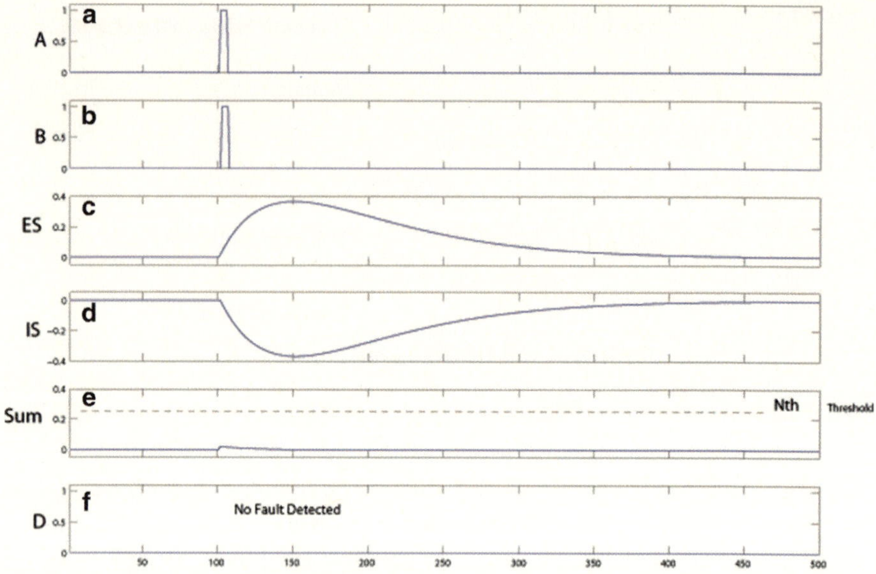

Fig. 7.9 Single channel wire FDU simulation

(IS) response illustrated in subplot (d). As the initiation of the inhibitory response was sufficiently close to the excitatory response (1 nS delay) the summation of both synapse responses, S, does not reach its predefined fault threshold value, as shown in plot (e). Thus no fault is detected and the neuron output remains low as shown in plot (f). The fault threshold value, Nth, has been configured to allow a maximum 15 nS propagation delay between the initiation of excitatory and inhibitory responses before there is risk of false positives where a fault can be detected unintentionally. A propagation time less than 15 nS will not produce false-positives however the maximum propagation time can be re-configured to accommodate longer propagation times. Figure 7.10 depicts the simulation output for the same single channel wire Fault Detection Unit model when a fault is injected. Plots (a) and (b) of Fig. 7.10 represent the digital pulses detected at points A and B respectively in the FDU model (see Fig. 7.6). A digital pulse at point A initiates the ES response shown in plot (c) however, as no pulse is detected at point B (simulated fault), no inhibitory response is provide to counteract the excitatory response. Therefore, the summation of both responses matches the excitatory synapse response. At time step 100 the summed value starts to increment due to the ES response and this progressively raises the summation value, *S*, above the predefined fault threshold value shown in plot (e). At this point the neuron output starts to fire producing a pulsing output as shown in plot (f). As the ES response decays the value, *S*, falls below the defined fault threshold value and the pulsing output stops. The pulsing out signifies a fault has been detected on the line, i.e. the connection between point A and B is compromised.

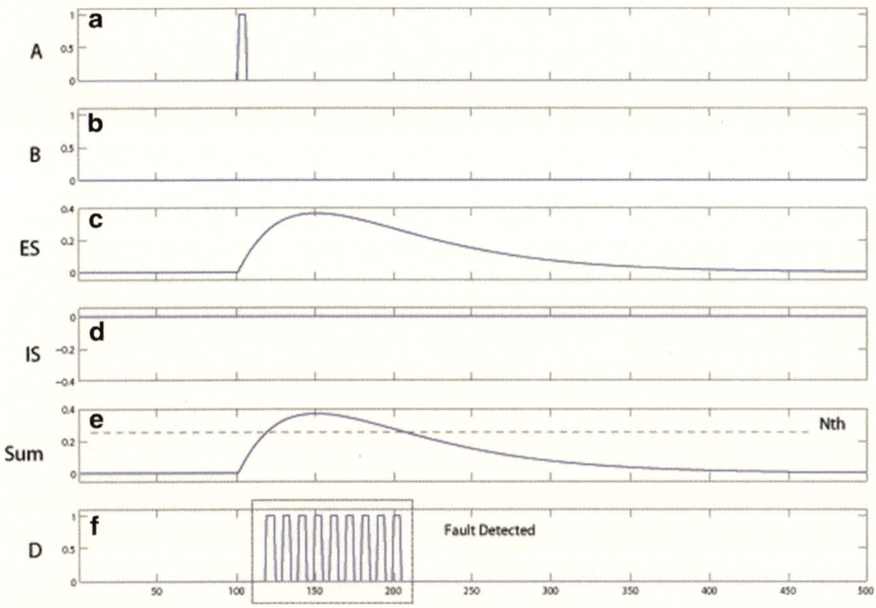

Fig. 7.10 Simulation of single channel wire under faulty conditions

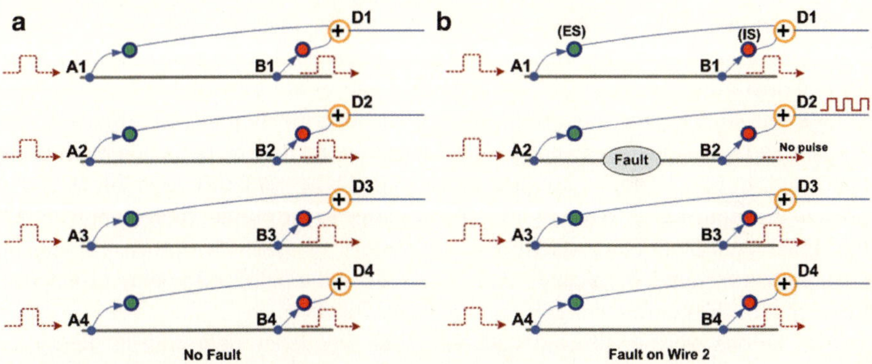

Fig. 7.11 Example of fault detection in multiple-wire channels

7.4.2 *Multiple Channel Wires*

Figure 7.11 illustrates how SMART can fault detect for a multiple wire channel structure (four lines are used as an example). In multiple wire channels each line uses a single fault detection unit, where the output of each FDU provides a status on an individual line. The example in Fig. 7.11 illustrates a scenario where no fault is detected, Fig. 7.11a, and a fault is injected on line 2 of the channel, Fig. 7.11b. Figure 7.11b illustrates the output from FDU #2 pulsing under the faulty condition.

Figure 7.12 represents the simulated output from the example 4-bit channel scenarios of Fig. 7.11. For clarity the inhibitory and excitatory synapse responses and the summation neuron responses have been removed from the simulation output. Figure 7.12a, b depicts the four input signals to each FDU (A1A4) and the summation neuron output signal associated with each FDU (D1D4). In Fig. 7.12a no fault is identified in the network (all excitatory responses in the FDU's were cancelled out by their inhibitory response), therefore no output neuron reached its threshold and did not cause the output to pulse. Alternatively, in Fig. 7.12b where a fault is simulated on wire #2, the same 4 bit input pulse train is used however, due to the fault on wire #2 no inhibitory response is generated and therefore the output from neuron D2 reaches its threshold after signal A2 is received, causing output D2 to fire its pulse train as shown in plot (f) of Fig. 7.12b. Furthermore, Fig. 7.13 illustrates the simulation output when multiple faults are injected in the channel. For example, plots (f) and (h) show neurons D2 and D4 pulsing, respectively, when the injected faults on wires #2 and #4 are detected.

7.4.3 Synchronous Registers Under Faulty Conditions

Figure 7.14 illustrates how a FDU can be applied across the input/output of a generic logic block such as the common D-type flip flop (FF). The Excitatory synapse is initiated by the pulse on the D*in* terminal of the FF, with the Inhibitory synapse connected to the D*out* output and initiated by clocked output signal from the FF. In normal operation, shown in Fig. 7.14a, both synapses are activated and cancel each other as the signal at the input of the FF is clocked to the output correctly; therefore no fault and subsequently no pulse output at the summation or detection neuron are generated. In Fig. 7.14b the scenario is illustrated where a faulty logic block could prevent the input signal to the FF from propagating to the output, thereby preventing the signal from D*out* from asserting the Inhibitory synapse. This process results in the summation value, S, exceeding the threshold and a pulse train being produce at the output of the neuron to signify a fault in the flip-flop.

Simulations were conducted with the clock frequency operating at the target speed of 100 MHz. Parameters for excitatory, inhibitory synapse responses and summation neuron thresholds have been configured to operate at this frequency. The plots (b) and (c) in Fig. 7.15 illustrate the function of the FF where data at D*in*, which is connected to the excitatory synapse of the FDU, appears at D*out* after one clock cycle. In this example the maximum propagation delay between D*in* and D*out* (FF delay) is 10 nS, which is exaggerated for illustration purposes and is normally in the order of pico seconds.

The simulated output of Fig. 7.16 illustrates a detected error across a flip flop as no D*out* signal pulse is generated. As no inhibitory response can then be initiated, the summation neuron threshold is reached and a pulse train evident at the Detect output indicating fault detection.

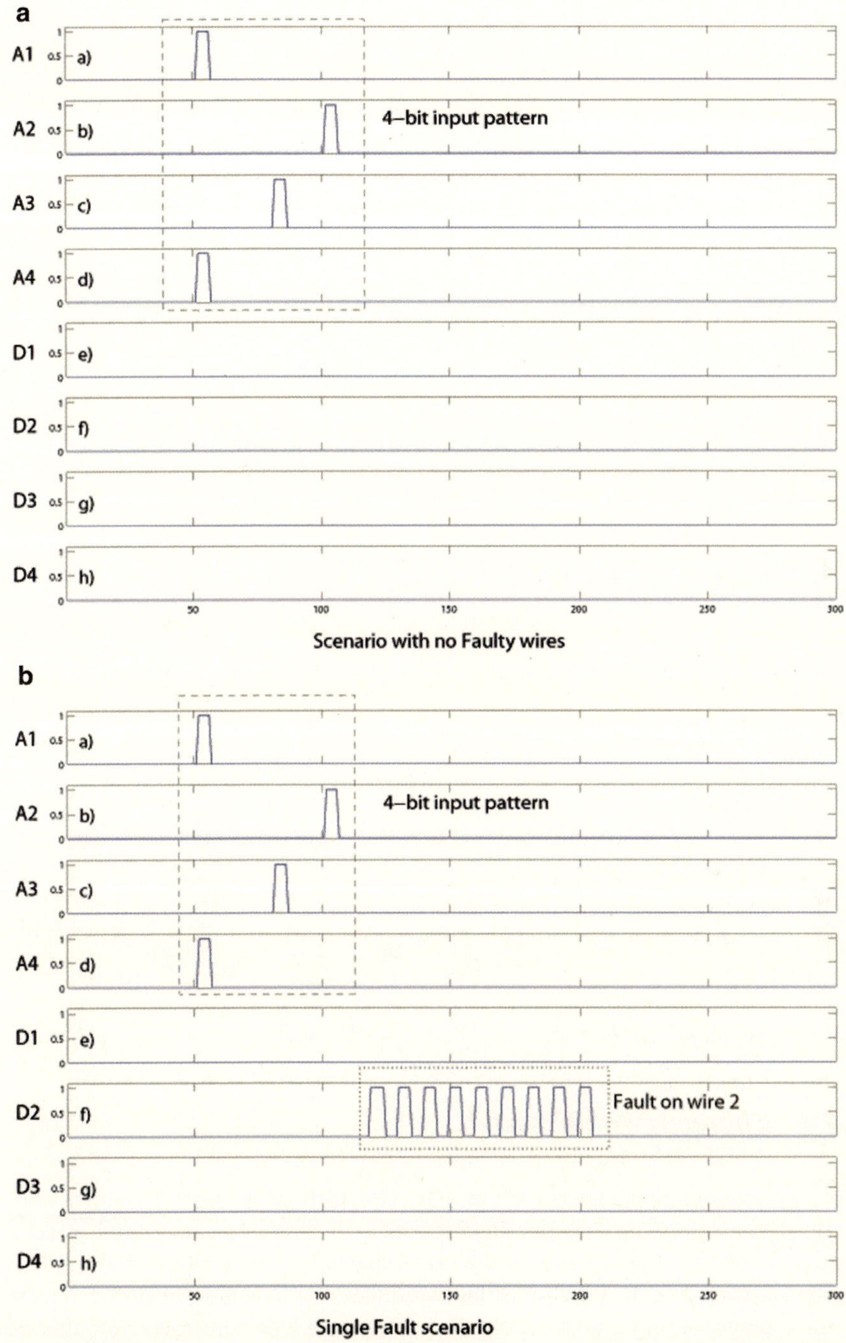

Fig. 7.12 FDU simulations with multiple channel wires

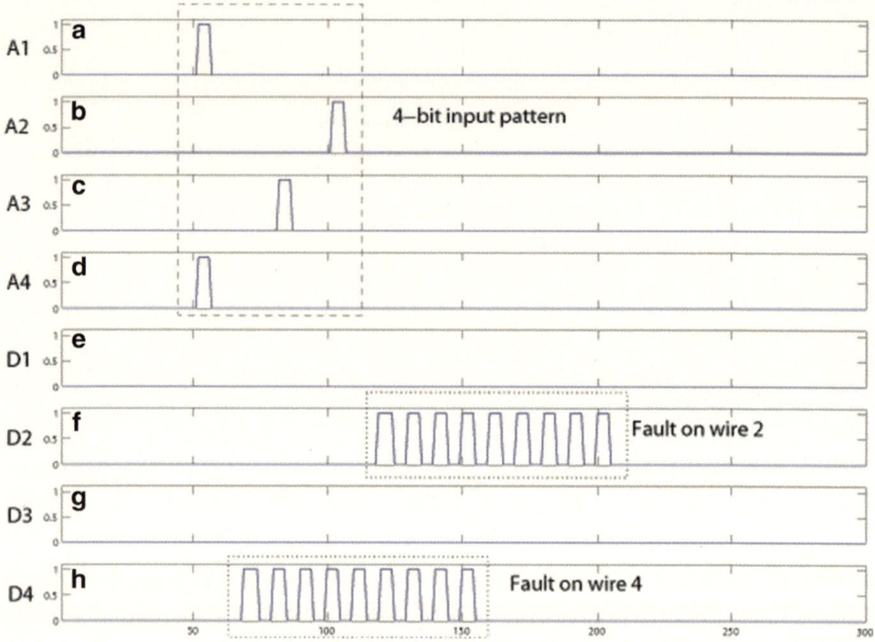

Fig. 7.13 Multiple channel FDU simulation – 2 faults

Fig. 7.14 FDU applied across D flip-flop (**a**) no fault (**b**) faulty logic block

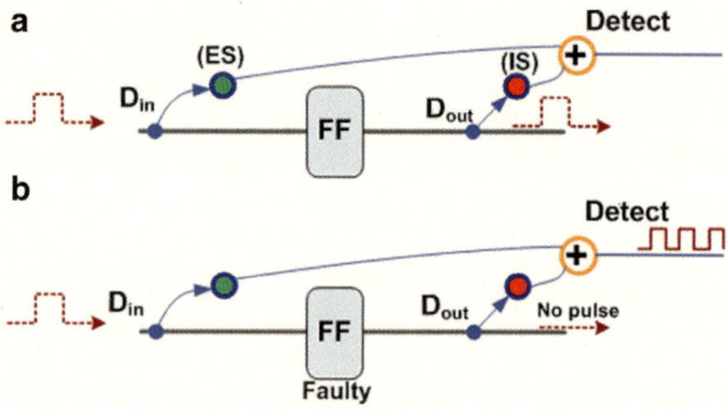

7.4.4 Channels with Crosstalk

Another potential application for the FDU is detection of crosstalk faults, where a signal transmitted on one line in a channel can induce an undesired effect or glitch on another adjacent line in the same channel. In its simplest form, a FDU implementation to detect crosstalk faults utilizes one detection neuron connected to three neighbouring lines in a channel. Figure 7.17 demonstrates how this can

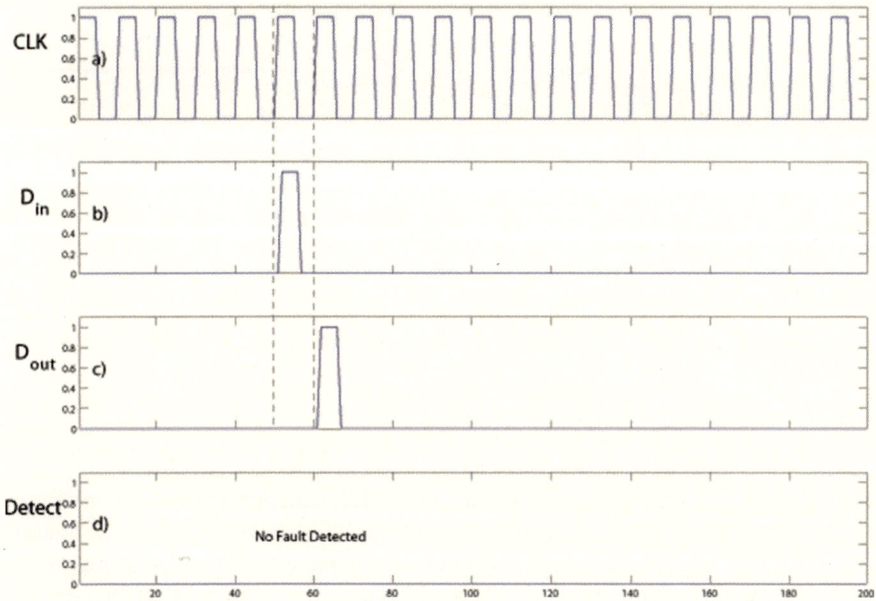

Fig. 7.15 FDU applied across D flip-flop no fault

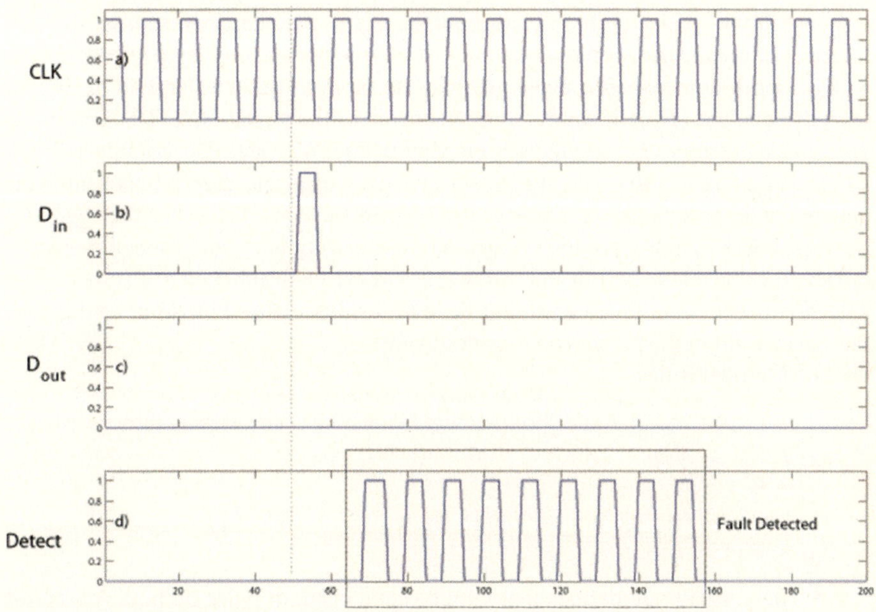

Fig. 7.16 FDU applied across D flip-flop – fault

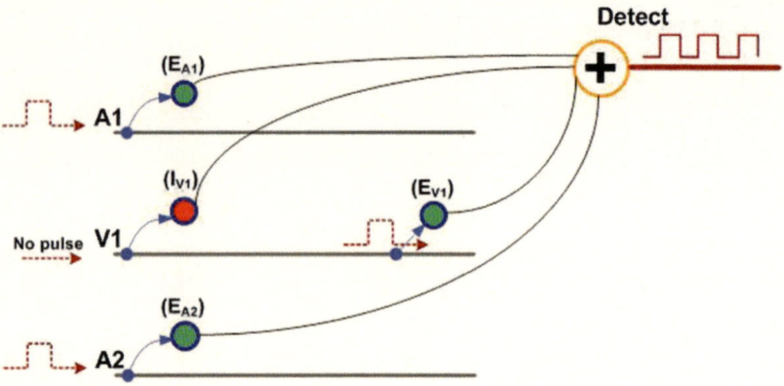

Fig. 7.17 FDU applied to a crosstalk model

be achieved using a sequence of Excitatory (ES) and Inhibitory (IS) synapses. In this scenario the detection neuron is sensing for crosstalk faults on the middle line, *V1*, know as the victim line. The lines A1 and A2 are in close proximity to the victim line and therefore are more likely to induce a crosstalk error on channel V1. These lines are identified as Aggressor 1 (*A1*) and Aggressor 2 (*A2*).

In previous FDU scenario when no fault is present the transmitted signal initiated a positive or excitatory synapse response followed by a negative or inhibitory response at the receive point to cancel out the positive excitatory response. In order to detect crosstalk errors on the victim line (V1) the response synapses are reversed, with an inhibitory response event (I_{V1}) at the input to the wire and an excitatory response evident (E_{V1}) if the input pulse traverses the signal path. The inhibitory, I_{V1}, and excitatory, E_{V1}, synapses on victim line, V1, are also configured with larger weight values than the Aggressor synapses, E_{A1} and E_{A2}. This ensures the magnitude of the synaptic response of the I_{V1} and E_{V1} are greater than the aggressor responses. Configuration of the synapse strengths (weighting) and Detection neuron threshold value, N_{TH}, permits the following rules to be formulated to ensure correct detection of crosstalk on a victim line. The four rules are used to aid in configuring the synapse strength and neuron threshold values.
Crosstalk model Rules:

1: $E_{A1} + E_{A2} < N_{TH}$ (sum of both strengths must be > the neuron threshold)
2: $E_{A1} + E_{V1} > N_{TH}$
3: $E_{A2} + E_{V1} > N_{TH}$
4: $|E_{V1}| \approx |I_{V1}|$ (strength of ES and IS on victim line must be balanced)
5: $E_{A1} < E_{V1} > E_{A2}$

(1) The derived rule defines that with a signal event or pulse on both Aggressor lines A1 and A2, the resultant summation in the detection neuron will not exceed the pre-defined threshold, N_{TH}.

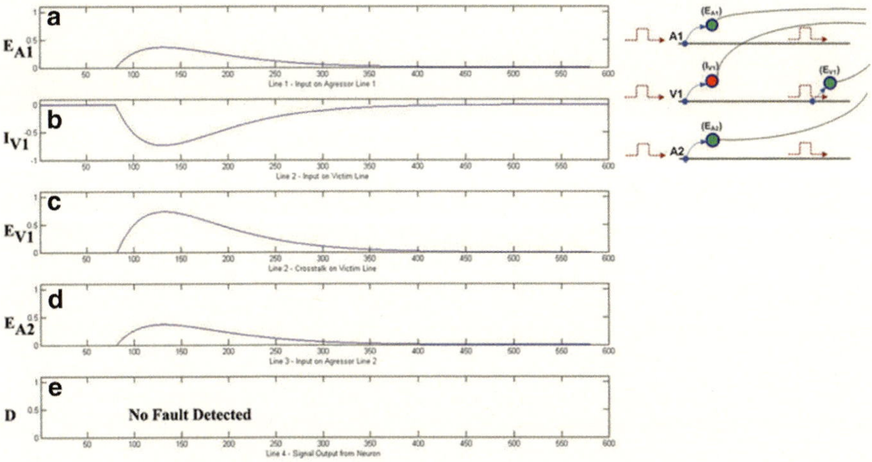

Fig. 7.18 Simulation of the FDU applied to the crosstalk model – no fault

(2–3) A signal event on either Aggressor line, A1 or A2, combined with a pulse or glitch on the Victim line, V1, will trigger the Excitatory response E_{V1}, which will exceed the neuron threshold N_{TH} and trigger a pulse train indicating a crosstalk error. This fault condition should only occur if no associated input pulse is detected on the Victim line, which would fire an inhibitory response (I_{V1}) of equal but opposite magnitude to the excitatory response E_{V1}.

(4) Thus to satisfy the crosstalk model the victim line synaptic responses I_{V1} and E_{V1} must be of equal magnitude (Eqn 1e) and this response magnitude must be greater than that of the aggressor line synaptic responses $E_{A1} + E_{A2}$. The weight parameters that affect these responses are configured in the existing synapse models.

To summarise, any pulse signal received on either aggressor input (E_{A1} or E_{A2}) combined with a pulse signal or glitch at on victim line at E_{V1}, without a pulse at the input I_{V1} will result in a genuine crosstalk error at the detection neuron. Figures 7.18 and 7.19 illustrate the output signals from the four synapses, shown in Fig. 7.17, that feed into the detection neuron. In Fig. 7.18 a pulse is detected at the input of all three lines, indicative of a synchronised bus switching multiple data lines at the same instance in time. In this case the larger excitatory response of E_{V1} is cancelled out by the inhibitory response from I_{V1}.

In Fig. 7.19 the inhibitory response of I_{V1} is not present indicating no input signal received, although the victim excitatory response is fired due to crosstalk coupling between the victim line and an adjacent aggressor line, A1. The magnitude of the summation in the detection neuron is now greater than the threshold value, and therefore initiates a fault detection pulse train.

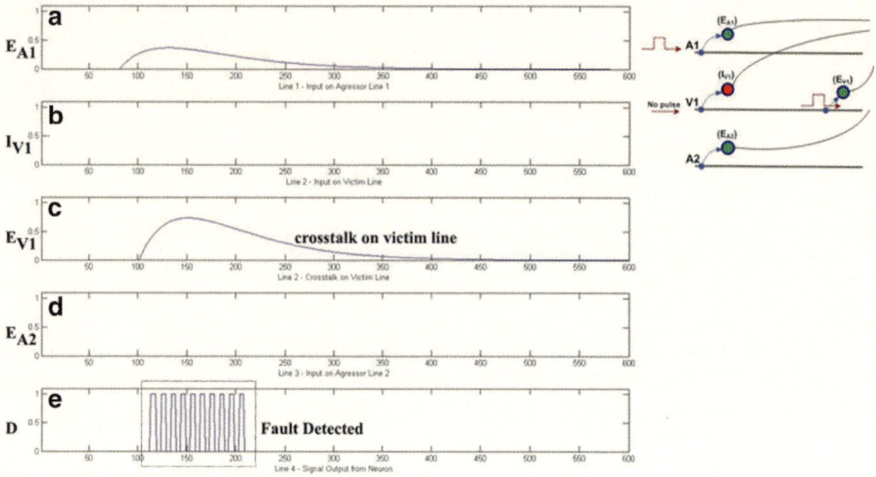

Fig. 7.19 Simulation of the FDU applied to the crosstalk model – fault induced

7.5 Hardware Implementation of the FDU

To validate the FDU model functionality in hardware, an Altera Cyclone 4 FPGA was targeted with an implementation of the circuit shown in Fig. 7.20. RTL hardware models of the excitatory synapse, inhibitory synapse, and integrate and fire neuron was used in the hardware demonstration of the FDU detection capability. The device under test (DUT) is an example synchronous Flip Flop (FF). This is used to underpin the core capability in detecting faults experienced in components and not just channel wires.

In this circuit configuration, an input train generator is used to generate a steady stream of pulses to the FF logic block. The generator is used to emulate digital traffic patterns feeding to the logic component and control the test conditions. During normal operation the output data, Q, of the FF would propagate the value present at the input data, D, on the positive edge of clock frequency. An excitatory synapse with facilitation response is connected to the input signal at point A, and an inhibitory synapse with depression response connected to the output signal at point B. The synapse responses are fed into the integrate and fire neuron. The neuron generates an output spike if the threshold level is reached, i.e. fault detected within the FF under test. Faults are injected into the FF path by using a switch to break the connection between the input and output of the FF. The switch is opened to emulate a fault in the FF, which prohibits the input from propagating to the output.

Figure 7.21 shows the plot from real-time data obtained via Altera's SignalTap on-chip analyser. The analyser output was obtained from the FPGA development board. Plot (a) in the figure shows the input train to the FF under test. Plot (b)

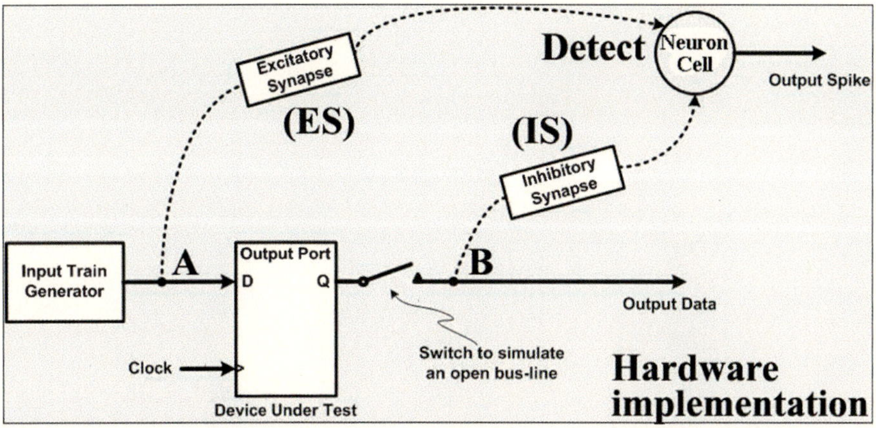

Fig. 7.20 Block diagram of the hardware model

illustrates the potential or voltage level of the excitatory synapse. From this plot it is evident that the synapse initially accumulates with each input pulse. It builds up to the maximum value that the synaptic hardware model can reach, and remains at this level due to the constant train of input pulses. Plot (c) shows the potential or voltage level of the inhibitory synapse. This also reaches a minimum level in response to the constant input pulses as all input pulses are propagating to the output of the FF, as shown in plot (f). This is shown between time 0 and 500 when the switch is closed (no fault present). Plot (e) presents the output from the neuron and between time step 0 and 500, and the plot illustrates there are no spike outputs as the switch is closed as the threshold level, Nth, has not been exceeded. However, when a fault is injected after time step 500 by opening the switch, the value of the inhibitory synapse starts to increase (becomes less negative in value and migrates towards 0 V). This causes the net value accumulated in the neuron to exceed the threshold level as the excitatory synapses is still receiving pulses at the input and maintaining its value. At this point the neuron starts to fire, generating pulses at the output as shown in plot (e). The value of the inhibitory synapse remains high (less negative) until the fault is removed by closing the switch again. For the duration of time that the pulses do not activate the inhibitory synapse, the neuron maintains its pulsing output. This provide data to a user that the fault is detected and active for the duration of time that the pulsing remains at the output of the neuron. Plot (d) shows the inhibitory synapse value decrease (become more negative) when the fault is removed by closing the switch again. This restores the balance in the accumulated value in the neuron to just below the threshold level and therefore stops the neuron from producing pulses at its output. This is shown in plot (e) just after time step 1,000. The remainder of the time steps in plots (a–f) show three other examples when a fault is injected and then removed.

This hardware implementation demonstrates that real-time fault detection can be achieved using the SMART strategy, where no active testing of the component

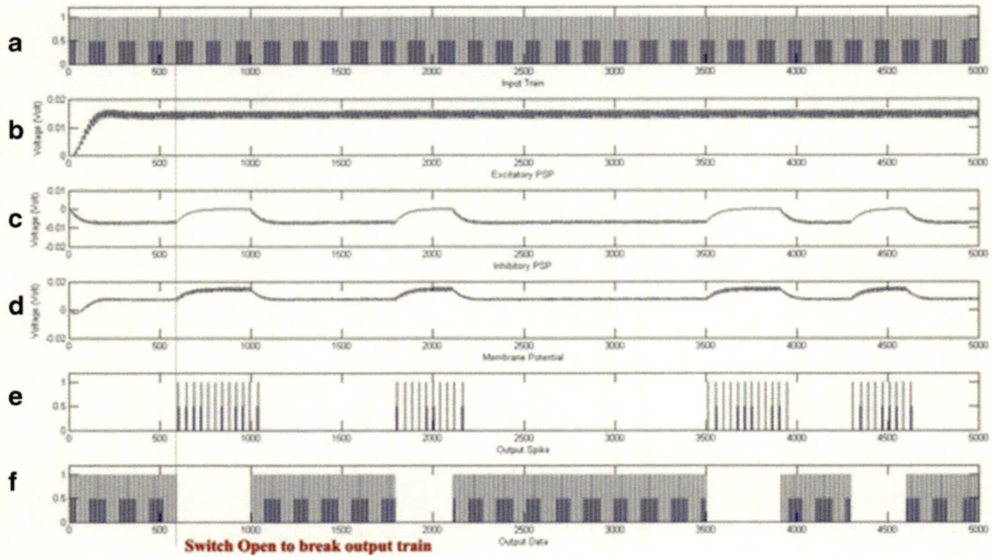

Fig. 7.21 Real time plot data from Altera's SignalTap on-chip analyser

or channel wires are required. Moreover, although digital implementations of the excitatory synapse, inhibitory synapse, and integrate and fire neuron has been used in this hardware verification experiment, highly optimised CMOS-based implementations have been developed for the synapses and neuron. For example, the area/power consumption [21] of the CMOS-based synapses ($24 \times 10^{-8}\,\text{mm}^2$ and power consumption/cycle of 1 nW) and neuron ($9 \times 10^{-6}\,\text{mm}^2$) are very low and therefore SMART is highly scalability. Using the example EMBRACE NoC router (discussed in Sect. 7.2.3.2), which exhibits a total area of $0.056\,\text{mm}^2$ with 48 lines per channel and four channel directions, the additional area overhead from implementing the hardware FDU, shown in Fig. 7.20, for all four channels is only $1.896 \times 10^{-3}\,\text{mm}^2$; this represents less than 4% increase in total router area which is significantly lower in cost than existing approaches [1, 12, 23]. Scalability is a critical aspect to any fault detection solution and this is particularly pertinent to NoC interconnect as any added functional benefit must not significantly increase its area and power costs. The SMART strategy clearly adheres to the scalability constraint while providing real-time fault detection.

7.6 Summary

Technology scaling has facilitated the integration of multiple processing cores on a single chip, with Network-on-chip (NoC) becoming an important strategy to alleviate the interconnect problem associated with communication between many processing elements. However, further advances in technology scaling are resulting in device components, including NoC components, that are more susceptible to faults. This chapter has provided an introduction to SMART, a novel online strategy for detecting faults in NoC interconnect by using biological synapses and neurons to identify temporal and spatial faults. Analysis of the SMART strategy has been provided which simulates the FDU monitoring signal wires in a number of different circuit configurations applicable to NoC architectures. These include single and multi-wire channels, and crosstalk between adjacent wires. To validate the SMART strategy in real time, a hardware model of the FDU has been developed and implemented on an Altera Cyclone 4 FPGA. A simple fault injection mechanism has been used to induce faults into the system and real-time fault detection using SMART has been demonstrated on the FPGA. In particular, analysis from synthesis results demonstrated that SMART provides a low area and power overhead, thereby, sustaining scalability criteria for large-scale NoC systems. Future work will explore the use of intelligent techniques in the classification of temporary and permanent faults using the temporal and spatial data provided by multiple FDUs across a complete NoC system.

References

1. A. Alaghi, N. Karimi, M. Sedghi, Z. Navabi, Online NoC switch fault detection and diagnosis using a high level fault model, in *22nd IEEE International Symposium on Defect and Fault Tolerance in VLSI Systems*, Rome, 2007
2. A. Benso, S. Di Carlo, G. Di Natale, P. Prinetto, A watchdog processor to detect data and control flow errors, in *IEEE Online Testing Symposium*, Greece, 2003
3. H. Paugam-Moisy and S.M. Bohte. Computing with Spiking Neuron Networks, In G. Rozenberg, T. BÅuck and J.N. Kok, Eds, *Handbook of Natural Computing*. Springer Verlag: Heidelberg, 2011
4. S. Carrillo, J. Harkin, L. McDaid, S. Pande, S. Cawley, F. Morgan, Adaptive routing strategies for large scale spiking neural network hardware implementations, in *21st International Conference on Artificial Neural Networks (ICANN)*, Finland, 2011
5. S. Carrillo, J. Harkin, L. McDaid, S. Pande, S. Cawley, B. McGinley, F. Morgan, Advancing interconnect density for spiking neural network hardware implementations using traffic-aware adaptive network-on-chip routers, in *Neural Networks* Elsevier Science, pp. 42–57 (2012)
6. C. Concatto, J. Almeida, G. Fachini, M. Herve, F. Kastensmidt, E. Cota, M. Lubaszewski, Improving the yield of NoC-based systems through fault diagnosis and adaptive routing. J. Parallel Distrib. Comput. **71**(5), 664–674 (2011)
7. A. Frantz, M. Cassel, F. Kastensmidt, E. Cota, L. Carro, Crosstalk- and SEU-aware networks on chips. IEEE Des. Test Comput. **24**, 340–350 (2007)
8. W. Gerstner, W. Kistler, *Spiking Neuron Models: Single Neurons, Populations, Plasticity* (Cambridge University Press, Cambridge, 2002)
9. A. Ghania, L. McDaid, A. Belatreche, S. Hall, S. Huang, J. Marsland, T. Dowrick, A. Smith, Evaluating the generalisation capability of a CMOS based synapse, Neurocomputing, 83, 188–197 (2012)
10. I. Golubev, R. Tsarev, T. Semenko, N-version software systems design. *Proceedings of the 11th International Scientific and Practical Conference of Students, Post-graduates and Young Scientists*, Boston, USA, 2007
11. O. Goloubeva, M. Rebaudengo, M. Reorda, M. Violante, Improved software-based processor control-flow errors detection technique, in *Reliability and Maintainability Symposium*, 2005
12. C. Grecu, A. Ivanov, R. Saleh, E. Sogomonyan, P. Pande, On-line fault detection and location for NoC interconnects, in *Proceedings of the 12th IEEE International Symposium on On-Line Testing (IOLTS)*, Spain, 2006
13. J. Harkin, P. Dempster, B. Cather, T. McGinnity, Fault detection for self repairing systems, in *IEE SMC*, UK, 2007
14. J. Harkin, F. Morgan, L. McDaid, S. Hall, B. McGinley, S. Cawley, A reconfigurable and biologically inspired paradigm for computation using network-on-chip and spiking neural networks. Int. J. Reconfigurable Comput. (2009)
15. F.Morgan, S.Cawley, P.McGinley, S. Pande, L. McDaid, B. Glackin and J. Harkin Exploring the Evolution of NoC-Based Spiking Neural Networks on FPGAs. In: IEEE Field Programmable Technology Conference, Sydney, Australia, pp. 24–27 (2009)
16. C. Hernandez, A. Roca, R. Flich, S. Duato, Characterizing the impact of process variation on 45 nm NoC-based CMPs. J. Parallel Distrib. Syst. **71**, 651–663 (2011)
17. M. Herve, E. Cota, F. Kastensmidt, M. Lubasewski, Diagnosis of interconnect shorts in mesh NoCs, in *3rd ACM/IEEE International Symposium on Networks-on-Chip*, San Diego, USA, 2009
18. M. Hosseinabady, A. Banaiyan, M. Bojnordi, Z. Navabi, A concurrent testing method for noc switches, in *IEEE Design, Automation and Test in Europe*, 2006
19. M. Kakoee, V. Bertacco, L. Benini, A distributed and topology-agnostic approach for on-line NoC testing, in *IEEE/ACM International Symposium on NoCs*, Pennsylvania, USA, 2011
20. A. Kohler, G. Schley, M. Radetzki, Fault tolerant network on-chip switching with graceful performance degradation. IEEE Trans. Comput. Aided Des. Integr. Circuit. Syst. **29**, 883–896 (2010)

21. L. McDaid, S. Hall, P. Kelly, A programmable facilitating synapse device, in *IEEE International Joint Conference on Neural Networks (IEEE World Congress on Computational Intelligence)*, Hong Kong, China, 2008
22. S. Mitra, E.J. McCloskey, Design of redundant systems protected against common-mode failures, in *Proceedings of the IEEE VLSI Test Symposium*, California, USA, 2001
23. E. Mourad, A. Nayak, Comparison-based system level fault diagnosis: a neural network approach. IEEE Trans. Parallel Distrib. Syst. **23**(6), 1047–1059 (2012)
24. S. Murali, G. De Micheli, L. Benini, T. Theocharides, N. Vijaykrishnan, M. Irwin, Analysis of error recovery schemes for networks on chips. IEEE Trans. Des. Test Comput. **22**(5), 434–442 (2005)
25. P. Reviriego, C. Argyrides, J. Maestro, D. Pradhan, Improving memory reliability against soft errors using block parity. IEEE Trans. Nucl. Sci. **58**(3), 981–986 (2011)
26. P. Yaghini, A. Eghbal, H. Pedram, H. Zarandi, Investigation of transient fault effects in an asynchronous NoC router. J. Syst. Archit. **57**(1), 61–68 (2011)
27. Nodoushan M, Miremadi S, and Ejlali A. Control Flow checking using Branch Instructions, IEEE International Conference on Embedded and Ubiquitous Computing, Vol 1, 66–72 (2008)
28. Giaconia, G.C., Di Stefano, A., Capponi, G. FPGA-based concurrent watchdog for real time control systems. Electronic Letters, **39**(10) 769–770 (2003)
29. Dai, L., Shang, D., Xia, F., Yakovlev, A. Monitoring circuit based on threshold for fault-tolerant NoC. Electronics Letters, **46** 984–5 (2010)

Chapter 8
Power-Efficient Fault-Tolerant Finite Field Multiplier

Jimson Mathew, A.M. Jabir, R.A. Shafik, and D.K. Pradhan

As integrated circuit density increases, digital circuits characterized by high operating frequencies and low voltage levels will be increasingly susceptible to faults. Furthermore, it has recently been shown that for many digital signature and identification schemes an attacker can inject faults into the hardware and the resulting incorrect outputs may completely expose their secrets. On-chip error masking techniques such as error correction could be one of the options to mitigate the above problems. To this end, this chapter presents a framework of techniques to design error correcting circuits. Fault attacks are based on injecting some faults into a cryptosystem and observing any leak of secret information, primarily by analyzing erroneous results produced by the cryptosystem due to the faults. For example, in [2] Boneh et al. presented the first fault-based attacks on public key cryptosystems, namely RSA and Rabin signature scheme. Since RSA is usually implemented using the Chinese Remainder theorem (CRT), having one correct signature and one faulty signature of the same massage can lead to the modulus factorization [3]. In order to avoid such fault-based attacks, the cryptosystem can be designed to detect errors and correct computations. Therefore, if we can correct errors the module will not produce any erroneous results as output.

Since, majority of the error correction techniques described in this chapter is based on finite field circuits, finite field arithmetic is reviewed. First, the basic definitions and properties of finite fields are introduced, which are relevant to the material treated later in this chapter. All the statements are given without proof because of space constraints. A detailed treatment of the subject can be

J. Mathew (✉) • R.A. Shafik • D.K. Pradhan
Department of Computer Science, University of Bristol, Bristol, UK
e-mail: jimson@cs.bris.ac.uk; Rishad.Shafik@bristol.ac.uk; pradhan@cs.bris.ac.uk

A.M. Jabir
Oxford Brookes University, Oxford, UK
e-mail: ajabir@brookes.ac.uk

J. Mathew et al. (eds.), *Energy-Efficient Fault-Tolerant Systems*,
DOI 10.1007/978-1-4614-4193-9_8, © Springer Science+Business Media New York 2014

found in [17]. Most of the structures proposed in this chapter are based on the Polynomial Basis (PB), therefore more emphasis is given on PB representation. Moreover, PB representation is by far the most versatile and offers suitable solutions to most computational problems. Here we discuss various design approaches for error correction finite field multiplier. Faults causing single error at the output bits of the multiplier is corrected on-line. Hence, on-line error correction results in more robust hardware modules against fault attacks. Also presented a systematic method for designing multiple detection and correction techniques for multiplier circuits for Galois fields over GF(2^m). We used multiple parity predictions to detect multiple errors applying popular error correcting codes. The expressions for the parity prediction are derived from the input operands, and are based on the primitive polynomials of the fields. For multiple bit error correction we use Reed Solomon codes. Comparison with traditional techniques, shows better area and power performance.

8.1 Introduction

This section provides the background material relevant to this chapter and the following chapter. Since, majority of the error correction techniques described here are based on finite field circuits, finite field arithmetic is reviewed.

8.1.1 Finite Field

Definition of fundamental algebraic structures are first introduced.

Definition 8.1 ([16]). A group is a set G together with a binary operation $\circ$ on G such that:

1. Binary operator $\circ$ is associative; i.e., for any $a,\ b,\ c \in G, a \circ (b \circ c) = (a \circ b) \circ c$
2. There is an identity (or unity) element e in G such that for all a $\in$ $G\ a \circ e\ =\ e \circ a =\ a.$
3. For each $a \in G$ there exits an inverse element $a^{-1} \in G$ such that $a \circ a^{-1} = e$

Definition 8.2 ([16]). A Ring (R, +, $\bullet$) is a set R together with two binary operation + on $\bullet$ such that:

1. R is abelian group with respect to +
2. Binary operator $\bullet$ is associative i.e. (a $\bullet$ b)$\bullet$ c = (a $\bullet$ b) $\bullet$ c for all $a,\ b,\ c \in R.$
3. The distribution law holds; that is, for all $a,\ b,\ c \in R$ we have a $\bullet$(b + c) = a $\bullet$ b + a $\bullet$ c and (b + c)$\bullet$ a= b$\bullet$ a + c $\bullet$ a.

Definition 8.3. A finite field F is a set F together with two binary operations denoted by + and $\bullet$ such that

1. F is a commutative ring under $+$ and $\bullet$.
2. Non-zero elements of F form a group under $\bullet$.

Definition 8.4. The order of a field is the number of its elements.

Theorem 8.1 ([16]). *The order q of a field must be a power of a prime;* $q = p^m$, *p is a prime.*

Theorem 8.2 ([16]). *There exists a unique field of order* p^m, *for any prime and any positive integer m.*

Definition 8.5. The smallest positive integer λ for which $\sum_{i=1}^{\lambda} 1 = 0$ in a field is called the field's characteristic.

The architectures proposed here are based on finite fields of characteristic two.

Definition 8.6. Let a be an element of GF(q). The smallest positive integer s for which $a^s = 1$ is called the order of the element.

Definition 8.7. Elements that have order s = q − 1 are called primitive elements.

It can be shown that elements with maximum order exists for every finite field. A primitive element and its powers generate the entire multiplicative group $\{1, \alpha, \alpha^2 \ldots, \alpha^{m-1}\}$ of a field.

8.1.2 *Polynomials Over Finite Fields*

Finite fields over the set GF(2^m) can be generated with monic irreducible polynomials of the form $P(x) = x^{m-1} + \sum_{i=0}^{m-2} c_i x^i$, where $c_i \in$ GF(2) [18]. It is conventional to represent the elements of GF(2^m) as a power of the primitive element denoted by α, where α is the root of $P(x)$, i.e. $P(\alpha) = 0$. The set $\{1, \alpha, \alpha^2 \ldots, \alpha^{m-1}\}$ is referred to as the polynomial basis. Each element $A \in$ GF(2^m) can be expressed with respect to the PB as a polynomial of degree m over GF(2), i.e. as $A(x) = \sum_{i=0}^{m-1} a_i x^i$, where $a_i \in$ GF(2). Given any two elements A and B over GF(2^m).

Example 8.1. Consider the Galois field GF(2^4) generated by the polynomial $P(x) = x^4 + x + 1$ with α as a primitive root. The Elements of $GF(2^4)$ generated by P(x) with primitive root α is shown in Table 8.1

Definition 8.8 ([13]). A polynomial A(x) is irreducible over GF(q) if A(x) is only divisible by c or cA(x) where $c \in GF(q)$

In this thesis, $a|b$ denotes "a divides b", where a and b can either be number or polynomial.

Definition 8.9. Let P(x) be a polynomial of degree m over GF(q) with P(0) = 0. The smallest positive integer s for which $P(x)(x^s + 1)$ is called the order of P(x).

Table 8.1 Field elements $GF(2^4)$ with $P(x) = x^4 + x + 1$

Element	Representation
0	0000
1	0001
α	0010
α^2	0100
α^3	1000
$\alpha^4 = \alpha + 1$	0011
$\alpha^5 = \alpha^2 + \alpha$	0110
$\alpha^6 = \alpha^3 + \alpha^2$	1100
$\alpha^7 = \alpha^3 + \alpha + 1$	1011
$\alpha^8 = \alpha^2 + 1$	0101
$\alpha^9 = \alpha^2 + 1$	0101
$\alpha^{10} = \alpha^3 + \alpha$	1010
$\alpha^{11} = \alpha^2 + \alpha + 1$	0111
$\alpha^{12} = \alpha^3 + \alpha^2 + \alpha$	1110
$\alpha^{13} = \alpha^3 + \alpha^2 + \alpha + 1$	1111
$\alpha^{13} = \alpha^3 + \alpha^2 + 1$	1101

Definition 8.10. A monic polynomial of degree m with maximum order $s = q^m - 1$ is said to be a primitive polynomial.

A primitive polynomial of degree m over GF(q) exist for any field GF(q).

8.1.3 Bases of Finite Field

In general, there are three different bases widely used in the literature which are standard, normal and dual. An extension of GF(q^m) of the field GF(q) can be viewed as m-dimensional vector space over GF(q). Each element of GF(q^m) can be represented as a linear combination of the m elements of the base $\beta_0, \beta_1, \cdots, \beta_{m-1}\}$.

Definition 8.11. The set $\{1, \alpha, \alpha^2, \ldots, \alpha^{m-1}\}$, where α is a root of the irreducible polynomial P(x) of degree m over GF(q), is called standard or canonical or polynomial basis.

Definition 8.12. The set $\{\alpha, \alpha^q, \alpha^{q^2}, \ldots, \alpha^{q^{m-1}}\}$, where α is a root of the irreducible polynomial P(x) of degree m over GF(q), is called normal base if the m elements are linearly independent.

Definition 8.13. Let B $= \beta_0, \beta_1 \cdots, \beta_{m-1}$ be a base of $GF(q^m)$ the dual base B is a base satisfying

$T_r(\beta_i \gamma_i) = \{^{1,\ if\ i=j}_{0,\ if\ i\neq j}$. It can be shown that there exits a dual base for every base

8.2 Polynomial Basis Addition and Multiplication

Consider a finite field of $K = GF(2)$ and its extension field $F = GF2^m$. Then F can be considered as m dimensional vector space over K and if $\{1, \alpha, \alpha^2, \ldots, \alpha^{m-1}\}$ is a basis of F over K, each element $A \in F$ can be uniquely represented in polynomial form as

$A(x) = a_0 + a_1x + a_2x^2, \ldots, +a_mx^{m-1}$, $a_i \in$ GF(2). Let $B(x) = b_0 + b_1x + b_2x^2, \ldots, +b_mx^{m-1}$, $b_i \in$ GF(2) be another element in F. Then addition or subtraction of A and B is given by

$A \pm B = a_0 \pm b_0 + a_1 \pm b_1x + a_2x^2, \ldots, +a_{m-1} \pm b_{m-1}x^{m-1}$

Addition can be easily realized using m EXOR gates.

A Polynomial Basis (PB) multiplication of A and B over GF(2^m) is defined as $W(x) = A(x) \cdot B(x)$ mod $P(x)$. It is nothing but a algebraic multiplication of the two polynomials and a modulo reduction with $P(x)$. The following example illustrates the basic polynomial multiplication.

Example 8.2. A multiplier structure over GF(2^4) defined by the primitive polynomial $P(x) = x^4 + x^3 + 1$ is illustrated as follows.

The two inputs of the multiplier are $A = (a_0, a_1, a_2, a_3)$ and $B = (b_0, b_1, b_2, b_3)$. The polynomial form of these elements are: $A(x) = a_0 + a_1x + a_2x^2 + a_3x^3$, and $B(x) = b_0 + b_1x + b_2x^2 + b_3x^3$, where $A, B \in$ GF(2^4). The product $C(x) = A(x) \cdot B(x)$.

Now, $C(x) = (a_0 + a_1x + a_2x^2 + a_3x^3) \cdot (b_0 + b_1x + b_2x^2 + b_3x^3) = a_0b_0 + (a_0b_1 + a_0b_1)x + (a_0b_2 + a_1b_1 + a_2b_0)x^2 + (a_0b_3 + a_1b_2 + a_2b_1 + a_3b_0)x^3 + (a_1b_3 + a_2b_2 + a_3b_1)x^4 + (a_2b_3 + a_3b_2)x^5 + a_3b_3x^6$.

Let us denote the lower order m coefficients as $d_0, d_1, \ldots, d_{m-1}$, and the higher order $m - 1$ coefficients as $e_0, e_1, \ldots, e_{m-2}$. Then, $C(x) = d_0 + d_1x + d_2x^2 + d_3x^3 + e_0x^4 + e_1x^5 + e_2x^6$. Here, we define product over the primitive polynomial $P(x) = x^4 + x^3 + 1$ as $W(x) = A(x) \cdot B(x)$ mod $P(x)$. Hence, we have, $x^4 = x^3 + 1$, $x^5 = x(x^3 + 1) = x^4 + x = x^3 + x + 1$, and $x^6 = x(x^5) = x(x^3 + x + 1) = x^4 + x^2 + x = x^3 + 1 + x^2 + x = x^3 + x^2 + x + 1$.

Substituting for x^4, x^5, and x^6 and then simplifying we get: $W(x) = C(x) = (d_0 + e_0 + e_1 + e_2) + (d_1 + e_1 + e_2)x + (d_2 + e_2)x^2 + (d_3 + e_0 + e_1 + e_2)x^3$. The general structure of the multiplier is shown in Fig. 8.1.

Mastrovito has proposed an algorithm, along with its hardware architecture, for PB multiplication [12], popularly known as the Mastrovito algorithm/multiplier. There are many reasons for choosing this architecture as representative for standard base multipliers. First, it has one of the lowest gate counts among traditional PB multipliers. Secondly, it gives a systematic method for design finite field multiplier for a given field and P(x).

The standard basic multiplication can be represented as

$A(x)B(x)modP(x) = (c_0 + c_1x + c_2x^2, \ldots, +c_mx^{m-1}) = (a_0 + a_1x + a_2x^2, \ldots, +a_mx^{m-1})(b_0 + b_1x + b_2x^2, \ldots, +b_mx^{m-1})modP(x)$

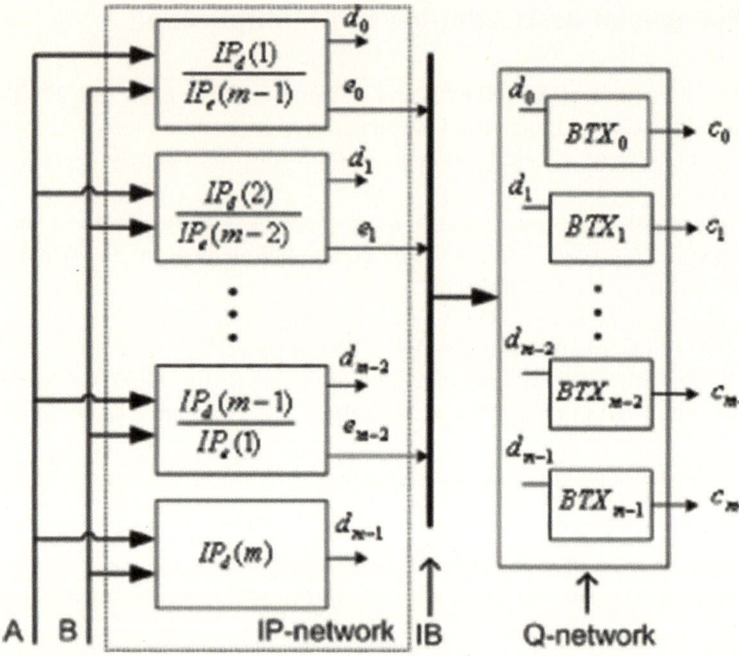

Fig. 8.1 Architecture of the BP multiplier over GF(2^m) [21]

The elements B(x) and C(x) can be represented as column vectors containing polynomial coefficients. By using the new Z matrix where Z = f(A(x), P(x)) the multiplication can be formulated as:

$$C = \begin{pmatrix} c_0 \\ c_1 \\ \vdots \\ c_{m-1} \end{pmatrix} = ZB = \begin{pmatrix} f_{0,0} & \cdots & f_{,-1} \\ \vdots & \ddots & \vdots \\ f_{m-1,0} & \cdots & f_{m-1,m-1} \end{pmatrix} \begin{pmatrix} b_0 \\ b_1 \\ \vdots \\ c_{m-1} \end{pmatrix} \tag{8.1}$$

The matrix Z is called product matrix. Its coefficients $f_{i,j} \in GF(2)$ depend on a_i and on the coefficients $q_{i,j}$ of the Q matrix which is given in Eq. 8.2

$$f_{i,j} = \begin{Bmatrix} a_i & ; & j = 0 \quad i = 0 \cdots n-1 \\ u(i-j)a_{i-j} + \sum_{i=0}^{j-1} q_{j-1-t,i} a_{m-1-t} & ; & j = 1 \cdots m-1 \; i = 0 \cdots n-1 \end{Bmatrix}$$

where the step function u is defined as

$$\mathrm{u}(\mu) = \begin{Bmatrix} 1 & \mu \geq 0 \\ 0 & \mu > 0. \end{Bmatrix}$$

The Equation 8.1 represents the entire Mastrovito multiplication. The Q matrix which is required to build Z is a function of the binary primitive polynomial P(x) of degree m, generating GF(2^m) Its binary entries $q_{i,j}$ are defined as

$$\begin{pmatrix} x^m \\ x^{m+1} \\ \vdots \\ x^{2m-2} \end{pmatrix} = \begin{pmatrix} q_{0,0} & q_{0,1} & \cdots & q_{0,m-1} \\ q_{1,0} & q_{1,1} & \cdots & \\ \vdots & \vdots & \ddots & \vdots \\ q_{m-2,0} & q_{m-2,1} & \cdots & q_{m-2,0} \end{pmatrix} \begin{pmatrix} 1 \\ x \\ \vdots \\ x^{m-1} \end{pmatrix} \bmod P(x) \tag{8.2}$$

The Q matrix describes the representation of the polynomials x^m, x^{m+1},$\cdots$, x^{2m-2} in the equivalence classes mod P(x), i.e. after the modulo P(x). The following example shows the construction of Q matrix and field multiplication by Mastrovito

Example 8.3. Let $P(x) = x^4 + x + 1$ be the primitive polynomial generating GF(2^4) Considering the equivalence classes mod P(x), the polynomial x^4, x^5, x^6 are as follows:

$$\begin{aligned} x^4 &\equiv (1 + x) \bmod x^4 + x + 1 \\ x^5 &\equiv (x + x^2) \bmod x^4 + x + 1 \\ x^6 &\equiv (x^2 + x^3) \bmod x^4 + x + 1 \end{aligned} \tag{8.3}$$

Equation 8.3 can be written in matrix form as

$$\begin{pmatrix} x^4 \\ x^5 \\ x^6 \end{pmatrix} = \begin{pmatrix} 1\,1\,0\,0 \\ 0\,1\,1\,0 \\ 0\,0\,1\,1 \end{pmatrix} \begin{pmatrix} 1 \\ x \\ x^2 \\ x^3 \end{pmatrix} \bmod P(x) \tag{8.4}$$

The final product can be written as

$$\mathrm{C} = \mathrm{ZB} = \begin{pmatrix} a_0 & a_1 & a_2 & a_3 \\ a_1 & a_0 + a_3 & a_2 + a_3 & a_1 + a_2 \\ a_2 & a_1 & a_0 + a_3 & a_2 + a_3 \\ a_3 & a_2 & a_1 & a_0 + a_3 \end{pmatrix} \begin{pmatrix} b_0 \\ b_1 \\ b_2 \\ b_3 \end{pmatrix}.$$

The implementation complexity of the above matrix vector product depends on the primitive polynomial P(x).

In [21], based on the Mastrovito algorithm, a new formulation for PB multiplication and generalized bit-parallel hardware architecture has been presented. Their formulation is summarized below for completeness, which we have used in the rest of the thesis.

Consider a multiplier with a and b inputs where $A = (\mathbf{a}) = [a_0, a_1, a_2, \ldots, a_{m-1}]$ and $B = (\mathbf{b}) = [b_0, b_1, b_2, \ldots, b_{m-1}]$. The a_i and b_i, where $0 \leq i \leq m-1$, are the coordinates of a and b respectively. The formulation is based on the three matrices: (i) an $m-1$ by m reduction matrix **Q**, (ii) the **L** matrix, and (iii) the **U** matrix. The **L** and **U** matrices are formed for implementation of this multiplication scheme. The **L** is a lower triangular matrix and the **U** is an upper triangular matrix. The outputs (d's and e's) of the IP-network are defined by the following two vectors, which are functions of A and B.

$$\mathbf{d} = \mathbf{L}\mathbf{b} \tag{8.5}$$

$$\mathbf{e} = \mathbf{U}\mathbf{b}, \tag{8.6}$$

where $\mathbf{b} = [b_0, b_1, b_2, \ldots, b_{m-1}]^T$, a vector column of the coordinates and x^T represents the x transpose. The matrices **L** and **U** are defined as follows.

$$\mathbf{L} = \begin{bmatrix} a_0 & 0 & 0 & \ldots & 0 \\ a_1 & a_0 & 0 & \ldots & 0 \\ a_2 & a_1 & a_0 & \ldots & 0 \\ \vdots & \vdots & \ddots & \vdots & \vdots \\ a_{m-2} & a_{m-3} & \ldots & a_0 & 0 \\ a_{m-1} & a_{m-2} & \ldots & a_1 & a_0 \end{bmatrix}$$

$$\mathbf{U} = \begin{bmatrix} 0 & a_{m-1} & a_{m-2} & \ldots & 0 & a_1 \\ 0 & 0 & a_{m-1} & \ldots & 0 & a_2 \\ \vdots & \vdots & \vdots & \ddots & \vdots & \vdots \\ 0 & 0 & \ldots & 0 & a_{m-1} & a_{m-2} \\ 0 & 0 & \ldots & 0 & 0 & a_{m-1} \end{bmatrix}.$$

The multiplication outputs are given by the equation:

$$\mathbf{c} = \mathbf{d} + \mathbf{Q}^T\mathbf{e}, \tag{8.7}$$

where the matrix **Q**, which is dependent on the irreducible polynomials, can be derived as shown in [21] and $\mathbf{c} = [c_0, c_1, c_2, \ldots, c_{m-1}]^T$ is the output bits.

The above polynomial multiplication and modulo reductions can be represented in the matrix form as follows.

Example 8.4. Consider the Example 8.2, using the above formulation we have

$$\begin{bmatrix} c_0 \\ c_1 \\ c_2 \\ c_3 \end{bmatrix} = \begin{bmatrix} d_0 \\ d_1 \\ d_2 \\ d_3 \end{bmatrix} + \begin{bmatrix} 1 & 1 & 1 \\ 0 & 1 & 1 \\ 0 & 0 & 1 \\ 1 & 1 & 1 \end{bmatrix} \begin{bmatrix} e_0 \\ e_1 \\ e_2 \end{bmatrix}$$

where $\mathbf{Q} = \begin{bmatrix} 1 & 0 & 0 & 1 \\ 1 & 1 & 0 & 1 \\ 1 & 1 & 1 & 1 \end{bmatrix}$.

$$e_0 = a_3b_1 + a_2b_2 + a_1b_3$$
$$e_1 = a_3b_2 + a_3b_2$$
$$e_2 = a_3b_3$$

$$\vec{d} = \begin{bmatrix} d_0 \\ d_1 \\ d_2 \\ d_3 \end{bmatrix} = \begin{bmatrix} a_0b_0 \\ a_1b_0 + a_0b_1 \\ a_2b_0 + a_1b_1 + a_0b_2 \\ a_3b_0 + a_2b_1 + a_1b_2 + a_0b_3 \end{bmatrix} \tag{8.8}$$

$$\vec{e} = \begin{bmatrix} e_0 \\ e_1 \\ e_2 \end{bmatrix} = \begin{bmatrix} a_3b_1 + a_2b_2 + a_1b_3 \\ a_3b_2 + a_2b_3 \\ a_3b_3 \end{bmatrix} \tag{8.9}$$

Definition 8.14. Hamming weight of $N^1(c)$ is defined as the number of non-zero components of c. The Hamming Distance between v and w is defined as the number of places in which they differ.

Definition 8.15. The parity check matrix **H** of a code consists of all the non-zero r-tuples as its columns. In the systematic form, the columns of **H** are arranged as $\mathbf{H} = [\mathbf{H_s H_p}]$, where $\mathbf{H_s}$ is the systematic part and $\mathbf{H_p}$ the parity part.

Lemma 8.1 ([22]). *Let A and B be two field elements and S be their multiplication without modulo reduction. Then the parity of S is given by $p_s = p_A \cdot p_B$, where p_A and p_B are the parity of A and B respectively.*

8.3 SEC Multiplier: Based on Hamming Code

In this section, a systematic method for designing single error correcting bit parallel *polynomial basis* (PB) multipliers over GF(2^m) is proposed. In [22] the authors have considered the detection of single stuck at faults in the PB multipliers over GF(2^m). They use simple parity prediction technique for error detection. The main problem with their approach is that for a low complexity bit parallel multiplier the delay overhead is 69.2%. For performance critical applications this delay overhead may be critical. Our technique fundamentally differs from this technique in two critical issues. Firstly, our technique addresses the problem of single error correction. Secondly, and more importantly, since our parity prediction circuit runs parallel with the multiplier, delay penalty comes only in the decoding and correction logic. While most of the previous work provides fault detection techniques, this work proposes error correction techniques and investigates the hardware cost and performance.

A number of approaches exist, e.g. [4]. One way to detect errors in finite field multipliers is to use parity prediction techniques [22]. The problem with this approach is that no error correction can be performed. Furthermore, we can not distinguish between error in multiplier and parity prediction logic. A second approach is to scale the inputs of the multiplier by a factor and at the end of the multiplication, the correctness of the result is checked by one or two divisions [6]. The main techniques that can be used for single error correction are (1) error

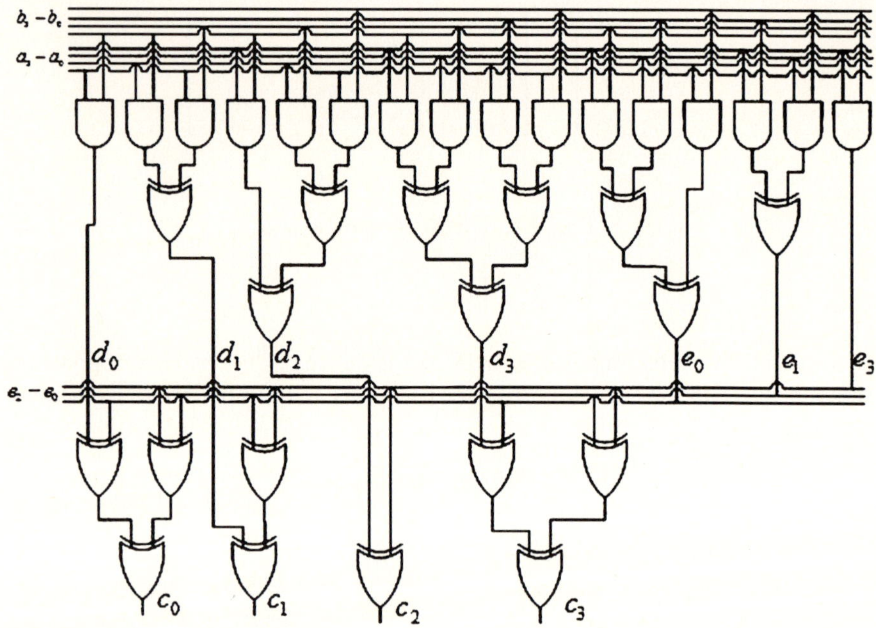

Fig. 8.2 A bit parallel GF(16) multiplier

detection and retry, and (2) error masking. Error detection and retry involves using concurrent error detection (CED) circuitry that monitors the outputs of a circuit for the occurrence of an error. If an error is detected, the system recovers through rollback and retry thereby preventing a failure. Error masking involves using circuitry that masks (i.e. *corrects*) errors using schemes such as the triple modular redundancy (TMR) [14].

Next, we present a novel technique for designing Single Error Correction (SEC) bit parallel multipliers over GF(2^m). The basic structure of the multiplier is shown in Fig. 8.1. The classical bit-parallel multiplier is designed by the method described in Sect. 8.2 (see e.g. Fig. 8.2). The modified single error correcting architecture is shown in Fig. 8.3. Apart from the functional unit of the multiplier, it consists of a parity prediction unit, output parity generation unit, and the comparison and decoding logic. For error detection and correction we use the parity prediction, which is based on Hamming's principles. Hamming codes are the simplest of a group of codes known as linear block codes [8]. The advantage of the Hamming codes is that the number of parity bits grows logarithmically as the number of output bits increases. However, the complexity of the encoding and decoding logic grows linearly as the number of data bits increases in memory based error correction. In the proposed approach, we consider parity prediction based encoders. For memory based Hamming encoders, the bits are encoded with a tree of exclusive-OR (XOR) operations. The principal difference between Hamming codes applied

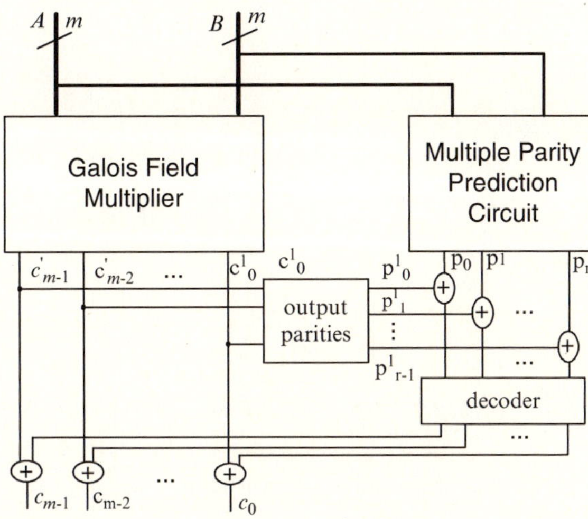

Fig. 8.3 Proposed SEC multiplier over GF(2^m)

to memory and our approach is that, in our approach, instead of encoders we have parity prediction circuits. The sizes of the parity prediction circuits depend on the number of input bits. At the output side the error correction steps include computing the syndrome, finding the error pattern and correcting error in the code word. Mathematically, the syndrome computation is similar to the one in [8]. In the implementation, the syndrome vector is generated by bit XOR operation of predicted parity bits and output generated parities. Error pattern is identified by the decoding the syndrome vectors. Finally, the correction is applied by the XOR operation of decoder output and output from the multiplier. These steps are shown in Fig. 8.3. Next, we derive the closed form expressions for the predicted parity bits.

Let $\vec{c}^{\,1} = [c_0^1, c_1^1, c_2^1, \ldots, c_{m-1}^1]^T$ be the output of the multiplier and $\vec{c} = [c_0, c_1, c_2, \ldots, c_{m-1}]^T$ the corrected output. Also let r be the number of parity bits, and $\vec{p} = [p_0, p_1, \ldots, p_{r-1}]^T$ and $\vec{p}^{\,1} = [p_0^1, p_1^1, \ldots, p_{r-1}^1]^T$ respectively be the predicted and the parity bits generated from the output bits. Let $\mathbf{H_p}$ be the parity check matrix associated with the proposed single error correction scheme.

Lemma 8.2. *Let A and B be two field elements and* $\mathbf{p}$ *be the predicted parity bits of their product. We have,*

$$\mathbf{p} = p_s + \mathbf{H_p}^{\mathbf{c}}\mathbf{d} + \mathbf{K}^{\mathbf{c}}\mathbf{e}, \tag{8.10}$$

where $\mathbf{K} = \mathbf{H_p}\mathbf{Q}^T$, *and the superscript* $\mathbf{c}$ *represents the complementary set.*

Proof. From the fundamentals we have the parity check equations

$$\mathbf{p} = \mathbf{H_p}\mathbf{c}. \tag{8.11}$$

Substituting **c** from Eq. (8.7) into Eq. (8.11),

$$\mathbf{p} = \mathbf{H_p d} + \mathbf{H_p Q}^T \mathbf{e} \tag{8.12}$$

Using Lemma 8.1, Eq. (8.11) can be rewritten as,

$$\mathbf{p} = p_s + \mathbf{H_p}^{\mathbf{c}}\mathbf{d} + \mathbf{K}^{\mathbf{c}}\mathbf{e}. \tag{8.13}$$

Hence the proof.

8.3.1 *Design Procedure Parity Prediction Bits*

- Determine the number of parity bits (r) required to satisfy the equation $m + r + 1 \leq 2^r$.
- Construct the **H** matrix, with $(m + r)$ non-zero r-bit column vectors. The dimension of the resulting matrix is $r \times (m + r)$.
- A column vector with a single 1 is assigned to parity P_i.
- The column vector with all 1s is assigned to output bit c_{m-1}.
- The remaining m columns are assigned the output bits c_i, without any constraints.
- Generate predicted parity expressions in terms of a_is and b_is from Eq. (8.10).

Example 8.5. Consider the multiplier structure over GF(2^4) constructed in Example 8.2. Here we have $m = 4$. Therefore, we need three parity bits to correct single errors. We have,

$$\mathbf{H} = \begin{array}{ccccccc} p_0 & p_1 & p_2 & c_0 & c_1 & c_2 & c_3 \\ 1 & 0 & 0 & 1 & 1 & 0 & 1 \\ 0 & 1 & 0 & 1 & 0 & 1 & 1 \\ 0 & 0 & 1 & 0 & 1 & 1 & 1. \end{array}$$

Therefore, $\mathbf{H_s} = \begin{array}{ccc} p_0 & p_1 & p_2 \\ 1 & 0 & 0 \\ 0 & 1 & 0 \\ 0 & 0 & 1 \end{array}$; $\mathbf{H_p} = \begin{bmatrix} 1\,1\,0\,1 \\ 1\,0\,1\,1 \\ 0\,1\,1\,1 \end{bmatrix}$; and $\mathbf{K} = \begin{bmatrix} 0\,1\,1 \\ 0\,0\,1 \\ 1\,0\,1 \end{bmatrix}$.

The predicted parity bit outputs based on Eq. (8.12), where $0 \leq i \leq 3$, are as follows. (Here, the '+' sign represents EXOR operation.)

$$p_0 = d_0 + d_1 + d_3 + e_1 + e_2 \tag{8.14}$$

$$p_1 = d_1 + d_2 + d_3 + e_1 \tag{8.15}$$

$$p_2 = d_1 + d_2 + d_3 + e_0 + e_2 \tag{8.16}$$

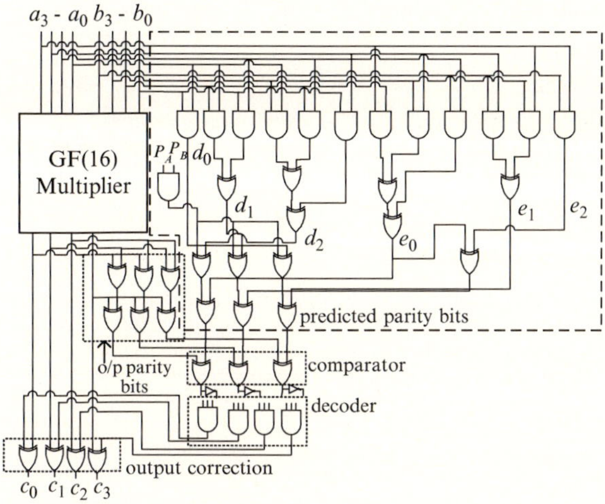

Fig. 8.4 SEC bit parallel multiplier over GF(2^4)

Applying Eq. (8.13) we have,

$$\mathbf{H_p^c} = \begin{bmatrix} 0\,0\,1\,0 \\ 0\,1\,0\,0 \\ 1\,0\,0\,0 \end{bmatrix}; \mathbf{K^c} = \begin{bmatrix} 1\,0\,0 \\ 1\,1\,0 \\ 0\,1\,0 \end{bmatrix}.$$

Moreover, we have $p_s = e_2 + e_1 + e_0 + d_3 + d_2 + d_1 + d_0$. Therefore, the final equations are

$$p_0 = p_s + d_2 + e_0 \tag{8.17}$$

$$p_1 = p_s + d_1 + e_0 + e_2 \tag{8.18}$$

$$p_2 = p_s + d_0 + e_1. \tag{8.19}$$

The **d** and **e** outputs of the network are given below.

$d_0 = a_0b_0$; $d_1 = a_1b_0 + a_0b_1$; $d_2 = a_2b_0 + a_1b_1 + a_0b_2$; $d_3 = a_3b_0 + a_2b_1 + a_1b_2 + a_0b_3$; $e_0 = a_3b_1 + a_2b_2 + a_1b_3$; $e_1 = a_3b_2 + a_2b_3$; $e_2 = a_3b_3$.

The implementation of this example is shown in Fig. 8.4. Here only the parity prediction and error correction logic units are shown. The idea of using the parity of input operands reduces the overall implementation complexity. For instance, in the above example to implement the final parity expression (Eqs. 8.14–8.16) we need 11 XOR gates whereas the modified expression (Eqs. 8.17–8.19) we need only 7 gates. This method also saves one XOR gate delay in the critical path.

Lemma 8.3. *Any single stuck-at faults in the parity prediction circuits, produce a non-zero syndrome vector with single one.*

Table 8.2 Comparison with other concurrent error detection schemes

Multiplier	[22]	[15]	Proposed
Structure	Bit parallel	Bit parallel systolic	Bit parallel
Basis	Polynomial	Dual	Polynomial
Area overhead	$(m) \cdot A_a$	$(2m^2 + m) \cdot A_a$	$(m^2 - m) \cdot A_a$
	$(4m - 2) \cdot A_x$	$(m^2 + 4m) \cdot A_x$	$[(m^2 - 3m + 2) + N^1(H^c) + N^1(H^c) + r \cdot 2^{r-1} + m] \cdot A_x$
		$(2m^2 - 2) \cdot A_L$	A_d
Delay overhead	$\lceil \log_2(m+1) \rceil T_x$	$T_a + 2T_x$	$(r+1) \cdot T_x + T_d$
Fault coverage (%)	100	100	100
Single error correction	No	No	Yes

Proof. Let r be the number of parity bits, and $\vec{p} = [p_0, p_1, \ldots, p_{r-1}]^T$ and $\vec{p}^{\,1} = [p_0^1, p_1^1, \ldots, p_{r-1}^1]^T$ respectively be the predicted and the parity bits generated from the output bits. The output syndrome vector is given by

$$\vec{S} = [p_0 + p_0^1, p_1 + p_1^1, \ldots, p_{r-1} + p_{r-1}^1]^T. \tag{8.20}$$

Here, the '+' sign represents EXOR operation. Assume that outputs are correct and hence the output generated parity bits $\vec{p}^{\,1}$. Now from the Eq. 8.20 it is clear than any error in predicted parity bit produce a non-zero syndrome with single one at the erroneous parity position. Hence the proof.

Example 8.6. Consider the Example 8.8, assume that the output bits are 0110; the output generated parity bits are $p_0^0 = 0 + 1 + 0 = 1$; $p_0^1 = 0 + 1 + 0 = 1$; $p_0^2 = 1 + 1 + 0 = 0$; If there is no error in the parity prediction circuit the predicted parity bit should be $\vec{p}_0 = [110]^T$. Let us assume that the third parity bit is in error, that is $\vec{p} = [111]^T$. Then from Eq. 8.20 $\vec{S} = [1 + 1, 1 + 1, 0 + 1]^T = [001]^T$. Hence the syndrome indicates third predicted parity bit is in error. On the other hand, if there is an error in the multiplier, it is clear from the parity check equation, two or more bits in the syndrome bits will indicate the error. Hence indicating the functional circuit is in error.

8.3.1.1 Comparisons with Other Approaches

Table 8.2 shows the comparison. Here, T_x, T_a and T_d denote the delays of the XOR, AND, and decoder circuits. Similarly, A_x, A_a and A_d denote their areas respectively. A_L represents the latch area. $N^1(\mathbf{H^c})$ denotes the Hamming weight of $\mathbf{H^c}$. Since in the proposed parity prediction function we are not implementing the d_{m-1} output, we save m AND gates. It is a well known fact that in a bit parallel multiplier

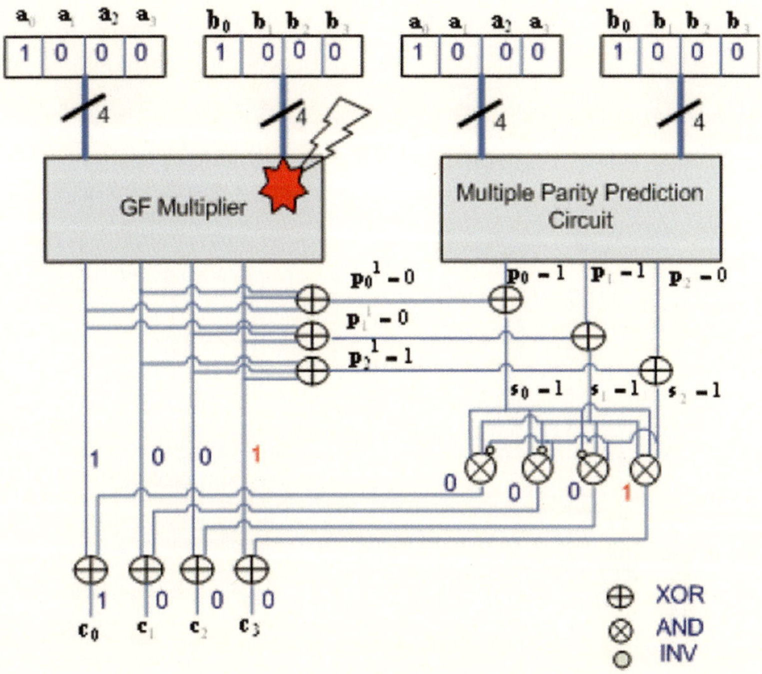

Fig. 8.5 Example showing permanent fault on multiplier and output bit is corrected by the proposed approach

the number of AND gates is m^2 [22]. The disjoint parity prediction circuits we considered here requires all the members of the vectors **e** and **d**, except d_{m-1}. Therefore the total number of AND gates is $m^2 - m$. For generating **e** and **d** from the inner products we require $m^2 - 3m + 2$ XOR gates. Apart from this, additional XOR gates are required to implement Eq. (8.10), which depend on $N^1(\mathbf{H^c})$ and $N^1(\mathbf{K^c})$. The delay overhead is contributed by the $r - 1$ XOR gates in the output parity generator plus one XOR gate delay for the correction, together with the decoder delay, T_d. Thus, the total delay overhead is $(r+1) \cdot T_x + T_d$. As we can see, the area of the parity prediction circuit is close to that of the multiplier. The delay penalty is mainly due to the output parity generation, decoder, and correction circuits.

8.3.2 Permanent Faults

Let us analyze various error scenarios on the proposed architecture and their effects. First, a permanent fault on the multiplier block is considered. We concentrate only on the multiplier block with the proposed error correction approach. Figure 8.5

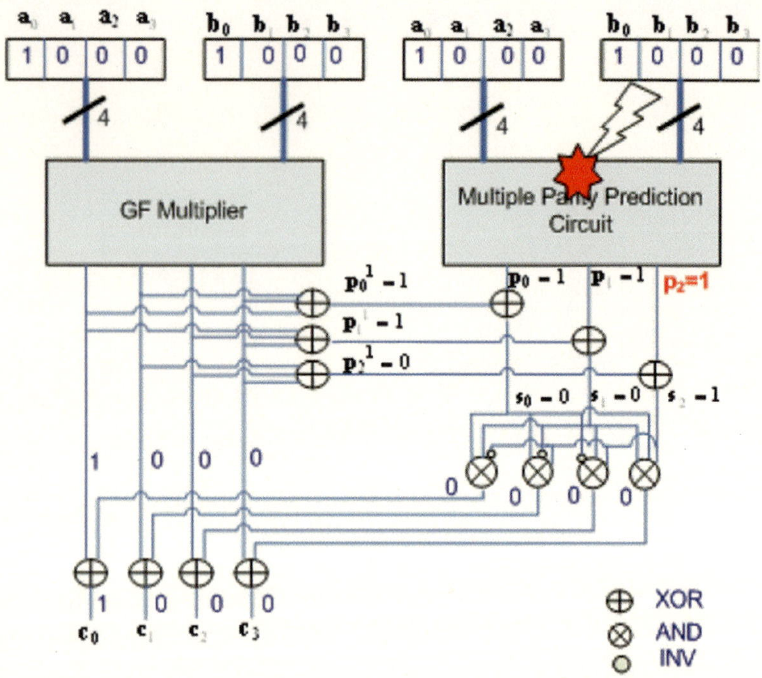

Fig. 8.6 Example showing permanent fault on PP. No correction required at the output as multiplier gives the right result

shows an error in the multiplier. Here we refer to Example 8.8 with the single error correcting **H** matrix. Let the two inputs of the multiplier be $A = (a_0, a_1, a_2, a_3) = (1000)$ and $B = (b_0, b_1, b_2, b_3) = (1000)$. The correct multiplier output bits are $C = (c_0, c_1, c_2, c_3) = (1000)$. Let us assume that, an error in the multiplier causes an erroneous output $C = (c_0, c_1, c_2, c_3) = (1001)$. However, the Parity Prediction(PP) part is not effected. Upon comparing the output parities and the PP bits gives the syndrome (111) and the bit corresponding to this syndrome is c_3 (see the **H** matrix). Therefore, the third bit gets automatically corrected as shown. Second, a permanent fault on the PP block is considered. Figure 8.6 shows an error in the parity prediction. In this case an attack/error on the PP block would cause no error in the functional output and the syndrome generated will be one of the following (001), (010), (100). Since these syndromes are not decoded and no correction is applied. With a similar argument any error in the input register that causes single bit error in the output can be corrected by the above technique, whereas in the errors in the PP register that causes signal error in the predicted parities are not corrected. Therefore, by introducing the PP logic will not compromise reliability.

8.4 Fault-Tolerant Multiplier Using LDPC Code

In this section, we present a novel technique for designing fault-tolerant bit parallel multipliers over GF(2^m). The classical bit-parallel multiplier is designed by the method described in [21]. The modified single error correcting architecture is shown in Fig. 8.3. Apart from the functional unit of the multiplier, it consists of a parity prediction unit, output parity generation unit, and the comparison and decoding logic. For error detection and correction we use multiple parity predictions, which is based on LDPC codes. Recently, LDPC codes have been received much attention because of their excellent performance and large degree of parallelism [5]. The advantage of the LDPC codes is that it has reduced decoding complexity. In Hamming code based correction the complexity of the encoding and decoding logic grows linearly as the number of data bits increases. In the proposed approach, we consider parity prediction based encoders. In conventional encoders, the bits are encoded with a tree of exclusive-OR (XOR) operations. The principal difference between LDPC code applied to other applications and our approach is that, in our approach instead of encoders we have parity prediction circuits. The sizes of the parity prediction circuits depend on the number of input bits and number of predicted parity bits. Next, a systematic method for designing fault-tolerant scheme using LDPC code is described.

Let $\vec{c}^{\,1} = [c_0^1, c_1^1, c_2^1, \ldots, c_{m-1}^1]^T$ be the output of the multiplier and $\vec{c} = [c_0, c_1, c_2, \ldots, c_{m-1}]^T$ the corrected output. Also let r be the number of parity bits, and $\vec{p} = [p_0, p_1, \ldots, p_{r-1}]^T$ and $\vec{p}^{\,1} = [p_0^1, p_1^1, \ldots, p_{r-1}^1]^T$ respectively be the predicted and the parity bits generated from the output bits. Let **H** be the parity check matrix associated with the proposed scheme.

8.4.1 Design Procedure

- Determine the number of parity bits (r) required to satisfy the given LDPC code.
- Construct the **H** matrix, with m non-zero r-bit column vectors with number of ones is 2 and r column vectors with a single 1. The dimension of the resulting matrix is $r \times (m + r)$.
- A column vector with a single 1 is assigned to parity p_i.
- The remaining m columns are assigned the output bits c_i, without any constraints.
- Generate predicted parity expressions in terms of a_is and b_is from the parity check equations.

Example 8.7. A multiplier structure over GF(2^3) defined by the primitive polynomial $P(x) = x^3 + x + 1$ is shown in Fig. 8.7.

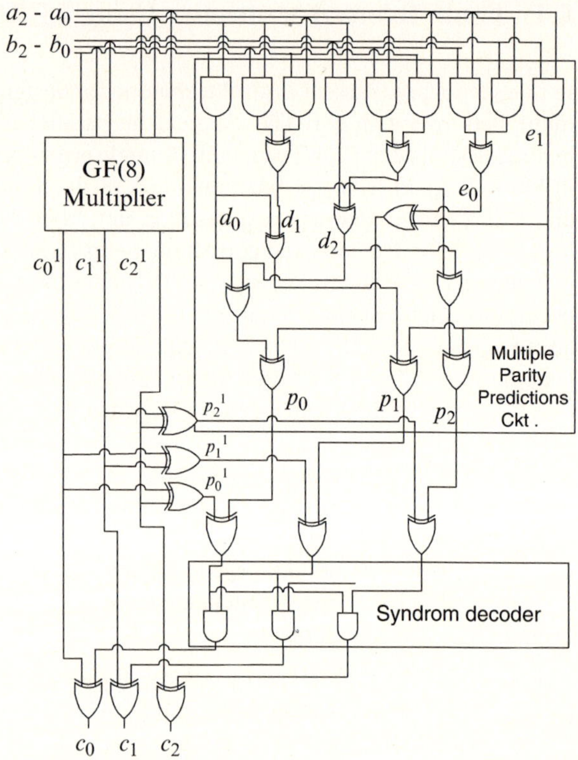

Fig. 8.7 Single error correcting GF(2^3) multiplier using LDPC code

The two inputs of the multiplier are $A = (a_0, a_1, a_2)$ and $B = (b_0, b_1, b_2)$. The polynomial form of these elements are: $A(x) = a_0 + a_1x + a_2x^2$, and $B(x) = b_0 + b_1x + b_2x^2$, where $A, B \in \mathrm{GF}(2^3)$. The product $C(x) = A(x) \cdot B(x)$.

Now, $C(x) = (a_0 + a_1x + a_2x^2) \cdot (b_0 + b_1x + b_2x^2) = a_0b_0 + (a_0b_1 + a_1b_0)x + (a_0b_2 + a_1b_1 + a_2b_0)x^2 + (a_1b_2 + a_2b_1)x^3 + a_2b_2x^4$.

Let us denote the lower order 3 coefficients as d_0, d_1 and d_2 and the higher order 2 coefficients as e_0 and e_1. Then, $C(x) = d_0 + d_1x + d_2x^2 + e_0x^3 + e_1x^4$. Here, we define product over the primitive polynomial $P(x) = x^3 + x + 1$ as $W(x) = A(x) \cdot B(x) \bmod P(x)$. Hence, we have, $x^3 = x + 1$ and $x^4 = x^2 + x$.

Substituting for x^3 and x^4, and then simplifying we get: $W(x) = C(x) = (d_0 + e_0) + (d_1 + e_0 + e_1)x + (d_2 + e_1)x^2$.

The above polynomial multiplication and modulo reductions can be represented in the matrix form as follows.

$$\mathbf{c} = \mathbf{d} + \mathbf{Q}^T\mathbf{e}, \tag{8.21}$$

where $\mathbf{Q} = \begin{bmatrix} 1 & 1 & 0 \\ 0 & 1 & 1 \end{bmatrix}$.

Example 8.8. Consider the multiplier structure over GF(2^3) constructed in Example 8.7. Here we have $m = 3$. Therefore, we need 3 parity bits to correct single errors. We have,

$$\mathbf{H} = \begin{array}{cccccc} p_0 & p_1 & p_2 & c_0 & c_1 & c_2 \\ 1 & 0 & 0 & 1 & 0 & 1 \\ 0 & 1 & 0 & 1 & 1 & 0 \\ 0 & 0 & 1 & 0 & 1 & 1. \end{array}$$

The predicted parity bit outputs based on the parity check matrix can be derived as follows: (Here, the '+' sign represents EXOR operation.)

$$p_0 = c_0 + c_2 = d_0 + d_2 + e_0 + e_1 \quad (8.22)$$

$$p_1 = c_0 + c_1 = d_0 + d_1 + e_1 \quad (8.23)$$

$$p_2 = c_1 + c_2 = d_1 + d_2 + e_1 \quad (8.24)$$

The **d** and **e** outputs of the network are given below.
$d_0 = a_0b_0$; $d_1 = a_1b_0 + a_0b_1$; $d_2 = a_2b_0 + a_1b_1 + a_0b_2$; $e_0 = a_2b_1 + a_1b_1$; $e_1 = a_2b_2$.

The implementation of this example is shown in Fig. 8.7. Here also the parity prediction and error correction logic units are shown. In the above example the delay overhead is 3 XOR delay plus one AND gate delay.

8.5 Experimental Results

The Four versions, classical, LDPC, Hamming and TMR multipliers over GF(2^m), have been designed and coded in VHDL. The analysis presented here are for multipliers over different primitive polynomials for the fields over GF(2^k), where $7 \leq k \leq 26$. However, the technique can be easily extended to higher order fields. The designs were simulated using Modelsim™. The designs were synthesized using the Synopsys tools in the UMC technology library, using the 0.18 μm CMOS technology. Synopsys Power Compiler™ was used to estimate the power consumptions. The area, delay and power estimates for the basic circuits are shown in Figs. 8.8–8.10. The x-axis shows the decimal representation of the primitive polynomials and the normalized hardware overheads are shown on the y-axis. For example, the primitive polynomial $P(x) = x^7 + x^1 + 1$ can be represented as 10000011 in binary and its decimal equivalent 131 is shown on the x-axis. As expected, the overhead varies depending on the primitive polynomials. The areas of multiplier and its parity prediction part are more or less the same. Therefore, the overhead comes to about 100%. Also, as the number of output bits increases, that is for larger multipliers, the overhead slightly goes down. Design examples show

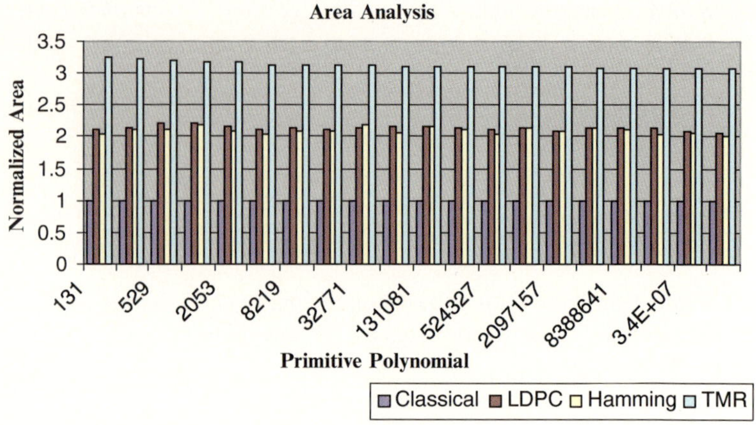

Fig. 8.8 Area analysis: classical, LDPC, Hamming and TMR

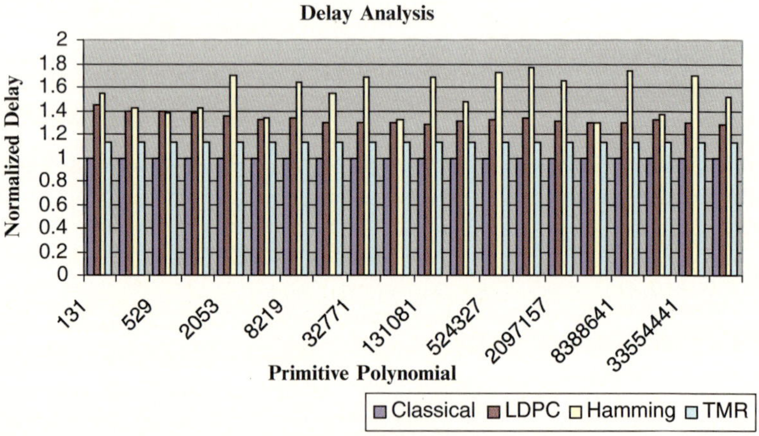

Fig. 8.9 Delay analysis: classical, LDPC, Hamming and TMR

that single error correction with Hamming based Multiple parity prediction requires an average area overhead of 109% accompanied by an increase of 54.5% in the critical path, whereas using LDPC the respective figures are 113.1 and 32.6%. Power analysis shows that there is on average 106.2% increase in power consumption in Hamming based technique and 109.2 for LDPC, whereas the corresponding figure for TMR based designs is 212.07%.

8.5.1 Comparisons Delay Overhead

The disjoint parity prediction circuits we considered here requires all the members of the vectors **e** and **d**. Therefore the total parity prediction logic comes to slightly

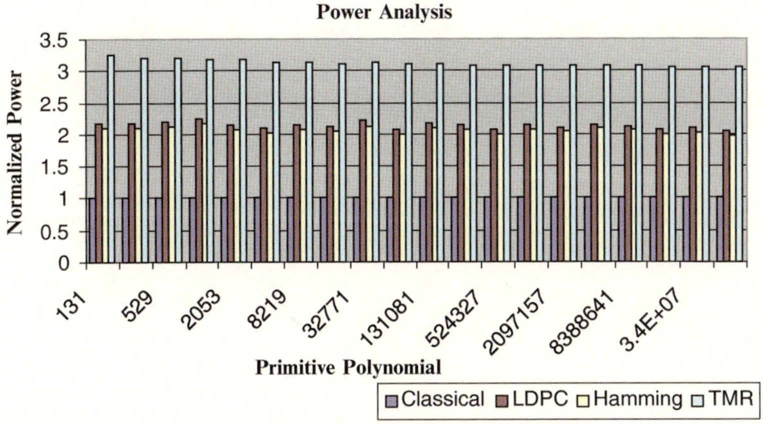

Fig. 8.10 Power analysis: classical, LDPC, Hamming and TMR

Table 8.3 Delay overhead in correction for various multiplier sizes

# of outputs	r LDPC	r Hamming	Delay Hamming	Delay LDPC
2–3	3	3	$2t_{xor} + 2t_{and} + t_{inv}$	$3t_{xor} + t_{and}$
4	4	3	$5t_{xor} + 2t_{and} + t_{inv}$	$4t_{xor} + t_{and}$
5–6	4	4	$5t_{xor} + 3t_{and} + t_{inv}$	$4t_{xor} + t_{and}$
7–10	5	4	$5t_{xor} + 3t_{and} + t_{inv}$	$5t_{xor} + t_{and}$
11	6	4	$5t_{xor} + 3t_{and} + t_{inv}$	$5t_{xor} + t_{and}$
12–15	6	5	$6t_{xor} + 4t_{and} + t_{inv}$	$5t_{xor} + t_{and}$
16–21	7	5	$6t_{xor} + 4t_{and} + t_{inv}$	$5t_{xor} + t_{and}$
22–26	8	5	$6t_{xor} + 4t_{and} + t_{inv}$	$5t_{xor} + t_{and}$

above 100%. Apart from this, additional XOR gates are required to implement the parity check equations, which depend on Hamming weight of the rows of **H**. The delay overhead is contributed by the output parity generator, syndrome decoding plus one XOR gate delay for the correction. Table 8.3 shows the comparison the proposed LDPC based scheme and Hamming based approach for various multiplier sizes. As we can see, the delay penalty is minimum in LDPC based design, this is mainly due to sparse matrix structure of LDPC, moreover only single AND is required for syndrome decoding.

8.6 SEC and DED Multiplier: Automated Synthesis

In this section, we present the proposed design of Single Error Correction (SEC) and Double Error Detection (DED) bit parallel multipliers over GF(2^m). The basic structure of the multiplier is shown in Fig. 8.11. The classical bit-parallel multiplier is designed by the method described in Sect. 8.2. Similar to previous proposals sizes

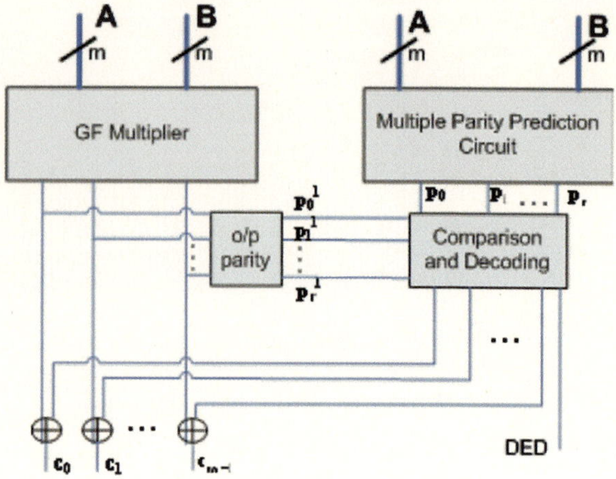

Fig. 8.11 Galois field bit parallel multiplier with SEC and DED

of the parity prediction circuits depend on the number of input bits. Next, we present the algorithm for designing the proposed scheme with an example.

Let $\vec{c}^{\,1} = [c_0^1, c_1^1, c_2^1, \ldots, c_{m-1}^1]^T$ be the output of the multiplier and $\vec{c} = [c_0, c_1, c_2, \ldots, c_{m-1}]^T$ the corrected output. Also let r be the number of parity bits, and $\vec{p} = [p_0, p_1, \ldots, p_{r-1}]^T$ and $\vec{p}^{\,1} = [p_0^1, p_1^1, \ldots, p_{r-1}^1]^T$ respectively be the predicted and the parity bits generated from the output bits. Let **H** be the parity check matrix associated with the proposed single error correction scheme.

Design Procedure:

- Determine the number of parity bits (r) required to satisfy the equation $m + r + 1 \leq 2^r$.
- Construct the **H** matrix, with $(m + r)$ non-zero r-bit column vectors. The dimension of the resulting matrix is $r \times (m + r)$.
- A column vector with a single 1 is assigned to parity P_i.
- The column vector with all 1s is assigned to output bit c_{m-1}.
- The remaining m columns are assigned the output bits c_i, without any constraints.
- Generate predicted parity expressions in terms of c_is. Next, substitute expressions for c_i and simplify to get the final expression in terms of a_is and b_is.
- For DED choose the parity check matrix such that the output bits are assigned to the columns with odd number of ones. In this case additional parity bits maybe required.
- Finally, combine the multiplier, PP, output encoder, decoder, and the correction logic as shown in Fig. 8.11.

The following example illustrates the above design procedure.

Example 8.9. Consider the multiplier structure over GF(2^4) constructed in Example 8.2. Here we have $m = 4$. Therefore, we need 3 parity bits to correct single errors. We have,

$$\mathbf{H} = \begin{array}{ccccccc} p_0 & p_1 & p_2 & c_0 & c_1 & c_2 & c_3 \\ 1 & 0 & 0 & 1 & 1 & 0 & 1 \\ 0 & 1 & 0 & 1 & 0 & 1 & 1 \\ 0 & 0 & 1 & 0 & 1 & 1 & 1 \end{array}.$$

Therefore, the parity check equations are: $p_0 = c_0 + c_1 + c_3$; $p_1 = c_0 + c_2 + c_3$; $p_2 = c_1 + c_2 + c_3$. Substituting for c_0, c_1, c_2, and c_3, the final predicted parity bits are: $p_0 = d_0 + d_1 + d_3 + e_1 + e_2$; $p_1 = d_1 + d_2 + d_3 + e_2$; $p_2 = d_1 + d_2 + d_3 + e_0 + e_2$.

The **d** and **e** are as follows: $d_0 = a_0 b_0$; $d_1 = a_1 b_0 + a_0 b_1$; $d_2 = a_2 b_0 + a_1 b_1 + a_0 b_2$; $d_3 = a_3 b_0 + a_2 b_1 + a_1 b_2 + a_0 b_3$; $e_0 = a_3 b_1 + a_2 b_2 + a_1 b_3$; $e_1 = a_3 b_2 + a_2 b_3$; $e_2 = a_3 b_3$.

The modified expression for double error detection is as follows.

Example 8.10. Consider the multiplier structure over GF(2^4) constructed in Example 8.2. The parity check matrix **H** that satisfies the double error detection condition is

$$\mathbf{H} = \begin{array}{cccccccc} p_0 & p_1 & p_2 & p_3 & c_0 & c_1 & c_2 & c_3 \\ 1 & 0 & 0 & 0 & 1 & 0 & 1 & 1 \\ 0 & 1 & 0 & 0 & 1 & 1 & 0 & 1 \\ 0 & 0 & 1 & 0 & 1 & 1 & 1 & 0 \\ 0 & 0 & 0 & 1 & 0 & 1 & 1 & 1 \end{array}.$$

Therefore, the parity check equations are: $p_0 = c_0 + c_2 + c_3$; $p_1 = c_0 + c_1 + c_3$; $p_2 = c_0 + c_1 + c_2$; $p_3 = c_1 + c_2 + c_3$. The final predicted parity bits for this case are: $p_0 = d_0 + d_2 + d_3 + e_2$; $p_1 = d_0 + d_1 + d_3 + e_1 + e_2$; $p_2 = c_0 + c_1 + c_2 + e_0 + e_2$; $p_3 = d_1 + d_2 + d_3 + e_0 + e_2$. In the DED architecture in some cases we need additional parity bits, as in the above case, which is associated with an increase in area.

8.6.1 Area Overhead

In the proposed parity prediction we need to generate the **d** and **e** from the input operands. Apart from this we have the decoding and correction circuitry. Hence, the total hardware overhead is greater than that of the multiplier hardware. The delay overhead is contributed by the EXOR gates in the output parity generator plus one EXOR gate delay for the correction, together with the decoder delay. As we can see, in the above structural approach we did not optimize the overall hardware requirement. Instead, we employ a synthesis tool, specifically targeted for the polynomials over GF(2^m), to optimize the hardware. We synthesized the multipliers and PP logic separately, and determined the area overhead of the PP logic to be less than 100%. The synthesis and optimization technique is presented in the following.

8.7 Synthesis and Optimization

We present a technique for synthesis and optimization of the multiple-output, multivariate polynomials over GF(2^m). The circuits with and without the error correction schemes have been represented in terms of these polynomials, which we have synthesized with this technique. The polynomials are represented as the Shared Galois Polynomial Decision Diagrams (SGPDDs) [10]. For example, Fig. 8.12a shows the SGPDD representation of the polynomial $f(a,b,c) = a + \beta bc^3 + \beta a^2c^3$ over GF(4), where $\{\alpha, \beta\} \in$ GF(4).

If the initial specification is not over finite fields, e.g. over Boolean or MIN-MAX post algebra, then the technique of [10] is applied for computing the coefficients of the polynomials and storing them as the SGPDDs.

Once the SGPDDs are obtained, circuits are synthesized by decomposing and factoring the SGPDDs based on finding *cuts* within the SGPDDs. A cut is a partitioning of the nodes in the SGPDD into two sets T and B, where T contains internal nodes and the root and B contains external, internal, and the last nodes, i.e. internal nodes which have the external nodes as their children. Effectively a cut can factorize an SGPDD realizing a function f in GF(2^m) as $f = D \cdot Q + R$. Cut based algorithms have been used for synthesis in the Boolean domain, e.g. [24]. In this approach we quickly factorize a polynomial over GF(2^m) based on cuts on their SGPDDs to construct an expression DAG based multiple output shared netlist. The netlist constitutes two types of nodes: internal nodes which can either be GF(2^m) adders or multipliers, or external nodes which can only be constants and variables over GF(2^m). The internal nodes can have two children. The netlists are further synthesized based on additional factorization and optimization.

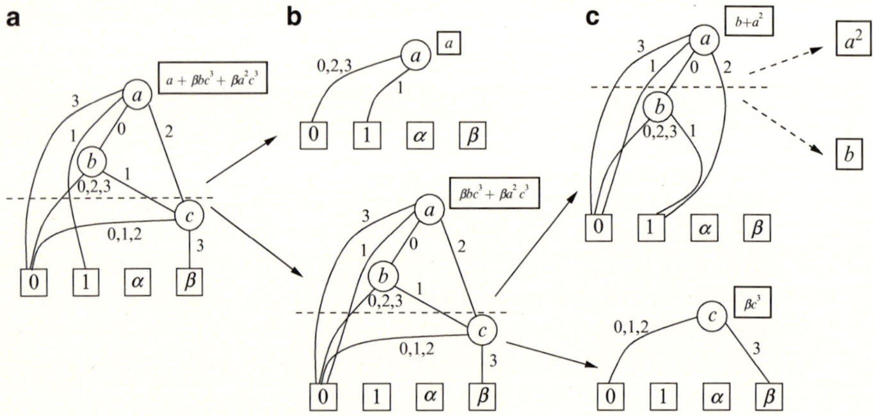

Fig. 8.12 Decomposition—an example

8.7.1 Decomposing from SGPDD

Given a variable x and a SGPDD for a function f, there are two types of decomposition possible: (i) multiplicative, which represents f as $f = X \times Y$, and (ii) additive, which represents f as $f = W + Z$. Here X, Y, W, and Z are SGPDDs. To perform a decomposition a cut is performed above the nodes representing x. Let v_x be a node representing x. To obtain the multiplicative decomposition all the paths in the original SGPDD from nodes above the cut leading to v_x are reconnected to the terminal node 1 and the result is reduced. This gives X. Y is simply the SGPDD rooted at v_x. To obtain the additive decomposition all the paths in the original SGPDD from nodes above the cut leading to v_x are reconnected to the terminal node 0 and the result is reduced. This gives W. Z is obtained by reconnecting all the paths from the nodes above the cut that do not pass through v_x to the terminal node 0 and reducing the result. The proof for this reasoning is straight forward and has been left out for brevity.

The netlists are obtained by decomposing the SGPDDs with a fast greedy heuristic algorithm based on the decomposition rules stated above. The algorithm recursively decomposes a function f, represented as an SGPDD rooted at f, as $f = D{\cdot}Q{+}R$, where D, Q, and R are SGPDDs also. Then each of the components, D, Q, and R are again decomposed until we reach a point where we cannot decompose the function further. The results are then added to the netlists, which are optimized with an efficient netlist optimization algorithm. Figure 8.12 shows an example. The cuts are shown with horizontal broken lines. Figure 8.12a performs an additive decomposition, Fig. 8.12b performs a multiplicative decomposition, and finally, Fig. 8.12c performs another additive decomposition to obtain the final result $((a) + ((\beta c^3) \times ((a^2) + (b))))$, which is added to the netlist. This requires one multiplier and two adders over GF(2^m). The terms with the exponents, i.e. βc^3 and a^2, can be implemented in two ways: either using *shared square and multiply*, or by using a 2-input 2-output look-up-table (LUT) for each of the terms since we are dealing with GF(4). Our technique can do either depending on which option is given. In general, if the LUT option is given, then the polynomials of the form $p(x) = \sum_{i=0}^{2^m-1} c_i x^i$ over GF(2^m), where $c_i \in$ GF(2^m), are generated as a single m-input m-output LUT.

8.7.2 Factorizing Netlists

Once the netlists are obtained, common factors are determined by walking through chains of multipliers following chains of adders. Figure 8.13a shows the netlist corresponding to a general structure of the form $Z = ((AX + Y) + BX)$. Clearly, X is factorizable. To factorize X, Z is restructured as $Z = ((AX + BX) + Y)$ (Fig. 8.13b) and then the factorization is carried out as $Z = ((X(A + B)) + Y)$ (Fig. 8.13c). The structure within the circle in Fig. 8.13b is the network of Fig. 8.13a

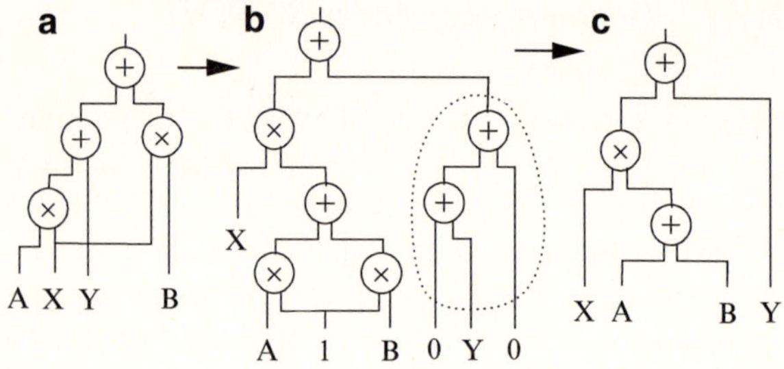

Fig. 8.13 Factorizing from netlist

after its pointers have been readjusted. After restructuring the netlists an efficient optimization technique, presented in the following, is applied to obtain the final optimized netlist.

The algorithm for factorization proceeds by trying out all the possible factorizations, and only stops when there are no more terms which can be factored out.

8.7.3 *Optimizing Netlists*

Netlists are optimized by processing them recursively from the external (i.e. variable or constant) nodes towards the root node. Each node is visited exactly once. For each node u the following is done. If u is already processed then its reference is returned so that it can be shared; otherwise its information is stored in a hash table for sharing. If u is an internal (i.e. GF(2^m) multiplier/adder) node, then let v_0 and v_1 be its two children. If both v_0 and v_1 are constants, then replace u with v_0 op v_1 where op is either addition or multiplication over GF(2^m), i.e. perform *constant propagation*. Replace u with v_i $(i \in \{0, 1\})$, if $u = v_i \times 1$ or $u = v_i + 0$. Replace u with 0, if $u = v_i \times 0$ or $u = v_i + v_i$. This algorithm can be shown to be optimal w.r.t. two input addition and multiplication over GF(2^m) under a fixed netlist transformation.

For example, given the netlist of Fig. 8.13b this algorithm will yield the netlist of Fig. 8.13c.

8.8 Experimental Analysis

Our experiences suggest that the industrial tools such as the Synopsys compilers seem to be incapable of efficiently optimizing the circuits over GF(2^m). This has motivated us in considering our own automatic synthesis and optimization techniques for the multipliers and PP logic. Therefore, in this section we include comparison with the Synopsys tools as well.

The techniques presented here have been implemented in Gnu C++ 3.2.2-5 on a computer with 640 MB RAM and a 600 MHz Athlon processor running

Table 8.4 Multipliers over GF(2^8) for all the 16 primitive polynomials

Prim Poly	Synopsys only (area, delay, power) (μm^2, ns, mW)	(a,m)	Proposed technique (area, delay, power) (μm^2, ns, mW)
285	(141,689.3, 18.8, 130.9)	(87,64)	(2,628.9, 1.6, 4.2)
299	(152,603.3, 14.6, 125.7)	(85,64)	(2,609.5, 1.7, 4.1)
301	(132,595.5, 15.0, 125.4)	(84,64)	(2,580.5, 1.5, 4.0)
333	(140,896.3, 15.2, 129.3)	(84,64)	(2,574.0, 2.0, 4.1)
351	(135,107.6, 17.9, 121.5)	(100,64)	(2,757.9, 1.5, 4.3)
355	(143,195.4, 17.4, 126.0)	(88,64)	(2,612.7, 1.8, 4.2)
357	(144,472.7, 14.8, 133.5)	(84,64)	(2,548.2, 1.5, 4.0)
361	(135,227.9, 16.3, 125.6)	(88,64)	(2,628.9, 1.7, 4.2)
369	(132,405.8, 17.6, 130.2)	(89,64)	(2,722.4, 1.9, 4.3)
391	(130,577.4, 15.9, 126.5)	(79,64)	(2,441.8, 2.2, 3.8)
397	(139,992.8, 14.4, 122.3)	(87,64)	(2,532.1, 2.0, 4.0)
425	(141,879.6, 15.6, 118.9)	(94,64)	(2,690.2, 2.0, 4.3)
451	(139,612.7, 17.6, 123.9)	(79,64)	(2,493.4, 1.8, 3.9)
463	(141,228.2, 17.8, 122.9)	(90,64)	(2,661.1, 1.9, 4.1)
487	(150,703.6, 16.7, 138.8)	(92,64)	(2,632.1, 1.9, 4.1)
501	(129,935.6, 16.0, 119.2)	(104,64)	(2,906.3, 2.3, 4.7)

RedHat Linux with kernel-2.4.20-43.9. The Synopsys design compiler was run on a dual processor Pentium 4 Linux machine with 2 GB RAM and kernel-2.4.21-20.EL. The benchmarks were stored as two-level AND-OR PLAs to enable us to determine how effective the proposed technique is in optimizing area, power, and delay. Also the Synopsys design compiler understands this format. After the synthesis and optimization the results were saved in the VHDL format, which were passed to the Synopsys design compiler (power was estimated with the Synopsys power compiler). The PLAs were also passed directly to the Synopsys compiler for optimizing without the aid of the proposed technique.

We have minimized multipliers over GF(2^m) ($2 \leq m \leq 8$) for all the 51 primitive polynomials in 0.18 μm CMOS technology. Table 8.4 shows results for the 8-bit multipliers. Column 1 represents the primitive polynomials, while Column 2 represents the area, delay, and power reported by the Synopsys design compiler without the aid of the proposed technique. The column with the heading "Proposed Technique" shows the result of applying the proposed technique first, and then applying the Synopsys compiler on the resulting VHDL files. The letters 'a' and 'm' represent 2-input EXOR and AND gates respectively. Here area, delay, and power are in 10^{-6} mm^2, nano seconds, and mW respectively. Power was estimated at 1.8 V. Significant area, delay, and power reduction is observable. Clearly the number of AND gates is m^2 for all the cases. The number of EXOR gates is $m^2 + k$, where k is a constant. As compared with [9], which reports $2m^2 - 1$ 2-input AND gates and $2m^2 - 3m + 1$ EXOR gates, the proposed technique produced better results. For the 8-bit multipliers this technique reports maximum area of 0.002906 mm^2, whereas [9] reported 0.0128 mm^2, i.e. about 4.4 times better

Table 8.5 Parity prediction over $GF(2^8)$ for all the 16 primitive polynomials

Prim Poly	Synopsys only (area, delay, power) (μm^2, ns, mW)	(a, m)	Proposed technique (area, delay, power) (μm^2, ns, mW)
285	(52,925.58, 14.41, 58.32)	(90,32)	(1,986.96, 1.70, 3.47)
299	(52,623.49, 13.40, 56.23)	(89,32)	(1,974.09, 1.69, 3.33)
301	(49,006.48, 11.54, 56.62)	(98,32)	(2,003.09, 2.02, 3.36)
333	(49,680.62, 9.14, 53.52)	(99,32)	(2,077.28, 1.58, 3.61)
351	(55,173.80, 10.93, 61.20)	(104,32)	(2,119.21, 1.80, 3.75)
355	(50,380.56, 10.10, 55.11)	(95,32)	(1,948.26, 1.82, 3.42)
357	(48,406.52, 11.25, 54.37)	(96,32)	(2,025.67, 1.61, 3.59)
361	(52,715.91, 11.86, 58.40)	(101,32)	(2,212.75, 1.57, 3.80)
369	(52,622.36, 10.76, 64.88)	(84,32)	(1,851.49, 1.74, 3.09)
391	(53,164.29, 14.66, 62.55)	(96,32)	(1,799.80, 1.58, 3.03)
397	(53,661.00, 12.11, 62.41)	(89,32)	(1,957.93, 1.56, 3.24)
425	(47,367.86, 9.29, 52.96)	(89,32)	(1,935.35, 1.65, 3.33)
451	(50,512.83, 10.86, 55.84)	(100,32)	(2,151.47, 1.66, 3.76)
463	(53,261.04, 12.45, 59.48)	(88,32)	(2,025.67, 1.62, 3.43)
487	(51,354.72, 13.45, 57.15)	(99,32)	(2,077.28, 2.00, 3.73)
501	(51,045.23, 13.30, 56.12)	(101,32)	(2,067.61, 1.83, 3.74)

(5.2 times better for polynomial 391). Also, it reports about two times reduced delay. For the 4-bit case the proposed technique reports about 0.000522 mm^2 area with the primitive polynomial 25, i.e. over five times better. It is not possible to directly compare this technique with [7,21] because these techniques have reported only theoretical results, with m^2 2-input AND gates and approximately $m^2 + k$ EXOR gates depending on the number of terms in the polynomials and their positions, without any implementation. However, clearly our technique is capable of closely matching the theoretical limits reported by these techniques, despite being a heuristic synthesis technique for the polynomials over $GF(2^m)$, where as techniques such as [7,21] are designed only for hand synthesizing the multipliers over $GF(2^m)$.

Table 8.5 shows the parity prediction counterpart of Table 8.4. Table 8.5 reports more or less the same performance improvements for the PP logic (area, delay, power) compared to Table 8.4. Figures 8.14–8.16 show the comparison between the multipliers and PP logic. Mostly, the area of the PP logic is less than that of the multipliers. On an average, for the designs considered here based on the proposed technique, the total area overhead is about 98.5%. As opposed to this, the structural designs required slightly over 100% area overhead. The improvement, we believe, could be because of the highly effective optimization done by the proposed technique. The delay penalty is about 55% of that of the multipliers. The power overhead is approximately 103%. Compared to the traditional techniques such as the TMR, which is associated with an overhead of more than 200%, the proposed technique is much better. Figure 8.17 shows the error correction scheme for the serial architecture.

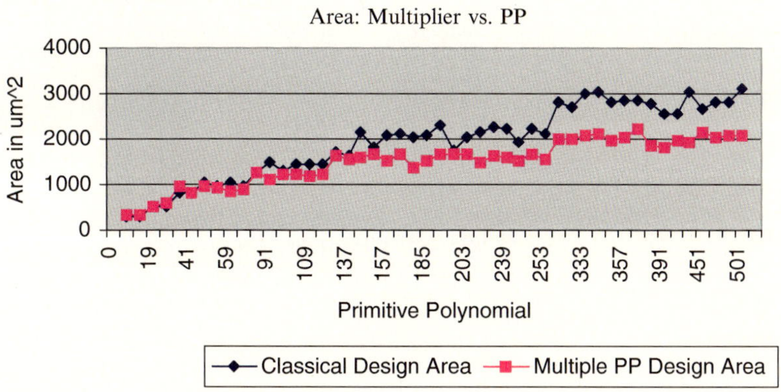

Fig. 8.14 Area: bit parallel multiplier vs. PP

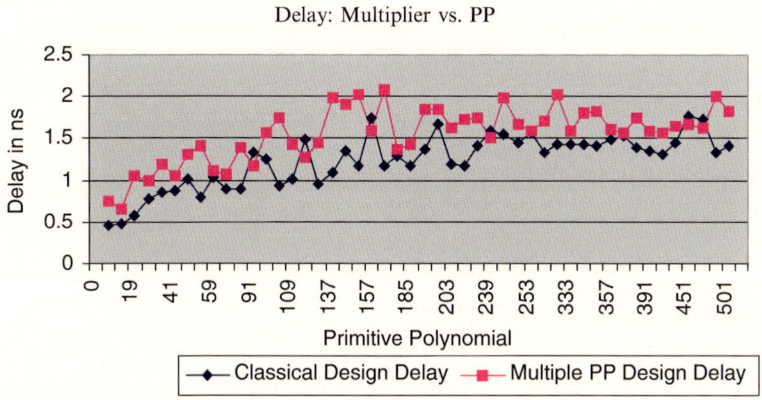

Fig. 8.15 Delay: bit parallel multiplier vs. PP

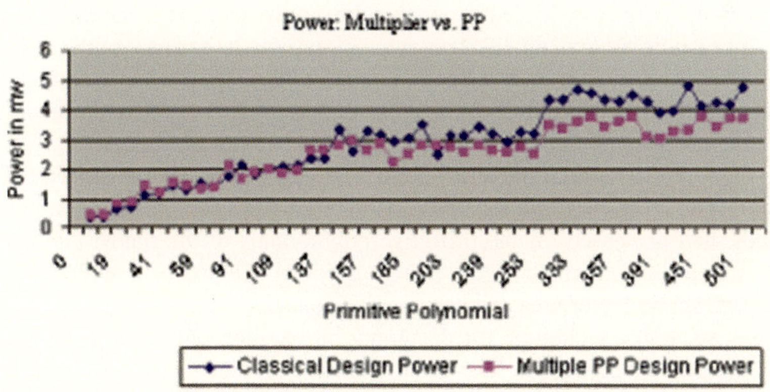

Fig. 8.16 Power: bit parallel multiplier vs. PP

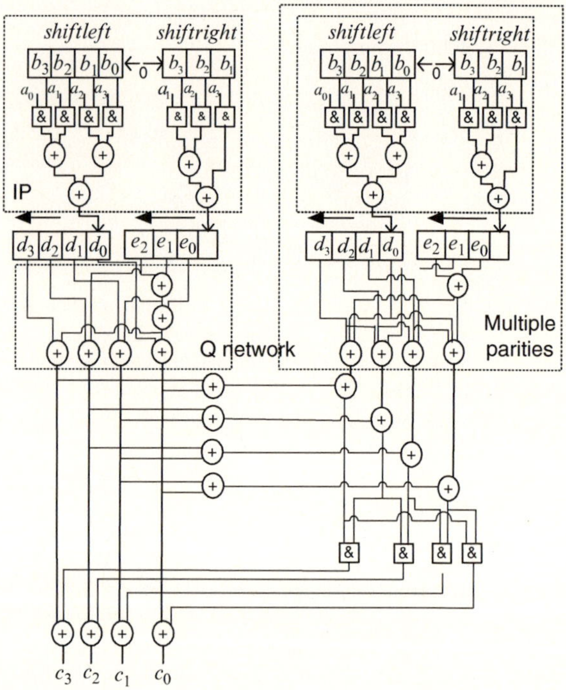

Fig. 8.17 Error correction in serial architecture

8.9 Multiple Error Detection

In this section, a systematic method for designing multiple bit error detection and correcting bit parallel *polynomial basis* (PB) multipliers over GF(2^m) is proposed. One of the problems in the approaches presented in the previous section is that the design can not correct multiple bit errors at the output. Due to the nature of faults in the cryptographic context and their potential effects on the security of the system, in certain cases multiple bit error detection is more important than correction. Fault-tolerance in finite filed multiplier have been addressed in [1, 20]. However, these techniques are specific to one particular multiplier structure.

First, presented a multiple error detection using Low Density Parity Check Codes (LDPC). The expressions for the parity prediction are derived from the input operands, and are based on the primitive polynomials of the fields. For multiple bit error correction we use Reed Solomon codes. Comparison with traditional techniques, shows better area and power performance.

In this section, a technique for designing bit parallel multipliers with multiple error detection is proposed. The basic structure of the multiplier is shown in Fig. 8.1. The classical bit-parallel multiplier is designed by the method described in Sect. 8.2. The modified single error correcting architecture is shown in Fig. 8.18. Apart from the functional unit of the multiplier, it consists of a parity prediction unit, output parity generation unit, and the comparison and error detection logic.

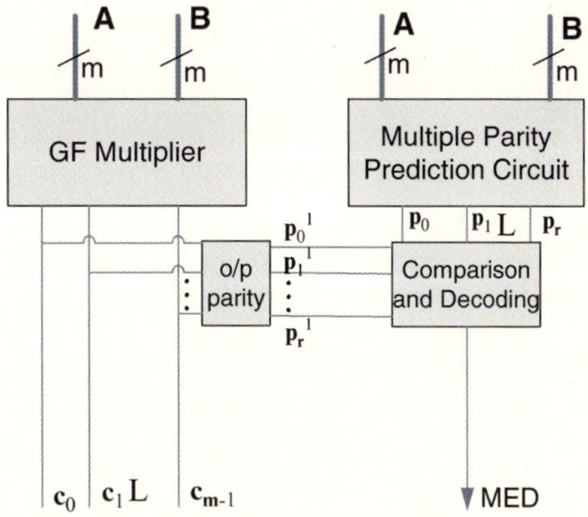

Fig. 8.18 Proposed multiple error detection multiplier over GF(2^m)

For error Multiple Error Detection (MED), we use the parity prediction, which is based on LDPC code. The advantage of the LDPC codes is that the complexity of the encoding and decoding logic is simple. We give more emphasis on the error detection, rather than correction, of fault. The reason is that the number of faults detectable (d − 1) is larger than the number of correctable faults ($\lfloor \frac{d}{2} \rfloor$), where d is the minimum distance of the error correction code used. Furthermore, due to the nature of faults in the cryptographic context and their potential effects on the security of the system, the detection of the errors is more important than correction. The sizes of the parity prediction circuits depend on the number of input bits. Next, we derive the closed form expressions for the predicted parity bits.

For the LDPC codes proposed in the previous section the hamming distance is 3. Now if we add an overall parity the overall hamming distance will be 4. Hence we could detect any three errors. Let $\vec{c} = [c_0, c_1, c_2, \ldots, c_{m-1}]^T$ be the output of the multiplier and the corrected output. Also let r be the number of parity bits, and $\vec{p} = [p_0, p_1, \ldots, p_{r-1}]^T$ and $\vec{p}^{\,1} = [p_0^1, p_1^1, \ldots, p_{r-1}^1]^T$ respectively be the predicted and the parity bits generated from the output bits. Let **H** be the parity check matrix associated with the proposed single error correction scheme.

8.9.1 Design Procedure

- Determine the number of parity bits (r) required to satisfy the required hamming distance and hence the number of detectable errors.
- Construct the **H** matrix based on the approach proposed in the previous section with an overall parity check bit added.
- A column vector with a single 1 is assigned to parity P_i.
- The remaining m columns are assigned the output bits c_i, without any constraints.
- Generate predicted parity expressions in terms of a_is and b_is.

Example 8.11. Consider the multiplier structure over GF(2^6), with primitive polynomial. Here we have $m = 6$. Therefore, we need 4 parity bits to detect three errors. We have,

$$\mathbf{H} = \begin{bmatrix} c_0 & c_1 & c_2 & c_3 & c_4 & c_5 & p_0 & p_1 & p_2 & p_3 & p_4 \\ 1 & 1 & 1 & 1 & 1 & 1 & 1 & 0 & 0 & 0 & 0 \\ 1 & 0 & 0 & 1 & 0 & 1 & 0 & 1 & 0 & 0 & 0 \\ 1 & 1 & 0 & 0 & 1 & 0 & 0 & 0 & 1 & 0 & 0 \\ 0 & 1 & 1 & 1 & 0 & 0 & 0 & 0 & 0 & 1 & 0 \\ 0 & 0 & 1 & 0 & 1 & 1 & 0 & 0 & 0 & 0 & 1 \end{bmatrix}$$

The predicted parity of outputs based on the above H matrix is given by

$$p_0 = c_0 + c_1 + c_3 + c_4 + c_5 \tag{8.25}$$

$$p_1 = c_0 + c_3 + c_4 \tag{8.26}$$

$$p_2 = c_0 + c_1 + c_4 \tag{8.27}$$

$$p_3 = c_1 + c_2 + c_3 + c_1 \tag{8.28}$$

$$p_4 = c_2 + c_4 + c_5 \tag{8.29}$$

(Here, the '+' sign represents XOR operation.)

$$p_0 = d_0 + d_1 + d_3 + d_4 + d_5 \tag{8.30}$$

$$p_1 = d_0 + e_0 + d_3 \;+\; e_2 + d_4 + e_4 \tag{8.31}$$

$$p_2 = d_0 + d_1 + e_1 + d_4 + e_4 + e_3 \tag{8.32}$$

$$p_3 = c_1 + c_2 + c_3 + c_1 \tag{8.33}$$

$$p_4 = d_2 + e_2 + e_1 + d_4 + e_3 + d_5 \tag{8.34}$$

where, the **d** and **e** outputs of the network are given below.

$d_0 = a_0b_0$; $d_1 = a_1b_0 + a_0b_1$; $d_2 = a_2b_0 + a_1b_1 + a_0b_2$; $d_3 = a_3b_0 + a_2b_1 + a_1b_2 + a_0b_3$; $d_4 = a_4b_0 + a_3b_1 + a_2b_2 + a_1b_3 + a_0b_4$; $d_5 = a_5b_0 + a_4b_1 + a_3b_2 + a_2b_3 + a_1b_4 + a_0b_5$;

$e_0 = a_5b_1 + a_4b_2 + a_3b_3 + a_2b_4 + a_1b_5$; $e_1 = a_5b_2 + a_4b_3 + a_3b_4 + a_2b_5$; $e_2 = a_5b_3 + a_4b_4 + a_3b_5$; $e_3 = a_5b_4 + a_4b_5$; $e_5 = a_5b_5$.

Table 8.6 shows the comparison of the area overhead for various multiplier sizes. On average the overhead is about 100%.

8.10 Multiple Bit Error Correction

In this sect a systematic method for designing multiple detection and correction techniques for multiplier circuits for Galois fields over GF(2^m) which is one of the key building blocks in many crypto architectures is considered. We used multiple parity predictions to detect multiple errors applying popular error correcting codes.

Table 8.6 Details of hardware overhead for different multipliers for multiple error detection

Field size	Multiplier area in μm²	PP and other circuitry area in μm²	% Overhead
$GF(2^{10})$	3,844.20	4,154.448	108.0
$GF(2^{11})$	4,656.90	4,941.5	106.1
$GF(2^{12})$	5,547.00	5,805.968	104.6
$GF(2^{13})$	6,514.50	6,747.852	103.5
$GF(2^{14})$	7,559.40	7,767.152	102.7
$GF(2^{15})$	8,681.70	8,863.868	102.0

8.10.1 Reed-Solomon Code

Reed-Solomon codes belong to a family of Forward Error Correction (FEC) codes known as linear block codes. Linear block codes encode message as a block and the redundant information is unique per block i.e. the whole message block is passed into the encoder one block at a time and the encoder has no memory of any information from the previous block. This is different to convolutional codes which encode continuously. One can think of this style of encoding as a window sliding over the information bits. In conventional Reed Solomon encoder, the bits inside the window are encoded, and therefore the encoded bits depend on previous bits, i.e. the encoder has memory. In the proposed technique a different approach is taken, that is a parallel encoder which does not have any memory is used.

Reed-Solomon (RS) codes were first proposed by Reed and Solomon in 1960 [19]. They are known to be very efficient algebraic codes, i.e. can correct a large number of errors with a low overhead. By the very nature of their structure, RS codes are well suited to FEC in bursty noise environments [11, 17, 23]. The codes have the power to correct errors that occur in a cluster. Decoding RS codes is a non-trivial task. There are two fundamental types of decoding. Hard Decision Decoding (HDD) first thresholds the received data, effectively making a hard decision for each bit (0 or 1). Then, using the properties of the code, the decoder detects and corrects the bits that are in error. In contrast, Soft Decision Decoding (SDD) uses all the information received from the channel, i.e. no thresholding of the received data.

8.10.2 Parallel Encoder

Reed-Solomon codes operate over an extended binary field $GF(2^m)$ and each symbol in a RS code word is an element from the corresponding Galois field. RS codes have the following parameters: $n = 2^m - 1$ symbols in a code word $k =$ the number of message symbols $n - k = 2t$ redundant symbols. $t =$ the number of errors to be corrected.

$$d_{min} = 2t + 1$$

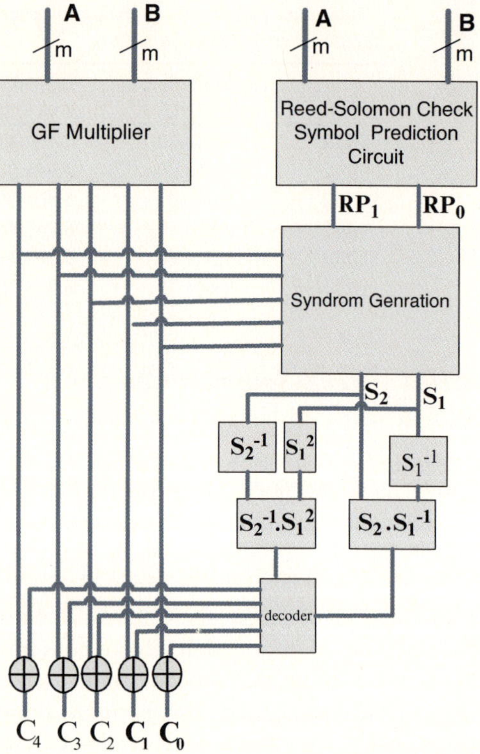

Fig. 8.19 Proposed multiple error correcting multiplier over GF(2^{15})

To construct the generator for a Reed Solomon code, we need to select the appropriate finite field and choose roots. Let the roots be β^i to β^{i+2t-1}, the generator polynomial will be g(X) = $(X+\beta^i)(X + \beta^{i+1})\cdots(X + \beta^{i+2t-2})(X + \beta^{i+2t-1})$ where t is number of symbol errors that can be corrected.

In this section, a systematic method for designing multiple bit error correcting bit parallel *polynomial basis* (PB) multipliers over GF(2^m) is proposed. Here we use Reed Solomon based scheme for error correction. The basic structure of the multiplier is shown in Fig. 8.1. The classical bit-parallel multiplier is designed by the method described in Sect. 8.2. The modified multiple error correcting architecture for a $GF(2^{15})$ multiplier is shown in Fig. 8.19. Apart from the functional unit of the multiplier, it consists of a reed Solomon check bits generation, syndrome generator, and correction logic. The sizes of the parity prediction circuits depend on the number of input bits. Next, we derive the closed form expressions for the predicted parity bits.

The basic principle of multiple bit error correction is explained by considering the following motivating example. Here we consider a GF$(2)^{15}$ bit multiplier. Let the roots be β and β^2, the generator polynomial will be g(X) = $(X+\beta)(X+\beta^2)$. i.e. g(X) = $X^2 + \beta^4 X + \beta^3$ The symbols encoded with above g(x) could correct one symbol error.

Table 8.7 Field elements GF(8) with $P(x) = x^3 + x + 1$

Element	Representation
0	000
1	001
β	010
β^2	100
$\beta^3 = \beta + 1$	011
$\beta^4 = \beta^2 + \beta$	110
$\beta^5 = \beta^2 + \beta + 1$	111
$\beta^6 = \beta^2 + 0 + 1$	101
$\beta^7 = 1$	001

Let $\vec{c} = [c_0, c_1, c_2, \ldots, c_{14}]^T$ be the output of the multiplier. The 15 bit output is divided into 5 three bit symbols over GF(2^3). Table 8.7 shows the field elements of GF(2^3) with $P(x) = x^3 + x + 1$.

Let RP_1 and RP_0 denotes the two Reed Solomon check symbols which is generated from the input operand.

$\mathrm{RP}_0 = \beta C_4 + \beta C_3 + \beta^3 C_2 + C_1 + \beta^3 C_0$

$\mathrm{RP}_1 = \beta^4 C_4 + \beta^5 C_3 + \beta^5 C_2 + C_1 + \beta^4 C_0$

$\mathrm{C}_4 = (c_{14},\ c_{13},\ c_{12})$

$\mathrm{C}_3 = (c_{11},\ c_{10},\ c_9)$

$\mathrm{C}_2 = (c_8,\ c_7,\ c_6)$

$\mathrm{C}_1 = (c_5,\ c_4,\ c_3)$

$\mathrm{C}_0 = (c_2,\ c_1,\ c_0)$

$\beta C_4 = (c_{13},\ c_{14} + c_{12},\ c_{14})$

$\beta C_3 = (c_{10},\ c_{11} + c_9,\ c_{11})$

$\beta^3 C_2 = (c_8 + c_7,\ c_8 + c_7 + c_6,\ c_8 + c_6)$

$\mathrm{C}_1 = (c_5,\ c_4,\ c_3)$

$\beta^4 C_0 = (c_2 + c_1 + c_0,\ c_1 + c_0,\ c_2 + c_1)$

$\mathrm{RP}_0 = (rp_{02}, rp_{01}, rp_{00})$

$rp_{02} = c_{13} + c_{10} + c_8 + c_7 + c_5 + c_2 + c_1 + c_0$ $rp_{01} = c_{14} + c_{12} + c_{11} + c_9 + c_8 + c_7 + c_6 + c_4 + c_1 + c_0$ $rp_{00} = c_{14} + c_{11} + c_8 + c_6 + c_3 + c_2 + c_1$

$rp_{02} = d_{13} + e_{13} + e_{12} + d_{10} + e_{10} + e_9 + d_8 + e_8 + d_7 + e_6 + d_5 + e_5 + e_4 + d_2 + d_1 + d_0 + e_0$

$rp_{01} = d_{14} + e_{13} + d_{11} + e_{11} + e_{10} + d_9 + e_9 + d_8 + + d_7 + d_6 + e_5 + d_4 + e_4 + e_3 + d_1 + e_1 + d_0$

$rp_{00} = d_{14} + e_{14} + e_{13} + d_{11} + e_{11} + e_{10} + d_8 + e_8 + e_7 + d_6 + e_6 + e_5 + d_3 + e_3 + d_2 + e_1 + d_1 + e_1 + e_0$

Deriving the bit level details of the RP_1, we have $\mathrm{RP}_1 = \beta^4 C_4 + \beta^5 C_3 + \beta^5 C_2 + C_1 + \beta^4 C_0$

$\beta^4 C_4 = (c_{14} + c_{13} + c_{12},\ c_{13} + c_{12},\ c_{14} + c_{13})$

$\beta^5 C_3 = (c_{10} + c_9, c_9\ ,\ c_{11} + c_{10} + c_9)$

$\beta^5 C_2 = (c_8 + c_7,\ c_6\ ,\ c_8,\ c_7 + c_6)$

$\mathrm{C}_1 = (c_5,\ c_4,\ c_3)$

$\beta^4 C_0 = (c_2 + c_1 + c_0\ ,\ c_1 + c_0,\ c_2 + c_1)$

Table 8.8 Hardware overhead for GF(2^{15}) multiplier with multiple error correction

Field size	Multiplier area in μm^2	PP and other circuitry area in μm^2	% Overhead
GF(2^{15})	8,681.70	15,958.868	184.82

$\mathrm{RP}_1 = (rp_{12}, rp_{11}, rp_{10})$
$rp_{12} = c_{14}+c_{13}+c_{12}+c_{10}+c_9+c_8+c_7+c_5+(c_2+c_1+c_0\ rp_{11} = c_{13}+c_{12}+c_9+c_6+c_4+c_1+c_0\ rp_{10} = c_{14}+c_{13}+c_{11}+c_{10}+c_9+c_8+c_7+c_6+c_3+c_2+c_1)$
$rp_{12} = d_{14}+e_{13}+d_{12}+e_{10}+e_9+d_9+e_8+e_7+d_7+e_6+e_7+d_5+e_4++e_5d_2+d_1+d_0+e_0\ rp_{11} = d_{13}+e_{13}+e_{12}+d_{12}+e_{10}+e_9+d_9+e_8+e_7+d_6+e_6+e_5+d_4+e_4++e_3d_1+d_1+d_0\ rp_{10} = d_{14}+e_{14}+d_{13}+e_{12}+d_{11}+e_{11}+d_{10}+d_9+e_8+d_7+e_7+e_6+d_3+e_3+e_2+d_1+e_1+e_0$

Let the output of the multiplier be

$C(X) = C_4X^6 + C_3X^5 + C_2X^4 + C_1X^3 + C_0X^2 + RP_1X + RP_0$

Syndrome $S_1 = (s_{12}, s_{11}, s_{10})$ is generated by

$S_1 = C_4\beta^6 + C_3\beta^5 + C_2\beta^4 + C_1\beta^3 + C_0\beta^2 + RP_1\beta + RP_0$
$\beta^6C_4 = (c_{12}, c_{14}, c_{13}+c_{12})\ \beta^5C_3 = (c_{10}+c_9, c_9, c_{11}+c_{10}+c_9)\ \beta^4C_2 = (c_8+c_7+c_6, c_7+c_6, c_8+c_7)\ \beta^3C_1 = (c_5+c_4, c_5+c_4+c_3, c_5+c_3)\ \beta^2C_0 = (c_2+c_0, c_1+c_0, c_1)$

$\beta^1RP_1 = (rp_{11}, rp_{12}+rp_{10}, rp_{12})$
$RP_0 = (rp_{02}, rp_{01}, rp_{00})$
$s_{12} = c_{12}+c_{10}+c_8+c_7+c_6+c_5+c_4+c_2+c_0+rp_{11}+rp_{02})\ s_{11} = c_{14}+c_9+c_7+c_6+c_5+c_4+c_3+c_1+c_0+rp_{12}+rp_{10}+rp_{01})\ s_{10} = c_{13}+c_{12}+c_{11}+c_{10}+c_9+c_8+c_7+c_5+c_3+rp_{10}+rp_{12}+rp_{00})$
Syndrome $S_2 = (s_{22},\ s_{21},\ s_{20})$ is generated by evaluating the output polynomial at β^2

$S_2 = C_4\beta^12 + C_3\beta^10 + C_2\beta^8 + C_1\beta^6 + C_0\beta^4 + RP_1\beta^2 + RP_0$
$S_2 = C_4\beta^5 + C_3\beta^3 + C_2\beta^1 + C_1\beta^6 + C_0\beta^4 + RP_1\beta^2 + RP_0$
$\beta^5C_4 = (c_{13}+c_{12}, c_{12}, c_{14}+c_{13}+c_{12})\ \beta^3C_3 = (c_{11}+c_{10}, c_{11}+c_{10}+c_9, c_{11}+c_9)\ \beta^1C_2 = (c_7, c_8+c_6, c_8)\ \beta^6C_1 = (c_2, c_5, c_4+c_3)\ \beta^4C_0 = (c_2+c_1+c_0, c_1+c_0, c_2+c_1)\ \beta^2RP_1 = (rp_{12}+rp_{10}, rp_{12}+rp_{11}, rp_{11})\ RP_0 = (rp_{02}, rp_{01}, rp_{00})$
$s_{22} = c_{13}+c_{12}+c_{11}+c_{10}+c_7+c_2+c_1+rp_{12}+rp_{10}+rp_{02})\ s_{21} = c_{12}+c_{11}+c_{10}+c_9+c_8+c_6+c_5+c_1+c_0+rp_{12}+rp_{11}+rp_{01}\ s_{20} = c_{14}+c_{13}+c_{12}+c_{11}+c_9+c_8+c_4+c_3+c_2+c_1+rp_{11}+rp_{00}$

The above check symbols are used for generating the syndrome. Table 8.8 shows area analysis of the above design example. The overhead is about 185%, however it is much less than a N modular error correction. The above design can correct three errors (one symbol error).

8.10.3 Improvement in Robustness

As discussed previously, for security reasons, especially to provide resistance against fault-based attacks, it is very important to verify the correctness of computations in cyptographic computations. Error detection may be a countermeasure for many security applications. However, error correction (in other words fault-tolerant characteristic) enables a module to perform its normal operation in spite of faults. Therefore, the above error detection and correction will result in more reliable modules.

8.11 Chapter Summary

Error correction is an effective way to mitigate fault related attacks in cryptographic hardware. Commonly, higher level mechanisms are adapted to protect the architecture. This chapter proposes an alternative hardware architecture compared to the existing approaches. Furthermore, the proposed technique can also be applied to tackle the problem of soft errors in logic circuits. The experimental results suggest that there is approximately, on an average, slightly over 100% area/power overhead and 54.5% delay penalty over the conventional designs. The delay penalty is mainly due to the output parity generation, decoding, and correction circuitry. Also presented in this chapter an automatic synthesis approach for designing SEC and DED finite field multipliers. Also presented a heuristic gate- as well as word-level synthesis and optimization technique for the polynomials over GF(2^m) for designing the SEC and DED multipliers. The experimental results suggest that this technique can significantly reduce area, delay, and power compared to the industrial tools and also closely match the theoretical limits. The performance figures also closely match those of the structural technique presented in this chapter. Therefore, we can conclude that using our technique near optimal SEC and DED multiplier circuits can be designed for the polynomials over GF(2^m).

References

1. S. Bayat-Sarmadi, M.A. Hasan, On concurrent detection of errors in polynomial basis multiplication. IEEE Trans. Very Large Scale Integr. VLSI Syst. **15**(4), 413–426 (2007)
2. D. Boneh, R. Demillo, R. Lipton, On the improtance of checking cryptographic protocols for faults, in *International Conference on the Theory and Applications of Cryptographic Techniques (Eurocrypt)*, Konstanz, 1997, pp. 37–51
3. D. Boneh, R.A. DeMillo, R.J. Lipton, On the importance of eliminating errors in cryptographic computations. J. Cryptol. **14**(2), 101–120 (2001)
4. S. Fenn, M. Gossel, M. Benaissa, D. Taylor, Online error dection for bitseial multipliers in GF(2^m). J. Electron. Test. Theory Appl. **13**, 29–40 (1998)
5. R. Gallager, *Low-Density Parity-Check Codes* (MIT, Cambridge, 1963)

6. G. Gaubatz, B. Sunar, Robust finite field arithmetic for fault-tolerant public-key cryptography, in *2nd Workshop on Fault Tolerance and Diagnosis in Cryptography (FTDC)*, Edinburgh, UK, 2005, pp. 196–207
7. A. Halbutogullari, C.K. Koc, Mastrovito multiplier for general irreducible polynomials. IEEE Trans. Comput. **49**(5), 503–518 (2000)
8. W. Hamming, Error detecting and error correcting codes. Bell Syst. Tech. J. **29**, 147–160 (1950)
9. N. Iliev, J.E. Stine, N. Jachimiec, Parallel programmable finite field GF(2^m) multipliers, in *Proceedings of the IEEE Computer Society Annual Symposium on VLSI Emerging Trends (ISVLSI'04)*, Tampa, Feb 2004, pp. 299–302
10. A. Jabir, D. Pradhan, A graph-based unified technique for computing and representing coefficients over finite fields. IEEE Trans. Comp. **56**(8), 1119–1132 (2007)
11. Mario Blaum, A course on error correcting codes. IBM Research Division IBM Corp, 1997
12. E.D. Mastrovito, VLSI architectures for computation in Galois fields. PhD thesis, Linkoping University, Linkoping, 1991
13. R.J. McEliece, *Finite Fields for Computer Scientists and Engineers* (Kluwer, Boston, 1987)
14. M. Nicolaidis, Y. Zorian, Online testing for VLSI a compendium of approaches. J. Electron. Test. Theory Appl. **11**, 7–10 (1998)
15. C.Y. Lee, C.W. Chiou, J.M. Lin, Concurrent error detection in a bitparallel systolic mulitplier for dual basis of GF(2^m). J. Electron. Test. Theory Appl. **21**, 539–549 (2005)
16. R. Lidl, H. Niederreiter, *Finite Fields* (Addison-Wesley, Reading, 1983)
17. S. Lin, D.J Costello, *Error Control Coding: Fundamentals and Applications* (Prentice-Hall, Englewood Cliffs, 1983)
18. D.K. Pradhan, A theory of Galois switching functions. IEEE Trans. Comp. **27**(3), 239–249 (1978)
19. S. Reed, G. Solomon, Polynomial codes over certain finite fields. SIAM J. Appl. Math. **8**, 300–304 (1960)
20. A. Reyhani-Masoleh, M.A. Hasan, Towards fault-tolerant cryptographic computations over finite fields. ACM Trans. Embed. Comput. Syst. **3**(3), 593–613 (2004)
21. A. Reyhani-Masoleh, M.A. Hasan, Low complexity bit parallel architectures for polynomial basis multiplication over gf(2m). IEEE Trans. Comput. **53**(8), 945–959 (2004)
22. A. Reyhani-Masoleh, M.A. Hasan, Fault detection architectures for field multiplication using polynomial bases. IEEE Trans. Comput. **55**(9), 1089–1103 (2006)
23. A. Vardy, Y. Beery, Bit level soft-decision decoding for reed-solomon codes. IEEE Trans. Commun. **39**, 440–444 (1991)
24. C. Wang, V. Singal, M. Ciesielski, BDD decomposition for efficient logic synthesis, in *International Conference on Computer Aided Design Aided Design (ICCAD)*, San Jose, 1999, pp. 626–631

Chapter 9
Low Cost C-Testable Finite Field Multiplier Architectures

Jimson Mathew, H. Rahaman, and D.K. Pradhan

9.1 Motivation

Design for Test (DFT) techniques attempt to improve access to the internal state of the crypto hardware either by improving control of internal nodes from the primary inputs or by improving observation capability of values on internal nodes at the primary outputs or both. Scan-based test is a powerful Design-For-Testability (DFT) technique. Scan DFT technique improves controllability and observability of internal circuit nodes. In scan DFT some internal registers and flip-flops are tied together into one or more scan chains and connected possibly to a five-pin serial JTAG boundary scan interface for external test [1]. In the JTAG interface, TCK is the test clock signal while TMS selects normal mode or test mode. TRST is the reset signal for test controller. During testing, test vectors can be scanned in via the TDI input pin and internal registers can be scanned out via TDO output pin [17]. Scan chains are typically automatically inserted into the design by test synthesis tools. A scan chain is classically organized according to the physical positions of the flip-flops. During chip packaging, scan chains are either connected to the external JTAG interface pins to provide on-chip debug and maintenance in field capabilities [6] or left unbound to prevent access. However, unbound scan chains can still be accessed as discussed [19]. Recently scan-based testing has been demonstrated to assist in non-invasive attacks to steal important information such as intellectual property (IP) and/or secret keys [5, 8, 15, 20]. Such scan testing based attacks have added to an already growing customer concern of hardware security [7, 13, 18]. As more information security measures are implemented on chip additional security measures must be implemented to defend from the multitude of intrusive and side-channel attacks.

J. Mathew (✉) • H. Rahaman • D.K. Pradhan
Department of Computer Science, University of Bristol, Bristol, UK
e-mail: jimson@cs.bris.ac.uk; hafizur@cs.bris.ac.uk; pradhan@cs.bris.ac.uk

J. Mathew et al. (eds.), *Energy-Efficient Fault-Tolerant Systems*,
DOI 10.1007/978-1-4614-4193-9_9, © Springer Science+Business Media New York 2014

Based on the test terminology used above, security of a chip can be defined as the extent to which the controllability and observability of its internal registers are restricted. Clearly, a scan based testability approach compromises the security of the internal hardware, which can lie in the scan data paths. Consequently, from a security perspective, cryptographic hardware should minimize controllability and observability of the internal state to a minimum. In this chapter we will investigate security aware DFT techniques which tests block on-chip. In particular, finite field multiplier which is one of the key block, is considered. Different testability approaches are investigated.

9.2 Introduction

First, a testability approach without any hardware modification of multiplier structure is considered. Second, two approaches with extra control pins are illustrated. C testability is important in cryptographic architectures because traditional scan based techniques are prone to side channel attacks [13, 19, 20].

To date, the testability issues of these multipliers have not been fully explored, despite their applications in critical areas such cryptography, error control and reliability, etc. First a structural approach is presented. Second, a C-testable designs of PB multipliers over GF(2^m) is considered. For an m-bit multiplier, a constant test set of length 8 is sufficient to detect all the single stuck-at faults. This method requires 3 control inputs. We also present another method which requires fewer control inputs and constant 7 vectors, but at the cost of about 33% extra hardware. We have observed that this test length is much lower than that required by the Automatic Test Pattern Generation (ATPG) techniques of the industrial tools such as the *Synopsys*^{TM} tools. The gate counts and the associated hardware, area, delay, and power of the proposed testable multipliers over different values of m are also analyzed. Finally, a simple Built In Self Test (BIST) architecture is proposed for generating the 8 constant test vectors. The area and delay of proposed testable and BIST circuits using a UMC 0.18 μm CMOS technology library have been presented. This test set provides 100% fault coverage and also detects single bit error in the Test Pattern Generator (TPG) itself. Owing to the possible applications of these circuits in sensitive areas such as the ECC systems, the BIST structure can also provide an added level of security.

Definition 9.1. A Boolean AND-EXOR function $F(x_1, x_2, \ldots, x_n)$ is in the Positive Polarity Reed Muller (PPRM) if only positive polarities are allowed for each input variable. For example $F_1 = x_1x_2 + x_2x_3$ is a PPRM. Several testable techniques for AND-EXOR circuits have appeared in [2, 10, 11, 14].

Definition 9.2. A circuit is constant (or C)-testable if it can be tested with a constant number of vectors independent of the circuit's complexity.

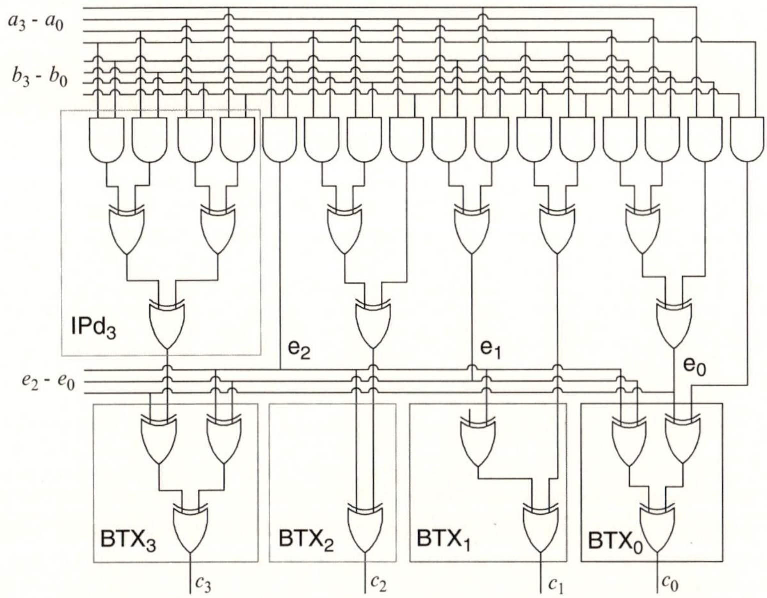

Fig. 9.1 Architecture of the polynomial basis multiplier over GF(2^4) with $P(x) = x^4 + x^3 + 1$

A general structure of the multiplier is shown in Fig. 8.1. An example over $GF(2^4)$ is shown in Fig. 9.1.

The proposed technique presented in the next section requires $(2m + 1)$ vectors for detecting all the single stuck-at faults in the AND part and multiple stuck-at faults in the EXOR part of the multipliers without incorporating any extra hardware. The test set is generated from the expressions of the inner product variables of the multiplication directly without the aid of any ATPG.

9.3 Fault Detection Technique

Although the multipliers over GF(2^m) are multiple output PPRM circuits, these are not conventional PPRM circuits. We can derive the test vectors without GF(2^m) any extra hardware. The fault detection technique is also different from those of the conventional PPRM circuits. We can derive directly from inner product variable expressions. As this is the most important part as far as testability is concerned, we first consider the EXOR part. The EXOR part is implemented with the tree structure resulting in high-speed operations. This technique detects multiple stuck-at faults in the EXOR part and single stuck-at faults in the AND part of the multiplier circuits [1].

Fig. 9.2 Test vectors and responses in an EXOR-tree

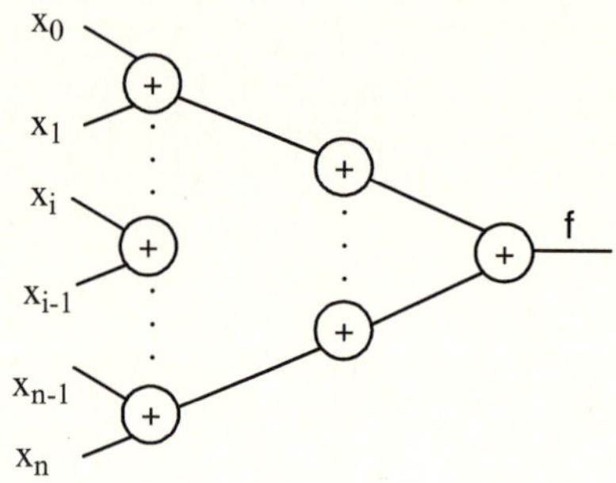

9.3.1 Tests for the EXOR Part

Deriving a testing technique with reduced universal test sets in AND-EXOR circuits having tree based XOR parts is an open problem. In this section we derive the test sets for the detection of the faults from the U, L, and Q matrices of the multipliers (explained in Sect. 8.2). We have applied the multiple fault detection assumption [2] in the EXOR part for testing the EXOR part of the multiplier circuits. It is shown that the (n + 1) test vectors are sufficient to detect multiple stuck-at fault in n input single output parity circuit as shown in Fig. 9.2.

Theorem 9.1 ([2]). *The vectors $t_0, t_1, t_2 \ldots t_n$ is a test for detecting multiple stuck-at faults in an EXOR network realizing a parity function $f = x_1 + x_2 + x_3 + \ldots x_n$, where*

$$\begin{bmatrix} t_0 \\ t_1 \\ t_2 \\ \vdots \\ t_n \end{bmatrix} = \begin{bmatrix} x_0 & x_1 & \cdots & x_n \\ 0 & 0 & \cdots & 0 \\ 1 & 0 & 0 & 0 \\ \vdots & \vdots & \ddots & \vdots \\ 0 & 0 & 0 & 1 \end{bmatrix} \tag{9.1}$$

Example 9.1. For the AND-EXOR circuit realizing $f = x_1 + x_2 + x_3 + x_4$ the test for detecting multiple stuck-at faults in the EXOR part is (0, 0, 0, 0), (1, 0, 0, 0), (0, 1, 0, 0), (0, 0, 1, 0), (0, 0, 0, 1). The multiple stuck-at faults in the EXOR tree are detected by this test.

This method is simple and directly applicable to linear EXOR circuits, but this concept cannot be applied directly for testing the multiplier circuits. For example consider the inner product circuit IPd(3) of the IP net-work as shown in Fig. 9.3. The input vectors as described in Example 9.1 cannot be applied to the EXOR part directly, since AND gates exist between the input terminals and the EXOR part. It is also shown in [14] that if $[t_1, t_2, \ldots, t_n]^T$ is a non-singular matrix, then this test can

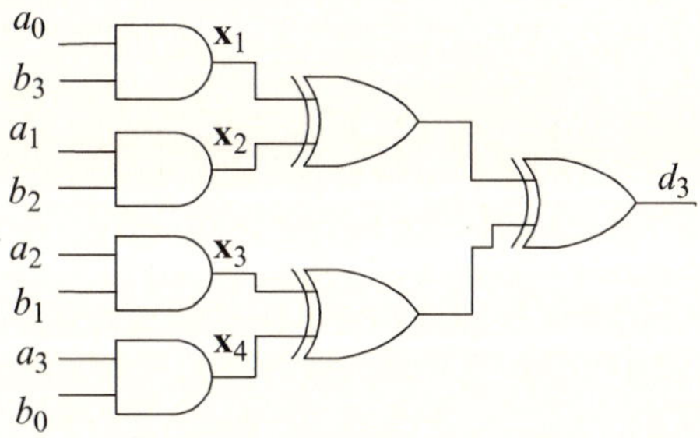

Fig. 9.3 IPd(3) part of GF(16) multiplier

detect the multiple stuck-at faults in the EXOR part of the AND-EXOR circuits. If the input set T is applied to the inputs of the circuit of Fig. 9.3, the output vectors of

$$T = \begin{bmatrix} t_0 \\ t_1 \\ t_2 \\ t_3 \\ t_4 \end{bmatrix} = \begin{bmatrix} a_0 & a_1 & a_2 & a_3 & b_0 & b_1 & b_2 & b_3 \\ 0 & 0 & 0 & 0 & 0 & 0 & 0 & 0 \\ 1 & 0 & 0 & 0 & 0 & 0 & 0 & 1 \\ 0 & 1 & 0 & 0 & 0 & 0 & 1 & 0 \\ 0 & 0 & 1 & 0 & 0 & 1 & 0 & 0 \\ 0 & 0 & 0 & 1 & 1 & 0 & 0 & 0 \end{bmatrix} \begin{bmatrix} g_0 \\ g_1 \\ g_2 \\ g_3 \\ g_4 \end{bmatrix} = \begin{bmatrix} x_1 & x_2 & x_3 & x_4 \\ 0 & 0 & 0 & 0 \\ 1 & 0 & 0 & 0 \\ 0 & 1 & 0 & 0 \\ 0 & 0 & 1 & 0 \\ 0 & 0 & 0 & 1 \end{bmatrix} \tag{9.2}$$

The IP-network has m number of d outputs and (m-1) number of e outputs. The d outputs are independent to each other, i.e. no two d outputs are feeding as inputs to the same c output in the Q-network. The faults in the IPd(j) block will be observed by the respective c_j output where $(0 \le j \le m-1)$. Each IPd block behaves as a single output AND-EXOR circuit provided that all the e out-puts are zeroes. From Eq. 8.5 we can derive the **d** and **e** outputs as follows.

$$\vec{d} = \begin{bmatrix} d_0 \\ d_1 \\ d_2 \\ \vdots \\ d_{m-2} \\ d_{m-1} \end{bmatrix} = \begin{bmatrix} a_0b_0 \\ a_1b_0 + a_0b_1 \\ a_2b_0 + a_1b_1 + a_0b_2 \\ \vdots \quad \vdots \quad \ddots \\ a_{m-2}b_0 + a_{m-3}b_1 + \cdots + a_0b_{m-2} \\ a_{m-1}b_0 + a_{m-2}b_1 + \cdots \cdots + a_0b_{m-1} \end{bmatrix} \tag{9.3}$$

$$\vec{e} = \begin{bmatrix} e_0 \\ e_1 \\ e_2 \\ \vdots \\ e_{m-2} \end{bmatrix} = \begin{bmatrix} a_{m-1}b_1 + a_{m-2}b_2 + \cdots + a_1b_{m-1} \\ a_{m-2}b_1 + \cdots + a_1b_{m-2} \\ a_{m-3}b_1 + \cdots + a_1b_{m-3} \\ \vdots \quad \vdots \\ a_{m-1}b_{m-1} \end{bmatrix} \tag{9.4}$$

$$T_d = \begin{bmatrix} t_0 \\ t_1 \\ t_2 \\ \vdots \\ t_i \\ \vdots \\ t_n \end{bmatrix} = \begin{bmatrix} a_0\ a_1 \cdots a_i \cdots a_{m-2}\ a_{m-1}\ b_0\ b_1 \cdots b_i \cdots b_{m-2}\ b_{m-1} \\ 0\ 0 \cdots 0 \cdots 0\ \ 0\ \ 0\ 0 \cdots 0 \cdots 0\ \ 0 \\ 1\ 1 \cdots 1 \cdots 1\ \ 1\ \ 1\ 0 \cdots 0 \cdots 0\ \ 0 \\ 1\ 1 \cdots 1 \cdots 1\ \ 0\ \ 0\ 1\ 0\ 0\ 0\ \ 0\ \ 0 \\ \vdots \cdots \vdots \quad \vdots \quad \vdots \cdots \vdots \\ 1\ 1 \cdots 1 \quad \cdots 0\ 1\ \ 0\ \ 0 \\ \vdots \cdots \vdots \quad \vdots\ \ 0\ 0\ 0\ \vdots \quad \vdots \\ 1\ 0\ 0\ 0\ 0\ \ 0\ \ 0\ \ 0\ 0\ 0\ 0\ 0\ \ 0\ \ 1 \end{bmatrix} \tag{9.5}$$

For the input vector t_1 in td, all the AND outputs in the first column in Eq. 9.3 produce 1, i.e. $a_0.b_0 = a_1.b_0 = a_2.b_0 = a_{m-2}b_0 = a_{m-1}b_0 = 1$. Similarly for the vector t_j, all the AND outputs in the jth column produce 1's. For the vector t_m, all the AND outputs in the mth column produce 1, i.e. $a_0.b_{m-1} = 1$. Each IPd block in the IP-network is independent of each other. Hence each IPd block is considered as a single output circuit. From Eq. 9.3, it is shown that the IPd(m-1) consists of a maximum number of AND terms in the expression. The expression for d_{m-1} is given as $d_{m-1} = a_{m-1}b_0 + a_{m-2}b_1 + a_{m-3}b_2 + \ldots + a_1b_{m-2} + a_0b_{m-1}$. The AND outputs of d_{m-1} for application of the Td inputs are derived in Gdm-1 which will be applied to the inputs of the XOR gates of IPd3 block. The $[g_1, g_2, g_3, \ldots, g_{m-1}, g_m]^T$ is a nonsingular matrix.

$$G_{d_{m-1}} = \begin{bmatrix} g_0 \\ g_1 \\ g_2 \\ \vdots \\ g_i \\ \\ g_m \end{bmatrix} = \begin{bmatrix} a_{m-1}b_0 & a_{m-2}b_1 & \cdots & a_{m-1}b_0 & a_{m-1}b_0 \\ 0 & 0 & \cdots & 0 & 0 \\ 1 & 0 & \cdots & 0 & 0 \\ 0 & 1 & \cdots & 0 & 0 \\ \vdots & \vdots & \ddots & \vdots & \vdots \\ 0 & 0 & \cdots & 1 & 0 \\ \vdots & \vdots & \cdots & \vdots & \vdots \\ 0 & 0 & \cdots & 0 & 1 \end{bmatrix} \tag{9.6}$$

The expression for d_{m-2} is given as $d_{m-2} = a_{m-2}b_0 + a_{m-3}b_1 + \ldots + a_1b_{m-3} + a_0b_{m-2}$. The AND outputs of d_{m-2} for application of the T_d inputs are given in Gd_{m-2} which is an non-singular matrix.

$$G_{d_{m-2}} = \begin{bmatrix} g_0 \\ g_1 \\ g_2 \\ \vdots \\ g_i \\ \\ g_m \end{bmatrix} = \begin{bmatrix} a_{m-2}b_0 & a_{m-3}b_1 & \cdots & a_{m-3}b_0 & a_{m-2}b_0 \\ 0 & 0 & \cdots & 0 & 0 \\ 1 & 0 & \cdots & 0 & 0 \\ 0 & 1 & \cdots & 0 & 0 \\ \vdots & \vdots & \ddots & \vdots & \vdots \\ 0 & 0 & \cdots & 1 & 0 \\ \vdots & \vdots & \cdots & \vdots & \vdots \\ 0 & 0 & \cdots & 0 & 1 \end{bmatrix} \tag{9.7}$$

Similarly the AND outputs for all other IPd blocks form a nonsingular matrix with the application of the T_d set. From Eqs. 9.3 and 9.4, it can be observed that when the **d** outputs receive a 1 for the vectors Td, all the **e** outputs remain 0. Hence the set Td is sufficient to detect the multiple stuck-at faults in the IPd blocks.

$$Te = \begin{bmatrix} te_0 \\ te_1 \\ \vdots \\ te_i \\ \vdots \\ te_{m-2} \end{bmatrix} = \begin{bmatrix} a_0 & a_1 & \cdots & a_i & \cdots & a_{m-2} & a_{m-1} & b_0 & b_1 & \cdots & b_i & \cdots & b_{m-2} & b_{m-1} \\ 0 & 0 & \cdots & 0 & \cdots & 0 & 1 & 0 & 1 & \cdots & 1 & \cdots & 1 & 1 \\ 0 & 0 & \cdots & 0 & \cdots & 1 & & 0 & 0 & \cdots & 1 & \cdots & 1 & 1 \\ & & \cdots & & \cdots & & & & & \cdots & 0 & 0 & \cdots & \\ 0 & 0 & & 1 & 0 & & & & & \cdots & & \cdots & 1 & 1 \\ & & \cdots & & \cdots & & & & & \cdots & & & \cdots & \\ & 1 & \cdots & 0 & \cdots & 0 & 0 & 0 & 0 & 0 & 0 & 0 & 0 & 1 \end{bmatrix} \tag{9.8}$$

$$G_{e0} = \begin{bmatrix} ge_0 \\ ge_1 \\ ge_2 \\ ge_3 \\ \vdots \\ ge_{m-2} \\ ge_{m-1} \end{bmatrix} = \begin{bmatrix} a_{m-1}b_1 & a_{m-2}b_2 & a_{m-3}b_3 & \cdots & a_2b_{m-2} & a_1b_{m-11} \\ 0 & 0 & 0 & \cdots & 0 & 0 \\ 1 & 0 & 0 & & 0 & 0 \\ 0 & 1 & 0 & \cdots & 0 & 0 \\ 0 & 0 & & & 0 & 0 \\ & \vdots & & \ddots & & \vdots \\ 0 & 0 & 0 & & 1 & \\ 0 & 0 & 0 & \cdots & 0 & 1 \end{bmatrix} \tag{9.9}$$

For the vector te_0 in T_e all the AND terms in the 1st column for the expressions corresponding to the e's in Eq. 9.4 receive a 1, i.e. $a_{m-1}b_1 = a_{m-1}b_2 = a_{m-1}b_3 = \ldots = a_{m-1}b_{m-2} = a_{m-1}b_{m-1} = 1$. Similarly for the vector t_j all the AND terms in the jth columns receive a 1. For the vector t_{m-2} all the AND. terms in the $(m-1)^{th}$ columns receive a 1, i.e. $a_1b_{m-1} = 1$. IPe(0) constitutes a large number of AND terms in the expression. The expression for e_0 is given as $e0 = a_{m-1}b_1 + a_{m-2}b_2 + \ldots + a_1b_{m-1}$. The AND outputs of the e_0 for T_e inputs are derived in the matrix Ge_0. The T_e is a $(m-1) \times 2m$ matrix. From the Eq. 9.4, it is shown that a_0 or b_0 is not part of any expression in the row. There is no need of any vector which sets either $a_0 = 1 \ or \ b_0 = 1$. The matrix $[ge_1, ge_2, ge_3, \ldots, ge_{m-2}, ge_{m-1}]^T$ is a nonsingular. From Eqs. 9.3 and 9.4 we observe that when the e outputs are 1 for the T_e inputs, all the d outputs remain 0. Hence the T_e set is sufficient to detect the multiple stuck-at faults in the IPe(0) block. Similarly the AND outputs for all other IPe blocks form a non-singular matrix for the T_e inputs. In all the IPe blocks the vector Te is sufficient to detect the multiple stuck-at faults.

9.3.2 EXOR Part in the Q-Network

The EXOR part in the Q-network is tested by the T_d and T_e test sets. When the d outputs are 1s and the e's are 0s, the EXOR gates, whose one of the inputs is connected to one of the d's outputs and the other input is connected to one of the e's out-puts, are receiving (1, 0) at their inputs. The other XOR gates are receiving (0,0) at their inputs. When the d outputs are 0s and e's are 1s, the XOR gates whose one input is connected with one of the d's and another input is connected with one of the e's outputs are receiving- (0, 1) signals at their inputs and other gates are receiving (1, 1) signal or (1, 0) at their inputs depending on the values of the e's. Due to the applications of the T_d and T_e test sets each BTX block in the Q-network receives the input vectors from the d and e outputs. As we see from Eq. 8.7, each c constitutes one d output and all or some of the e_j where $(0 \leq j \leq m-2)$. From Eq. 9.4 e0 contains the maximum number of AND terms. If we consider one of the c expressions, say $c_i = d_i + e_{m-2} + \ldots + e_0$, assuming c_i constitutes the maximum number of the e inputs, we can form the following test matrix for that BTX block, which is non-singular. Hence the vectors in this matrix detect multiple stuck-at faults in the BTX block realizing the c_i expressions. Similarly every BTX block receives from T_d and T_e input vectors, which forms a non-singular matrix.

$$X_Q = \begin{bmatrix} d_i & e_{m-2} & e_{m-3} & \cdots & e_i & \cdots & e_1 & e_0 \\ 0 & 0 & 0 & \cdots & 0 & \cdots & 0 & \\ 1 & 0 & 0 & \cdots & 0 & \cdots & 0 & 0 \\ 0 & 1 & 1 & \cdots & 1 & \cdots & 1 & 1 \\ 0 & 0 & 1 & \cdots & 1 & \cdots & 1 & 1 \\ \vdots & & & \cdots & & \cdots & & \\ 0 & 0 & 0 & \cdots & 1 & \cdots & 1 & 1 \\ \vdots & & & \vdots & & \vdots & & \\ 0 & 0 & 0 & \cdots & 0 & \cdots & 1 & 1 \\ 0 & 0 & 0 & \cdots & 0 & \cdots & 0 & 1 \end{bmatrix} \tag{9.10}$$

Lemma 9.1. *The multiple stuck-at faults in XOR part of the multipliers are detected by the test set TXOR, where* $TXOR = T_d \cup T_e$

Proof. Follows from Sect. 9.3.1

9.3.3 Test Set for the AND-Part of IP-Network

The multiplier network is multiple output positive polarity AND-EXOR network. Here,a conventional walking zero sequence is not required to detect stuck-at fault in the input/output of the AND gates. To test a stuck-at-1 fault at an input of an AND gate, we set it to 0 and set the other line of this gate to 1. Similarly, to test a stuck-at-0 fault in these gates, we set the other lines accordingly to 1. Hence Lemma 9.2 follows:

Lemma 9.2. *All single stuck-at faults in the AND-part of the IP network of the multiplier circuits are testable at the functional outputs by* $T_{and} = T_d \cup T_e \cup T_{a0b0}$, *where* $T_{a0b0} = a_0 a1 \ldots a_j \ldots a_{m-2} a_{m-1} b_0 b_1 \ldots b_j \ldots b_{m-2} b_{m-1} = 00\ldots 0\ldots 0010\ldots 0\ldots 00$

Proof. The AND part is tested for single stuck-at-1 faults by $T_{and} = T_d \cup T_e$. Due to the test vectors in T_d (excluding t_0 vector), all the 2-input AND gates receive the (1, 0) combination at the inputs. Again for test vectors in T_e (excluding t_0 vector) all the 2-input AND gates receive the (0, 1) combination at the inputs. From T_{a0b0} test set the AND term $(a_0 b_0)$ receives (0, 1). Hence, any single stuck-at-1 fault in the input/output of any AND gate is detected by the combination (0, 1) or (1, 0). It can be verified from Eqs. 9.3 and 9.4 that all the AND gates receive (1, 1) at their two inputs by T_d and T_e. Any single stuck-at-0 fault at the input/output of any AND gate will propagate to the functional outputs and will be detected. Hence the complete test set T_and (excluding the t_0 vector) detects all the single stuck faults at the primary inputs/outputs of the AND gates in the IP-network.

Theorem 9.2. *Any single stuck-at fault in the AND part and multiple stuck-at faults in the EXOR part in the multiplier circuits is testable by the function independent test set T of length (2m + 1), where* $T = T_d \cup T_e \cup T_{a0b0}$. *Proof: Follows from Lemmas 9.1 and 9.2.*

Example 9.2. The complete test set of length 9 for GF(2^4) multiplier with $P(x) = x^4 + x^3 + 1$ is formulated as follows. T =

$$\begin{bmatrix} a_0 & a_1 & a_2 & a_3 & b_0 & b_1 & b_2 & b_3 \\ 0&0&0&0&0&0&0&0 \\ 1&1&1&1&1&0&0&0 \\ 1&1&1&0&0&1&0&0 \\ 1&1&0&0&0&0&1&0 \\ 1&0&0&0&0&0&0&1 \\ 0&0&0&1&0&1&1&1 \\ 0&0&1&0&0&0&1&1 \\ 0&1&0&0&0&0&0&1 \\ 0&0&0&0&1&0&0&0 \end{bmatrix}$$

9.4 Experimental Analysis

Table 9.1 shows experimental results. As our algebraic test set is dependent only on the primitive polynomial, this scheme eliminates the need for test generation programs, e.g. the ATPG tool. Table 9.1 gives the number of test vectors obtained from different schemes for detecting single stuck-at faults to achieve 100% fault

Table 9.1 Number of tests required for achieving 100% coverage

	Number of tests required			
Size (m)	ATPG	SIS	[12]	Proposed
4	13	16	12	9
6	20	23	16	13
7	24	26	18	15
8	27	29	20	17
9	30	32	22	18
16	51	51	36	33
20	63	66	44	41
24	75	77	52	49
32	99	101	68	65

coverage although our algebraic test set detects single stuck-at faults in the AND part. The table compares our test scheme with ATPG based test generation and with algorithmic test generation schemes (SIS tool [16]). *Synopsys*[TM] (TetraMAX) tool is used to generate ATPG based test patterns. Clearly from the above table both the ATPG-based test generation and algorithmic test generation schemes require more test patterns compared to the proposed schemes for achieving 100% fault coverage.

9.5 C-Testable Scheme

The proposed testable design of PB multiplier over GF(2^m) is shown in Fig. 9.4. Basically it consists of two parts: IP-network and Q-network. The IP-network constitutes AND-parts followed by trees of EXOR gates. The Q-network constitutes trees of EXOR gates only. To achieve 100% testability, the IP-network has been augmented as shown in Fig. 9.4. AND parts of the IP-network are modified with three control lines k_0, k_1 *and* k_2. All two inputs AND gates have been replaced by three input AND gates.

9.5.1 *Testability in Single Output EXOR Tree*

Testing of single stuck-at fault in single output general EXOR tree can be performed by exactly 4 tests, which exhaustively applies all the 4 input combinations (00, 01, 10, 11) to each of 2-inputs EXOR gate. In this design, we need three control inputs k_0, k_1 *and* k_2 to achieve this. This is based on the following observation: in Fig. 9.5, the inputs to the last EXOR gate require 00, 11, 10, 01 to generate the output sequence s: 0011. Thus, its two inputs should receive the sequence q: 0110 and r: 0101. Similarly, q: 0110 and s: 0011 arriving at the two inputs of an EXOR gate will generate the output sequence r: 0101. Again, input sequences r: 0101 and s: 0011 will generate q: 0110 as the output sequence. There exist the following

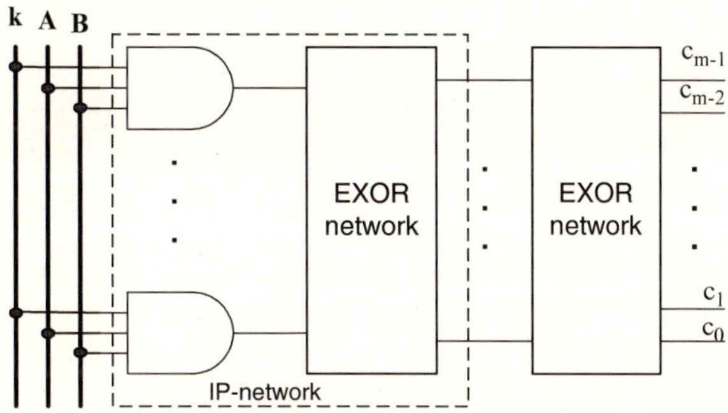

Fig. 9.4 Block Schematic of C-testable GF multiplier

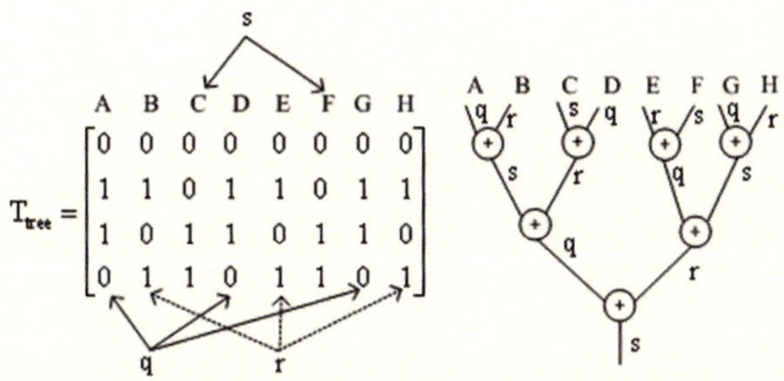

Fig. 9.5 Test vectors and responses in an EXOR-tree

relations among the vectors (q, r, s): $q \oplus r =$ s, $q \oplus s = r$, $r \oplus s = q$. Hence, by applying the three sequences (q, r, s) to the inputs of a tree, any one of the above three combinations can be applied to the inputs of each EXOR gate in the tree

Example 9.3. In the EXOR tree shown in Fig. 9.2, we assign sequence vectors q, r, s, q, r, s, q, r, …(i.e. by repeating the pattern (q, r, s)) to the inputs of the EXOR tree from left-to right until all of them are assigned. The outputs of the first level are propagated down to the root, i.e. the final output of the tree. Thus, each EXOR gate in the tree receives the desired input combination from the above three combinations. Three constant test vectors that are to be applied to the inputs of the tree of Fig. 9.2 are shown as a matrix Ttree. This matrix has four (constant) rows and y columns, where y is the number of leaf nodes of the tree, and is equal to the number of AND outputs (m^2) in the multiplier circuits. The columns of the matrix, if seen from left-to-right, will correspond to the sequence vectors: q, r, s, q, r, s, q, r, and so on. The number of distinct columns in the matrix is only four (constant), regardless of the size of the tree. Since EXOR-trees are embedded in the overall designs of the single output AND-EXOR circuits, the inputs of the trees are not directly accessible. In the IP network of Fig. 9.2, each AND output feeds an EXOR input. Hence, by applying the following four test vectors $v_1, v_2, v_3,$ *and* v_4, to the primary inputs of Fig. 9.2, all the three sequences q, r, s can be produced at the outputs of the AND-part. Note that in Fig. 9.6 the original function can be obtained by setting the three control inputs to 1 (k_0, k_1 *and* k_2). In this design, the AND outputs are partitioned into three groups based on the sequence vectors q, r,

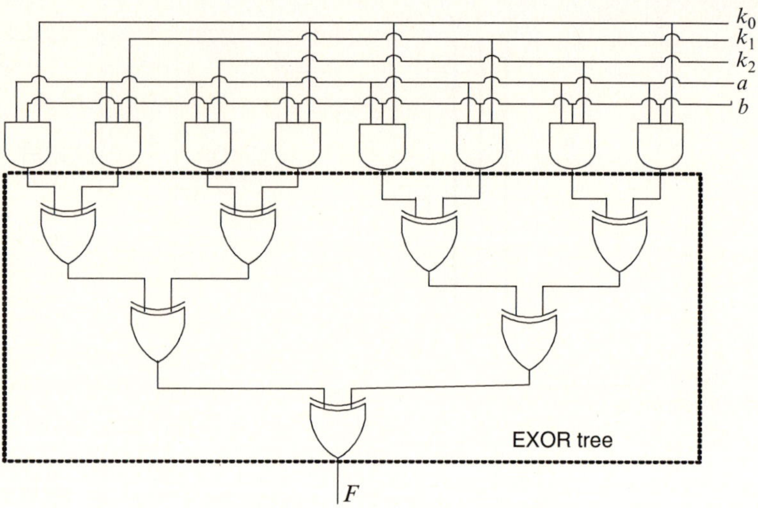

Fig. 9.6 EXOR-tree with a control level

and s. The output lines of the AND gates connected with k_0, k_1 *and* k_2 control lines receive the sequence vectors q: 0110, r:0101, and s:0011 respectively. Note that only the two input AND gates in the IP-networks have been replaced by three input AND gates and no additional hardware is required to gain C-testability.

$$\begin{array}{lccccccccccc} & k_0 & k_1 & k_2 & a_0 & a_1 & \cdots & a_{m-1} & b_0 & b_1 & \cdots & b_{m-1} \\ v_1 = & 0 & 0 & 0 & 0 & 0 & \cdots & 0 & 0 & 0 & \cdots & 0 \\ v_2 = & 1 & 1 & 0 & 1 & 1 & \cdots & 1 & 1 & 1 & \cdots & 1 \\ v_3 = & 1 & 0 & 1 & 1 & 1 & \cdots & 1 & 1 & 1 & \cdots & 1 \\ v_4 = & 1 & 1 & 1 & 1 & 1 & \cdots & 1 & 1 & 1 & \cdots & 1 \end{array}$$

9.5.2 *Embedding EXOR-Tree in Multi-output Multiplier Circuit*

The technique we have discussed above is applicable to single output AND-EXOR circuits. In this section we extend this idea to multiple output AND-EXOR circuits. To achieve 100% testability in multiplier circuits, the inputs of the EXOR gates of the IP- and Q-networks will be properly mapped. We assume that the IP-network would generate the following sequence from left-to-right:$q, r, s, q, \ldots, q, r, s, q \ldots$ and so on at the outputs e_j, where $0 \leq j \leq m-2$. To propagate these e_j outputs of the IP-network at the outputs of the Q-network, the d_i outputs, where $0 \leq i \leq m$, will be properly mapped with the sequences q, r, and s. After assigning all the root nodes d_i, and e_j of the IP-network, the input nodes of each EXOR gate in the IP-networks will be activated from the AND outputs with the proper sequences so that

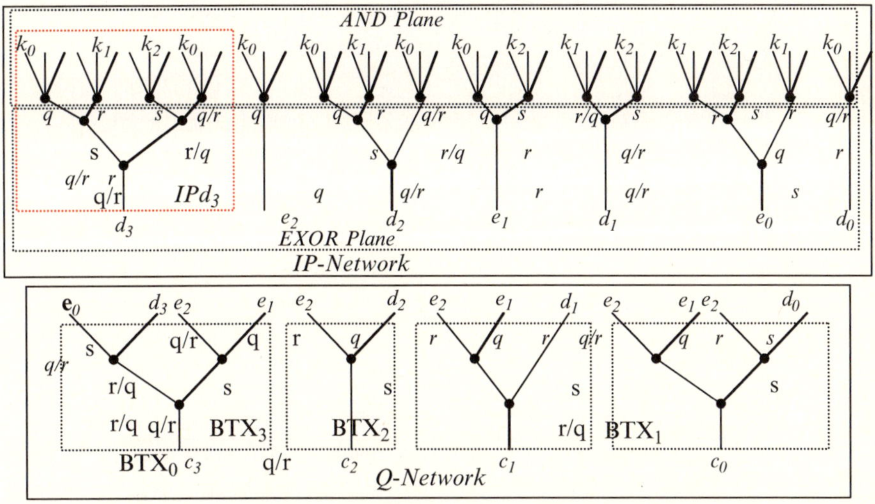

Fig. 9.7 Tree representation of nodes of Fig. 9.1

no two inputs of each EXOR gate receive the same sequence vector. Figure 9.7 shows the sequence in which the vectors are applied in the circuit of Example 8.4 and how the test signals propagate through to the primary output. The following algorithms outline this process.

Step-1: Assignment of the sequences q, r, and s to e_j, where $(0 \leq j \leq m-2)$.
Algorithm_seq_assignment_e

for $(j = 2; j \leq m; j++)$
{
$e_{(m-j)} = q;$
$e_{(m-(j+1))} = r;$
$e_{(m-(j+2))} = s$
}

Example 9.4. For the multiplier circuit over GF(2^4) shown in Fig. 9.8, the sequence values at e_j, where $0 \leq j \leq 2$ are assigned as follows: $e_2 = q$, $e_1 = r$, $e_0 = s$.

Step-2: Assignment of the sequences q, r, and s to d_i, where $(0 \leq i \leq m-1)$. After assigning the sequences at the e_i nodes in step-1, the sequences at di nodes, where $0 \leq i \leq m-1$, will be assigned in a such way that no two input nodes of each EXOR gate receive same sequence vector in the Q-networks.

Example 9.5. Consider the BTX3 block of the Q-network in Fig. 9.7. As shown in this figure, the nodes e_2, e_1, e_0 are assigned q, r, and s respectively. The nodes e_2 and e_1 produce s at the output node. To propagate the signal value at c_3 output node q or r will have to be assigned at other input node of the gate connected with c_3.

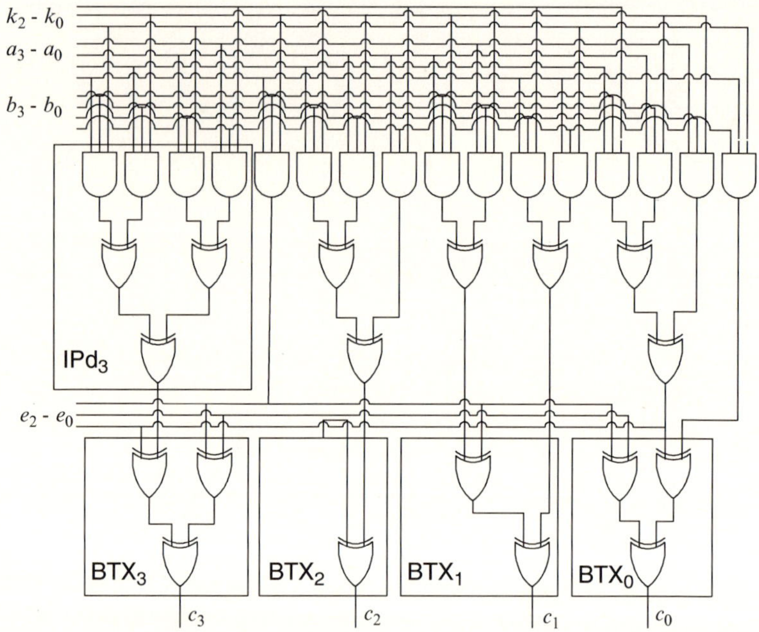

Fig. 9.8 Testable design of GF(16) multiplier

As e_0 is already assigned s, d_3 will have to be assigned either q or r to achieve either r or q at the output node of the gate connected to the inputs e_0 and d_3. In this way the sequences at the d_i nodes, where $(0 \leq i \leq m-1)$, are assigned as shown in Fig. 9.7.

Step-3: Assignment of the sequences q, r, and s to the internal nodes of the IP-network

After assigning the sequences at d_i's and $e_j s$, assign the input nodes of the EXOR gates in the IP-network with proper sequence vectors so that no two inputs of an EXOR gate receive the same sequence vector. To propagate the signal from the d_3 output to the final output c_3, the d_3 input is assigned either the sequence q or s. Similarly $d_2 =$ s or r; $d_1 =$ r or q, $d_0 =$ q or r. After assigning $e_j \, (0 \leq j \leq 2)$ and $d_i \, (0 \leq i \leq 3)$ outputs, every EXOR gate in the IP-network is mapped. If the test sequence t_1, t_2, t_3, t_4 is applied, then the output lines of the AND gates connected to control lines k_0, k_1 *and* k_2 receive the sequence vectors q: 0110, r: 0101, and s: 0011 respectively.

Example 9.6. Consider the IPd3 block in the IP-network of Example 9.3 (see Fig. 9.7). The d_3 node is already assigned with either q or r. If d_3 is assigned with q, then the input nodes of the associated gate with d_3 as the output node will have to be assigned with s and r respectively. If d_3 is assigned r, then the input nodes of the associated gate with d_3 as output will have to be assigned s and q respectively. In this way the input nodes of the EXOR gates of all the IPd blocks are assigned with the proper sequence vectors. After assigning the sequence vectors to each input node of the EXOR gates at the first level, we will have to determine the proper connection of control inputs $k_0,\ k_1,\ and\ k_2$. In the AND plane 3-input AND gates are used, instead of 2-input ones. One input of each AND gate is connected with one of the control inputs $k_0,\ k_1,\ and\ k_2$ and the other two inputs are connected with the A and B inputs respectively. One of the input nodes of each AND gate generating the q sequence will be connected to the control input k_0. Similarly one of the input nodes of each of AND gate generating the r sequence will be connected to the control input k_1. Again, one of the input node of each of AND gate generating the r sequence will be connected to the k_2 control input.

Example 9.7. The internal mapping of the interconnections in the IP and Q-networks of Example 9.7 is shown in Fig. 9.8, which is also the testable design of the multiplier designed from the irreducible polynomial $P(x) = x^4 + x^3 + 1$. The complete assignment mapping of the sequence vectors for this circuit appears graphically Fig. 9.7.

9.5.3 Constant Test Set

The output functions are the same as the original functions, i.e. the circuit will perform in its normal mode, when $k_0 = k_1 = k_2 = 1$. EXOR part: All single stuck-at faults can be tested by applying just four vectors at the inputs of Fig. 9.4, and by observing the circuit responses at the circuit outputs.

Lemma 9.3. *Any single stuck-at fault in the EXOR-part of the network is testable by v_1, v_2, v_3, v_4 tests.*

Proof. Follows from the discussions in Sect. 9.5.2

AND-part, primary inputs, and control inputs: To test for a s-a-1 fault at an input line of an AND gate, we set it to 0 and set the other two lines of this gate to 1 (each AND gate has 3 inputs). Similarly to test for a s-a-0 fault in these gates, we set all the inputs to 1.

Lemma 9.4. *All single stuck-at faults in the AND-part, at the primary inputs, are testable at the functional outputs of the multiplier circuit by the test set (v_5, v_6, v_7, v_8) of length 4 as follows, $k_0k_1k_2a_0a_1 \ldots a_{m-1}b_0b_1 \ldots b_{m-1} = (v_5 :$ $0001 \ldots 1 \ldots 111 \ldots 1,\ v_6\ :\ 1110 \ldots 0 \ldots 011 \ldots 1\ v_7\ :\ 1111 \ldots 1 \ldots 100 \ldots 0,$ $v_8 = 1111 \ldots 1 \ldots 111 \ldots 1)$.*

Proof. The AND part is tested for single stuck-at-1 faults by the first three vectors (v_5, v_6, v_7). Due to v_6, v_7, and v_5 all the 3-input AND gates receive the combinations (1, 0, 1), (1, 1, 0), and (0, 1, 1) respectively at their inputs. It can be seen that any single s-a-1 fault at any input and output of any AND gate will propagate to the functional outputs. The vector v_8, i.e. the all-1 vector, detects any single stuck-at-0

Table 9.2 Gate count for different polynomial basis

	Original implementation		Testable implementation	
Type of poly.	# of 2 inputs AND gate	# of 2 inputs EXOR gate	# of 3 inputs AND gate	# of 2 inputs EXOR gate
ESP	m^2	$m^2 - s$	m^2	$m^2 - s$
Trinomial	m^2	$m^2 - 1$	m^2	$m^2 - 1$
Pentanomial	m^2	$m^2 + m$	m^2	$m^2 + m$

fault at the inputs and output of any AND gate. Hence the test set (v_5, v_6, v_7, v_8) detects all the single stuck-at faults at the primary inputs and outputs of the AND gates, and control inputs of the IP-network.

Theorem 9.3. *Any single stuck-at fault in the proposed Multiplier network is testable by the constant test set T_s of length 8, where $T_s = (v_1, v_2, v_3, v_4, v_5, v_6, v_7, v_8)$.*

Proof. Follows from Lemmas 9.3 and 9.4.

Example 9.8. The constant test set of length 8 for the multiplier over GF(2^4) with the irreducible polynomial $P(x) = x^4 + x^3 + 1$ is given as follows:$k_0k_1k_2a_0a_1a_2a_3$ $b_0b_1b_2b_3$ = (v_1 : 00000000000, v_2 : 11011111111, v_3 : 10111111111, v_4 : 01111111111, v_5 : 00011111111, v_6 : 11100011111, v_7 : 11111100000, v_8 : 11111111111).

9.5.4 Gate Complexities

The gate complexities of the testable multiplier for different types of polynomials are given in the Table 9.2. In this table, the value of s is 1 for All One Polynomial (AOP) and $m/2$ for trinomial of the form $(x^m + x^k + 1)$ where $k = m/2$. For $k \neq m/2$, s = 1 for trinomial.

9.5.5 Experimental Analysis

We performed area, delay, power and test set size analysis on various GF(2^m) multipliers based on different polynomial bases. The area, power and delay analysis is based on UMC 0.18 μm CMOS technology library. All area measurements are expressed in cell units, excluding the interconnection wires. The two versions (classical and C- Testable Design) of Galois multipliers have been designed and coded in VHDL. The design was simulated using *Modelsim^{TM}* and was tested for functionality by giving various inputs. The designs were synthesized using the Synopsys tools. Synopsy's Power *Compiler^{TM}* was used to estimate the power consumption. The comparative analysis of area, delay and power is shown in

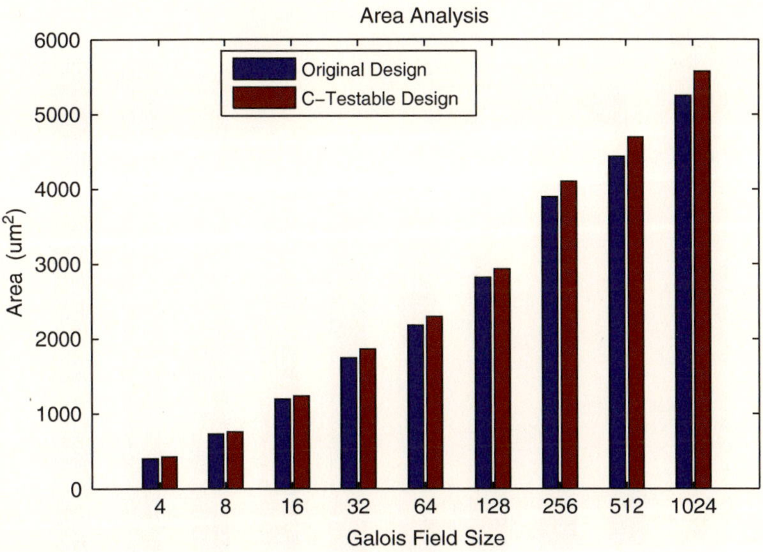

Fig. 9.9 Area analysis: original vs. C-testable version

Figs. 9.9–9.11 respectively. On an average there is 6% increase in area and power. The delay overhead is negligible, when the overall delay of the multiplier is considered.

The test set is constant of length 8, which eliminates the need for test generation programs. Table 9.3 gives the number of test vectors obtained from different schemes for single stuck-at faults to achieve 100% fault coverage. It shows that our test set is much smaller than either the pseudo-random or the ATPG based test patterns. Synopsys^{TM} tools are used to generate the ATPG based test patterns. For example, in both the schemes, the multiplier circuit over GF(2^{32}) requires more than 68 patterns of length 64 bits to achieve 100% fault coverage, whereas our scheme requires only 8 test patterns of length 64 bits. This also ensures reductions in test application time and the associated power consumption.

9.5.6 Reduction of Control Pins

To achieve C-testability in the multiplier circuits, our proposed scheme discussed in Sect. 9.5 requires three additional control pin. We can eliminate one of the pin by introducing the following scheme. This requires only 7 vectors for detecting all the single stuck-at faults, which is independent of the multiplier sizes, at the cost of at most $(2/3)m^2$ extra 2-input EXOR gates (33% extra hardware overhead) and two control inputs. This C-testable scheme for the GF(16) multiplier circuits is shown in Fig. 9.13. The EXOR part of the IP-network has been augmented

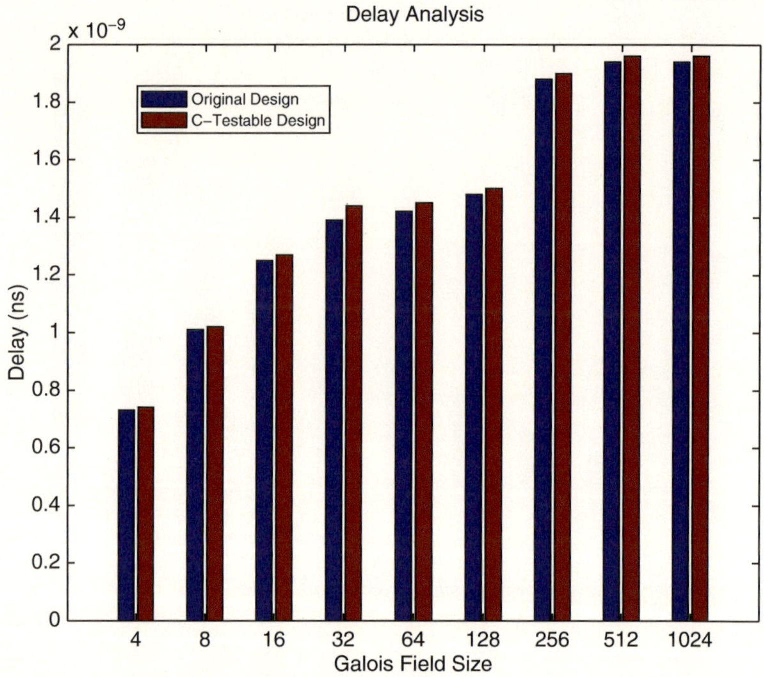

Fig. 9.10 Delay analysis: original vs. C-testable version

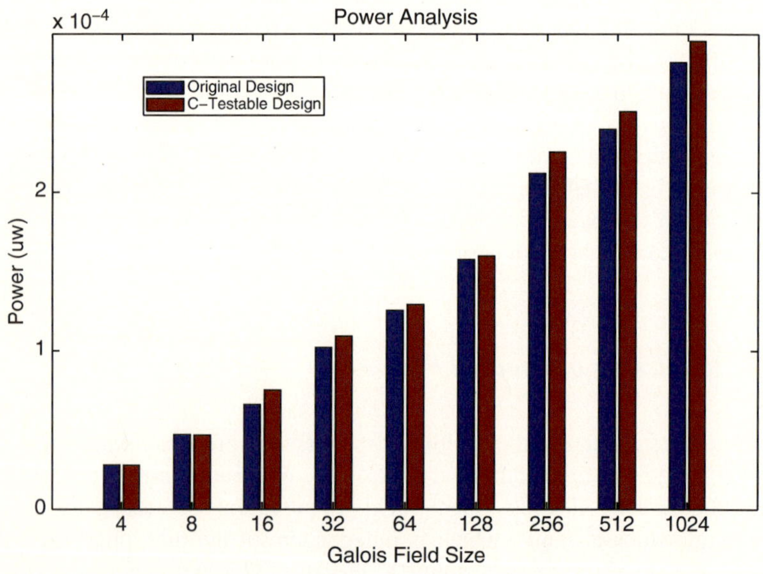

Fig. 9.11 Power analysis: original vs. C-testable version

Table 9.3 Number of tests required for achieving 100% fault coverage

GF Multiplier	#of i/ps,	ATPG (Synopsys)	Pseudo random	Proposed (C-testable)
GF(2^4)	8,4	12	14	8
GF(2^5)	10,5	14	17	8
GF(2^6)	12,6	15	20	8
GF(2^7)	14,7	18	22	8
GF(2^8)	16,8	20	24	8
GF(2^9)	18,9	23	26	8
GF(2^{16})	32,16	36	39	8
GF(2^{20})	36,18	44	47	8
GF(2^{24})	48,24	52	55	8
GF(2^{32})	64,32	68	71	8

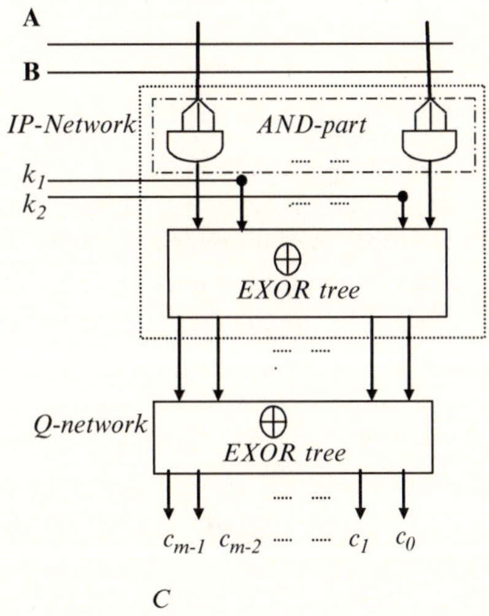

Fig. 9.12 Testable GF multiplier with two control inputs

with some additional EXOR gates and two control lines, k_1 and k_2. This scheme exhaustively applies all the 4 input combinations (00, 01, 10, 11) to each of 2-inputs EXOR gate. The generation of three-sequence vectors q, r, and s in single output EXOR tree was explained in [11]. We have extended this concept in this multi-output EXOR tree. The interconnection mapping of the IP- and Q-networks has been done using the mapping technique in Sect. 9.5.2. Figure 9.12 shows the general structure of the testing scheme with two control inputs. By applying a test $a_0a_1a_2a_3b_0b_1b_2b_3 = t_1 : 00000000(t_2 : 11111111)$ to the primary inputs of Fig. 9.13, an all-zero (all-one) vector can be produced at the outputs of the AND-part. If a test sequence t_1, t_2, t_1, t_2 is applied, then all the EXOR gates at the first level of the EXOR tree of the IP-network will receive the sequence q: 0110. To generate the other two sequence vectors r: 0101 and s: 0011, we use a control level with a few additional EXOR gates and two control inputs k_1 and k_2.

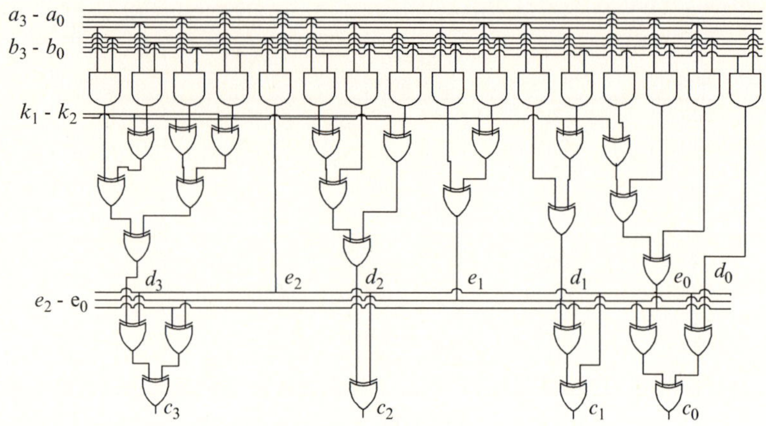

Fig. 9.13 Testable design of GF(16) multiplier with two control pins

In the EXOR tree shown in Fig. 9.5, we assign sequence vectors q, r, s, q, r, s, q, r,... (by repeating the pattern (q, r, s) to the inputs of the EXOR tree from left-to right until all of them are assigned. The outputs of the first level are propagated down to the root (final output of the tree). Thus, each EXOR gate in the tree receives the desired four combinations 00, 01, 10, 11 as an input. The four universal test vectors that are to be applied to the inputs of the tree are shown as a matrix Ttree in Fig. 9.5. This matrix has four (constant) rows and y columns, where y is the number of leaf nodes of the tree, which in our case, is equal to the number of AND outputs (m^2) in the multiplier circuit. The columns of the matrix, if seen from left-to-right, will correspond to the sequence vectors: q, r, s, q, r, s, q, r, and so on. The number of distinct columns in the matrix is only three (constant), regardless of the size of the tree.

9.5.7 Embedding an EXOR-Tree in the Single-Output AND-EXOR Circuit

Since the EXOR-tree is embedded in the overall design as in Fig. 9.12, the inputs of the tree are not directly accessible. In the IP network, each AND output feeds an EXOR input. Hence, by applying a test $a_0a_1 \ldots a_{m-1}b_0b_1 \ldots b_{m-1} = t_1 : 00\ldots000\ldots0$ ($t_2 : 11\ldots111\ldots1$) to the primary inputs of the Fig. 9.12, an all-zero (all-one) vector can be produced at the outputs of the AND-part. So, if a test sequence t_1, t_2, t_1, t_2 is applied, all the EXOR gates at the first level of the tree will receive the sequence q: 0110. To generate the other two sequence vectors r: 0101 and s: 0011, we use a control level with a few additional EXOR gates and two control inputs k_1 and k_2 as shown in Fig. 9.12. By setting these control inputs to 0, the original function can be obtained. In this design, the AND outputs are

Table 9.4 Control logic for the EXOR-tree

O/p of AND	Desired Tree o/p	Value of k_1	O/p of AND	Desired Tree o/p	Value of k_2
0	0	0	0	0	0
1	1	0	1	0	1
1	0	1	1	1	0
0	1	1	0	1	1

partitioned into 3 groups based on sequence vectors q, r, s. The output lines of the first group receiving the sequence vector q: 0110 are allowed to pass directly through the control level to reach inputs of the EXOR tree. The AND outputs for the other two groups (r and s) are passed through an additional EXOR gate, using control input k_1 and k_2 respectively. By controlling the values of $k_1(k_2)$ the desired sequence vector r(s) can be obtained from q (Table 9.4. The modified EXOR-tree (Fig. 9.12) requires $\leq \lceil(2y/3)\rceil$ additional EXOR gates for the control level.

9.5.8 Embedding XOR-Tree in Multi-output Multiplier Circuit

The above technique is applicable to single output AND-EXOR circuits. This concept has been extended to the multi-output AND-EXOR circuits. To get 100% testability, the inputs of the EXOR gates of the IP-network and Q-network will be properly mapped. The mapping of the interconnections of the IP-network and Q-network has been done on basis of the following steps.

- First, assign $e_{m-2} = \text{q}$, $e_{m-3} = \text{r}$, $e_{m-4} = \text{s}$, $e_m - 5 = \text{q}$, $e_{m-6} = \text{r}$, $e_{m-7} = \text{s}$, $e_{m-8} = \text{q}$. It is assumed that e_j, where $0 = j = m-2$ of IP-network (from left) will generate the sequences q, r, s, q, r, s, q, etc.
- Then map d_i, where 0 = i = m-1, with proper sequence in such way that no two inputs of each EXOR gate get the same sequence vector in the Q network.
- Assign inputs of EXOR gates in IP-network with proper sequence vectors so that no two inputs of each EXOR gate of IP-network receive the same sequence vector.
- The column vector with all 1s is assigned to output bit c_{m-1}

We assume that the IP-network would generate the following sequence from the left side: q, r, s, q,..., q, r, s, q and so on at the outputs e_j, where 0 = j = m-2. To propagate these e_j outputs of IP-network at the outputs of Q-network, the d_i outputs, where (0 = i = m-1) will be properly mapped with the sequences of q, r, s. After assigning all the root nodes of the IP-network d_i, and e_j, the inputs of each EXOR gate in the IP-network will be activated from the AND outputs with the proper sequences so that no two inputs of each EXOR gate receive the same sequence vector.

Table 9.5 Details of hardware for different polynomial basis

Type of Polynomial	# of AND	# of EXOR
ESP	m^2	$\lceil 1.67\,m^2 - s \rceil$
Trinomial	m^2	$\lceil 1.67\,m^2 - 1 \rceil$
Pentanomial	m^2	$\lceil 1.67\,m^2 + m \rceil$

9.5.8.1 Constant Test Set

If we put k1 = k2 = 0, then the output functions are the same as the original functions. All the single stuck-at faults can be tested by applying just following four vectors at the inputs of Fig. 9.11, and by observing the circuit responses at the functional outputs: $k_1k_2a_0a_1a_2a_3b_0b_1b_2b_3$ = (0000000000, 0111111111, 10 11111 111, 11 00000000). All the single stuck-at faults at the control inputs k_1 and k_2 are testable at the functional outputs of the multiplier circuits by the above test set. The multipliers are multiple output positive polarity AND-EXOR networks. To test a stuck-at-1 fault at an input line of an AND gate, we set it to 0 and set the other lines of this gate to 1s. Similarly to test a stuck-at-0 fault in these gates, we set all the inputs to 1s. All the single stuck-at faults in the AND-part at the primary inputs are testable at the functional outputs of the multiplier circuit by the following test set T2 of length 3: k1k2 a0a1a2a3 b0b1b2b3 = (000000 1111, 00 1111 0000 , 00 1111 1111). Any single stuck-at fault in the proposed Multiplier network is testable by the above constant test set of length 7. Gate Complexities: The gate complexities of the testable GF multiplier for different types of polynomials are given in Table 9.5. This C-testable design with two control pins requires approximately $(2m^2/3)$ EXOR gates i.e., the upper bound of the extra hardware overhead is approximately 33%.

9.5.9 *BIST Circuit*

In this section, we present a simple on-chip BIST scheme for our proposed C-testable design. The BIST scheme generates the required 8 vectors internally. This BIST structure provides two benefits: firstly it eliminates the need for the three control inputs necessary for fully testing the multipliers, and secondly it provides an added level of security. Figure 9.14 shows the basic schematic of the proposed BIST hardware. It constitutes 3 flip-flops and some combinational logic to generate 8 test vectors. Area, delay and power of the BIST hardware are shown in Table 9.6. The circuit is synthesized using the *Synopsys*^{TM} design compiler based on the 0.18 μm technology library from UMC. Note that for higher fields the logic will remain same because the pattern remains the same. Only word length varies depending upon m. For instance to generate the test set for GF(2^m), only two combinational logic block is necessary to generate a_i and b_i. In other words, all the logic blocks for generating a_i $(0 \leq i \leq m-1)$ remain the same, therefore the logic can be reused with additional buffers to drive the inputs a_i. Hence only 5 logic blocks (k_0, k_1, k_2, a_i *and* b_i) are necessary as shown in Fig. 9.14. As an additional

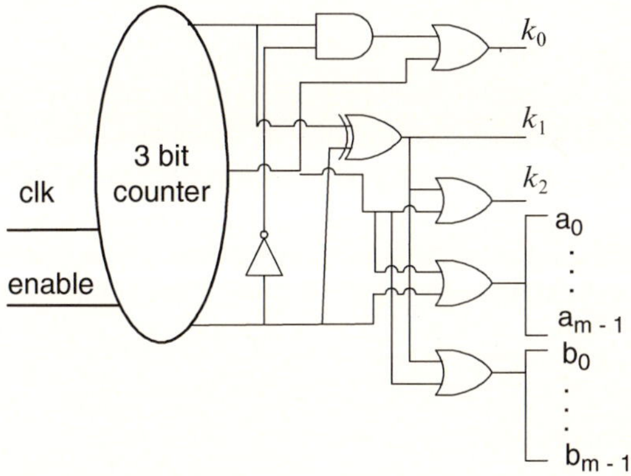

Fig. 9.14 Schematic of the proposed BIST hardware

Table 9.6 Details of area, delay and power for BIST hardware

Technology	Area	Delay	Power
0.18 μ	216 μm^2	0.61 ns	81.56 μW

advantage, instead of inserting additional buffers to drive a_i and b_i one could build additional a_i and b_i, and a simple comparison of the a_i (b_i) will detect 1 bit errors. The "enable" signal is used to activate the test process. The strength of the proposed scheme is that it facilitates the detection of any potential error in the TPG itself without compromising its performance in terms of fault coverage of the circuit.

Since we need 8 test vectors regardless of the size of the multiplier, both test and response data are very small. Then the collected response can be analyzed with a small on-chip hardware, or by a traditional Multiple-Input Signature Registers (MISRs) [3,4,9] which can provide with a simple true or false response depending on whether the circuits are faulty or not. Hence both the test pattern generation and response evaluation can be embedded within the systems, thus eliminating the ATE. Since the ATEs could require information on the internal structure of the systems, or at least a reasonable amount of probing, this could be eliminated completely with the added BIST scheme.

9.6 Chapter Summary

In this chapter, first, technique for generating test vectors for bit-parallel PB multiplier circuits over GF(2^m). For an m-bit multiplier circuit a function independent test set of length $(2m+1)$ is sufficient to detect all the single stuck-at faults in the AND part and multiple stuck-at faults in the EXOR part of multiplier circuits. All the test patterns can be determined readily from the corresponding algebraic forms, without any need of running an ATPG. The test set provides 100% single stuck-at fault coverage. The test set being very short in length, reduces test application time and test power. Also presented is a C-testable design of bit parallel polynomial basis multipliers over GF(2^m). The testable design requires approximately 6% extra hardware compared to original mulitplier design without testability and 3 control inputs. The only 8 constant test vectors are required for achieving 100% testability of stuck-at faults. Since these multipliers have found critical applications in cryptographic systems (e.g. Elliptic Curve Crypto (ECC) systems) and requires public-key secure testing, we also presented BIST structures for efficiently generating the required test vectors internally. The BIST structure eliminates the need for the three control inputs necessary for testing the multipliers.

References

1. M.L. Bushnell, V.D. Agrawal, *Essentials of Electronic Testing* (Kluwer, New York, 2000)
2. H. Fujiwara, On closedness and test complexity of logic circuits. IEEE Trans. Comput **30**(8), 556–562 (1981).
3. S.K. Gupta, D.K. Pradhan, A new framework for designing and analyzing bist techniques and zero aliasing compression. IEEE Trans. Comput. **40**(6), 743–763 (1991)
4. S.K. Gupta, M. Karpovsky, D.K. Pradhan, Aliasing probability for a multiple-input signature analyzer and a new compression technique. IEEE Trans. Comput. **39**(6), 586–591 (1990)
5. D. Hely, M.-L. Flottes, F. Bancel, B. Rouzeyre, N. Berard, M. Renovell, Scan design and secure chip, in *Proceedings of the 10th IEEE International On-Line Testing Symposium*, Funchal, 2004
6. D. Josephson, S. Poehhnan, Debug methodology for the mckinley processor, in *IEEE International Test Conference*, Baltimore, 2001, pp. 451–460
7. P. Kocher, R. Lee, G. McGraw, A. Raghunathan, S. Ravi, Security as a new dimension in embedded system design, in *Proceedings of the 41st Annual Conference on Design Automation*, San Diego, 2004, pp. 753–760
8. J. Lee, M. Tehranipoor, C. Patel, J. Plusquellic, Securing scan design using lock and key technique, in *IEEE International Symposium on Defect and Fault Tolerance in VLSI Systems*, Monterey, 2005
9. D.K. Pradhan, Glfsr: a novel method for test generation in bist environment. IEEE Trans. CAD 239–247 (1999)
10. H. Rahaman, D.K. Das, B.B. Bhattacharya, Easily testable realization of grm and esop networks for detecting stuck-at and bridging faults, in *International Conference on VLSI Design (VLSID 2004)*, Mumbai, Jan 2004
11. H. Rahaman, D.K. Das, B.B. Bhattacharya, Testable design of grm network with exor-tree for detecting stuck-at and bridging faults, in *ASPDAC 2004*, Yokohama, 2004, pp. 224–229
12. H. Rahaman, J. Mathew, A.M. Jabir, D.K. Pradhan, Easily testable implementation for bit parallel multipliers in GF(2^m), in *International Conference High Level Validation and Test (HLDVT)*, Monterey, Nov 2006
13. S. Ravi, A. Raghunathan, S. Chakradhar, Tamper resistance mechanisms for secure embedded systems, in *Proceedings of the 17th International Conference on VLSI Design*, Mumbai, 2004, pp. 605–611
14. T. Sasao, Easilytestable realizations for generalized reed-muller expressions. IEEE Trans. Comput. **46**(6), 709–716 (1997)
15. S. Scheiber, The best-laid boards, http://www.reed-electronics.com/tmworld/article/CA513261 (2005)
16. E.M. Sentovich, *Sis: A Sequential System for Sequential Circuit Synthesis*. Technical Report UCB/$ERLm92/41$, Electronic Research Laboratory, University of California, Berkeley, May 1992

17. Standard test access port and boundary-scan architecture. IEEE Standard, 1149.1–2001
18. K. Tiri, I. Verbauwhede, A vlsi design flow for secure side-channel attack resistant ics, in *Proceedings of the Design, Automation and Test in Europe*, Munich, 2005, pp. 58–63
19. B. Yang, K. Wu, R. Karri, Scan based side channel attack on dedicated hardware implementations of data encryption standard, in *International Test Conference*, Charlotte, 2004, pp. 339–344
20. B. Yang, K. Wu, R. Karri, Secure scan: a design-for-test architecture for crypto chips, in *Proceedings of the 42nd Annual Conference on Design Automation*, San Diego, 2005, pp. 135–140

Index

A
Advanced microprocessor bus architecture (AMBA), 172
Application task
  mapping, 180, 181, 183, 186, 188, 190–194, 196, 201, 202, 206, 217, 218, 221
  scheduling, 217, 218, 221
Approximate computing, 118, 121–123
Architecture allocation, 180, 202–205
ATPG. *See* Automatic test pattern generation (ATPG)
Automated synthesis, 289–291
Automatic test pattern generation (ATPG), 308, 309, 315, 316, 323, 325, 329
Availability, 2, 5, 7

B
Bio-inspired systems, 241–265
Biological synapses, 242, 250, 265
Birth and death process, 12, 31, 32, 60, 84
Bit parallel multiplier, 277, 278, 281, 282, 285, 289, 290, 297, 298
Bridging fault, 5
Built-In Self Test (BIST) circuit, 308, 328–329

C
Caches, 139, 141–143, 151, 157, 163, 164
CAM. *See* Content addressable memory (CAM)
CED. *See* Concurrent error detection (CED)
CFC. *See* Control flow checking (CFC)
Clock gating, 104, 107
Common-cause system, 77–81
Concurrent error detection (CED), 278, 282
Constant failure rate, 64, 82
Constant test set, 321–322, 328
Content addressable memory (CAM), 143–144, 147, 164
Control flow checking (CFC), 244, 246
Control pins, 308, 323–326
Core mapping, 217
Crosstalk, 242, 245, 253, 258–262, 265
C-testable, 307–330

D
DED. *See* Double error detection (DED)
Delay fault, 6–7
Delay model, 118
Design-for-test (DFT), 307, 308
Design optimization, 174–177, 181–206
Design-time synthesis, 218
Design trade-offs, 100, 101
Difference set (DS) codes, 151, 155–161, 163, 164
Digital signal processor (DSP), 170, 171, 178
Double error detection (DED), 289–291, 305
DPM. *See* Dynamic power management (DPM)
DRAM. *See* Dynamic random access memory (DRAM)
DS codes. *See* Difference set (DS) codes
DSP. *See* Digital signal processor (DSP)
DVFS. *See* Dynamic voltage frequency scaling (DVFS)
Dynamic power dissipation, 102–103
Dynamic power management (DPM), 3, 4
Dynamic random access memory (DRAM), 142–143, 163–164
Dynamic voltage frequency scaling (DVFS), 3, 4, 99, 112, 171, 175, 181

J. Mathew et al. (eds.), *Energy-Efficient Fault-Tolerant Systems*,
DOI 10.1007/978-1-4614-4193-9, © Springer Science+Business Media New York 2014

E

EG codes. *See* Euclidean geometry (EG) codes
Electronic system-level (ESL), 206
Embedded systems, 1
ESL. *See* Electronic system-level (ESL)
Euclidean geometry (EG) codes, 151, 153–156, 158, 159, 163, 164

F

Factorizing netlists, 293–294
Failures in time, 5
Fault
  correction, 137–165
  coverage, 242, 246, 247
  detection, 217, 222, 225, 229, 309–315
  detection techniques, 242–244
  injection, 181, 184–185, 198
  models, 5, 6
  tolerant design, 211–237
  tree analysis, 11, 12, 15, 20
Field programmable gate arrays (FPGAs), 211, 222–224, 228
  system, 262, 265
Finite field multipliers, 269–305, 307–330
Flash memory, 139, 144
Forward error correction (FED), 301
FPGAs. *See* Field programmable gate arrays (FPGAs)
Freund's model, 12, 16, 27–31

G

Galois field, 270, 271, 279, 290, 300, 301

H

Hamming codes, 147, 278
Hardware description language (HDL), 173
Hardware implementation, 242, 248, 262–265
Hardware redundancy, 2, 7
Hardware/software co-design, 177–179
HDL. *See* Hardware description language (HDL)
Heterogeneous cores, 213, 214

I

Information redundancy, 7–8
Instruction set architecture (ISA), 170, 171
Intelligent techniques, 265
ISA. *See* Instruction set architecture (ISA)

J

JTAG boundary, 307

K

k-out-of-n systems, 13, 31, 84

L

LDPC. *See* Low density parity codes (LDPC)
Load-sharing system, 12, 13, 77–81
Low density parity codes (LDPC), 285–289, 298, 299, 305
Low power design, 1

M

Many-core systems, 241
Markov chains, 12, 13, 31–41, 50–52, 58, 59, 65, 77, 82
Markov model, 11, 12, 31–60
Mean time-between-failures (MTBF), 5
Mean time-to-failure (MTTF), 4, 5, 99, 114, 115
Mean time-to-repair (MTTR), 5
MPSoC. *See* Multiprocessor system-on-chip (MPSoC)
Multi-bit error correction codes (ECCs), 150–152, 157, 158, 162–165
Multiple channel system, 255–258
Multiple error detection, 298–301
Multiple-input signature registers (MIARs), 329
Multiple V_{dd} design, 106
Multiplier architectures, 307–330
Multiprocessor system-on-chip (MPSoC), 169–177, 180–190, 192, 196–200, 202, 203, 205, 206

N

Network interfaces (NIs), 216–218, 225, 228
Network-on-chip (NoC), 212–224
Network topology, 212, 217
NIs. *See* Network interfaces (NIs)
NoC-based multiprocessor, 251
NoC interconnect, 241–265

O

OLS codes. *See* Orthogonal latin square (OLS) codes
On-chip communication architecture, 170–173, 175

OpenCL framework, 224
Orthogonal Latin square (OLS) codes, 151–154, 157–159, 162–164

P
Parallel encoder, 301–304
Parallel systems, 11, 21, 22
Parity prediction, 270, 277–285, 287, 288, 290, 291, 296, 298–300, 302
PB. *See* Polynomial basis (PB)
Permanent faults, 4, 7, 283–284
Platform efficiency, 219–220
Poisson process, 12, 23, 31–33, 35, 41–50, 54, 63, 67, 68, 72, 74, 77
Polynomial basis (PB), 270–277, 298, 302, 309, 322, 328, 330
Power gating, 104
Power minimization, 3, 99, 104–111, 114–118
Primitive polynomials, 270, 273–275, 286, 287, 295, 296, 298

R
Real-time systems, 242, 247, 248, 262, 263, 265
Reconfigurable systems, 211–237
Reed-Solomon (RS) codes, 143, 151, 163, 301
Registers, 137–142, 146, 147, 162, 164
Reliability, 2, 4, 5, 7
Reliability model, 12, 13, 23, 61–77
Reliable design, 2, 4–8
Routers, 212, 213, 216–218, 222–224, 226, 228–230, 233, 234
RS codes. *See* Reed-Solomon (RS) codes
Run time fault detection, 230

S
Scan chain, 307
SEC-DED codes. *See* Single error correction and double error detection (SEC-DED) codes
SER. *See* Soft error rate (SER)
Series systems, 17, 22–23
SEUs. *See* Single event-upsets (SEUs)
Single channel system, 251, 253–255
Single error correcting codes, 277, 278, 285, 286, 298
Single error correction and double error detection (SEC-DED) codes, 147–151, 155, 158, 161–165
Single event-upsets (SEUs), 4, 175, 184–188, 190–195, 198–205
Single parity check (SPC) codes, 146–147, 162, 164
SMART strategy, 251–253, 263, 265
SoC. *See* System-on-chip (SoC)
Soft error rate (SER), 5
Software reliability models, 12, 13, 61–77
SPC. *See* Single parity check (SPC) codes
Spiking neural networks, 241, 247–250
SRAM. *See* Static random access memory (SRAM)
Static power (P_{stat}) dissipation, 2, 3, 103–104
Static random access memory (SRAM), 142, 163, 164
Stochastic process, 11, 13, 31–33, 41, 50, 62, 67–69
Stuck-at fault, 5, 308–311, 313–316, 321–323, 328–330
Stuck open fault, 6
Synthesis, 289–296, 305, 307
System gap, 173–174
System-level technique, 8
System-on-chip (SoC), 1–3, 211–213, 215, 219, 222

T
Task graph, 217, 218, 220, 225, 234, 235, 237
Technology scaling, 1, 3, 6, 7, 104, 105, 114, 119, 265
Test
  set, 308–310, 314–316, 321–323, 328, 329
  vector, 307–310, 315, 317, 323, 326, 328–330
Time redundancy, 7
Timing error, 114, 117–121
Timing speculation (TS), 118–121
Transient fault, 4, 6, 7
TS. *See* Timing speculation (TS)

V
Voltage island(s), 170, 171
Voltage island design, 106
Voltage scaling, 181–183, 186, 188–190, 192, 194–200, 202, 205

W
Watchdog processor, 244, 246

Printed by Printforce, the Netherlands